Springer Series in Operations Research

Editor:
Peter Glynn

Springer
New York
Berlin
Heidelberg
Barcelona
Budapest
Hong Kong
London
Milan
Paris
Santa Clara
Singapore
Tokyo

Springer Series in Operations Research

Altiok: Performance Analysis of Manufacturing Systems
Dantzig and Thapa: Linear Programming: Introduction
Drezner (Editor): Facility Location: A Survey of Applications
 and Methods
Fishman: Monte Carlo: Concepts, Algorithms, and Applications
Olson: Decision Aids for Selection Problems
Yao (Editor): Stochastic Modeling and Analysis of Manufacturing
 Systems

Tayfur Altıok

Performance Analysis of Manufacturing Systems

With 140 Illustrations

 Springer

Tayfur Altiok
Department of Industrial Engineering
Rutgers University
P.O. Box 909
Piscataway, NJ 08855
USA

Series Editor:
Peter Glynn
Department of Operations Research
Stanford University
Stanford, CA 94305
USA

Library of Congress Cataloging-in-Publication Data
Altiok, Tayfur.
  Performance Analysis of Manufacturing Systems / Tayfur Altiok.
    p. cm. – (Springer series in operations research)
  Includes bibliographical references and index.
  ISBN 0-387-94773-6 (hard : alk. paper)
    1. Production Management–Evaluation. 2. Production management–
  Mathematical models. 3. Stochastic analysis. 4. Inventory
  control–Mathematical models. I. Title. II. Series.
  TS155.A42   1996
  658.5–dc20   96-14273

Printed on acid-free paper.

© 1997 Springer-Verlag New York Inc.
All rights reserved. This work may not be translated or copied in whole or in part without the written permission of the publisher (Springer-Verlag New York, Inc., 175 Fifth Avenue, New York, NY 10010, USA), except for brief excerpts in connection with reviews or scholarly analysis. Use in connection with any form of information storage and retrieval, electronic adaptation, computer software, or by similar or dissimilar methodology now known or here-after developed is forbidden.
The use of general descriptive names, trade names, trademarks, etc., in this publication, even if the former are not especially identified, is not to be taken as a sign that such names, as understood by the Trade Marks and Merchandise Marks Act, may accordingly be used freely by anyone.

Production managed by Frank Ganz; manufacturing supervised by Jeffrey Taub.
Typeset by The Bartlett Press, Marietta, GA.
Printed and bound by Braun Brumfield, Inc., Ann Arbor, MI.
Printed in the United States of America.

9 8 7 6 5 4 3 2 1

ISBN 0-387-94773-6 Springer-Verlag New York Berlin Heidelberg   SPIN 10537677

Preface

Manufacturing industries are devoted to producing high-quality products in the most economical and timely manner. Quality, economics, and time not only indicate the customer-satisfaction level, but also measure the manufacturing performance of a company. Today's manufacturing environments are becoming more and more complex, flexible, and information-intensive. Companies invest into the information technologies such as computers, communication networks, sensors, actuators, and other equipment that give them an abundance of information about their materials and resources. In the face of global competition, a manufacturing company's survival is becoming more dependent on how best this influx of information is utilized. Consequently, there evolves a great need for sophisticated tools of performance analysis that use this information to help decision makers in choosing the right course of action. These tools will have the capability of data analysis, modeling, computer simulation, and optimization for use in designing products and processes.

International competition also has had its impact on manufacturing education and the government's support of it in the US. We see more courses offered in this area in industrial engineering and manufacturing systems engineering departments, operations research programs, and business schools. In fact, we see an increasing number of manufacturing systems engineering departments and manufacturing research centers in universities not only in the US but also in Europe, Japan, and many developing countries. Moreover, the National Science Foundation in the US has recently increased its emphasis on manufacturing by forming a program in manufacturing systems, resulting in increased funding for research in this area.

During the past two decades, a large volume of research has accumulated in modeling production, manufacturing, and inventory systems with the purpose of designing and measuring their performance. Due to the natural existence of queueing phenomenon in the manufacturing systems, queueing theory and queueing network models have been a common approach. The current-state-of-the-art in the analysis of stochastic service systems has produced new tools to analyze complex manufacturing systems. Unfortunately, the published research is dispersed into a large number of journals because of the diverse backgrounds and interests of the researchers. Consequently, although modeling of manufacturing systems started in early 1950s, only recently the research appeared in a few books in this area.

The purpose of this book is to present the current state-of-the-art in stochastic modeling of manufacturing systems, with an emphasis on the production function together with the issues of inventory control. It will consist of exact, approximate, and numerical techniques to obtain various measures of performance. The ultimate goal is to put together a set of tools for design and performance analysis of a variety of complex manufacturing systems. Performance analysis in general is the first step in the design process of any system, and it is becoming crucially important for its contribution to effective industrial logistics.

Manufacturing systems is a very broad field, and an enormous amount of research is under way. A number of researchers in the US, France, Japan, England, Italy, and other countries are working in this area. It is not possible to write a book that covers everything. The contents in this book were chosen to produce a cohesive document in which topics are continuations of each other. Though many more books will be written in this fertile field, emphasizing various different aspects of manufacturing systems, the strength of this book is in its emphasis on the numerical aspects of the models presented. Phase-type distributions are quite useful in modeling data with a wide range of variability. Therefore, models involving phase-type random variables are highly versatile and quite powerful in representing real-life situations. The exact analysis of almost any manufacturing system turns out to be impossible. This book presents approximations for various complex systems, which not only help in understanding how these systems function but also provide acceptable results for important measures of performance.

It took a long time to complete this book, even though it is not very long. Emphasis on numerical aspects, examples, and exercises took an intense effort. Also, the process of connecting the book's contents with real-life manufacturing practices took its toll. The models and approaches I cover mainly use queueing-theoretic arguments. However, while doing that I stayed away from the bold reviews of queueing models and created a slightly different jargon for manufacturing systems than the one in queueing theory. I believe this approach appeals to the reader in the manufacturing field. We started our course "Performance Analysis of Manufacturing Systems" in the Department of Industrial Engineering at Rutgers University years ago. I have emphasized both theoretical issues as well as the practical aspects in modeling manufacturing systems. It has been very fruitful in the sense that the students worked on a number of projects, many of which ended up being their theses or dissertations. Many ideas have emerged out of this material, and some of them are already implemented in the industry.

In my quest for the solution in complex manufacturing performance problems, I had the pleasure of receiving the support of or being associated with many individuals and organizations. With my good friend and co-author Harry Perros, I developed many ideas and gave birth to many projects, most of which started over a discussion in a cafe in Paris and continued in a restaurant in San Francisco and in Denali, Alaska! We have worked very well together, and I hope to continue to do so. Sandy Stidham, with whom I had the pleasure of working in my dissertation, has been very supportive and constructive. E. A. Elsayed, my department chairman, always encouraged me by providing industrial problems. Our plant visits have been very helpful and enjoyable. My colleague and co-author Melike Gursoy helped by reviewing parts of the manuscript. My students Zu-Ling Kao, Raghav Ranjan, Mayur Amin, Goang-An Shiue and Esra Dogan contributed to certain chapters significantly. Süreyya Dipsar reviewed parts of the book. Muge Z. Avşar worked out many of the problems at the end of chapters.

Parts of the material in this book were developed while I had the support of the program in Production Systems at the National Science Foundation over several years. The support not only helped in theoretical developments but also enabled me to have a closer look at the pharmaceutical manufacturing industry. A large portion of the manuscript was rewritten and modified while I was visiting the ISEM Laboratory at Universite de Paris Sud in 1986, the MASI Laboratory at the Universite Pierre et Marie Curie in 1992 and 1993, and ENT at Evry, France in 1993. I am thankful to E. Gelenbe, G. Pujolle, S. Fdida, Y. Dallery, and T. Atmaca for making my excursions to Paris possible. I was awarded grants and a fellowship from the United Nations and the Fulbright Commission to give short courses and seminars on the same topic in industrial engineering departments at Middle East Technical and Bilkent universities in Ankara, Turkey. These visits were not only productive for the book but were also full of pleasant memories.

Finally, I want to mention my special thanks to my wife, Binnur, and to my parents, Mehmet and Nezihe Altiok, for their encouragement to finish this project and for all they have given me over the years that seem only to be too short.

Tayfur Altiok
February 1996

To Binnur, Eda, and Selen
for all the joy and happiness

Contents

Preface	v
1. Introduction to the Modeling of Manufacturing Systems	**1**
1.1. Typical Decision Problems in Manufacturing Systems	3
1.2. Performance Evaluation	3
1.3. Models of Manufacturing Systems	4
1.4. Design of Manufacturing Systems	9
References	11
2. Tools of Probability	**15**
2.1. Random Variables and Their Distributions	15
2.2. Joint Distributions	16
2.3. Conditional Distributions	16
2.4. Functions of Random Variables	17
2.5. Moments of Random Variables	18
2.6. Special Distributions	22
2.7. Stochastic Processes	26
2.8. Phase-Type Distributions	40
Related Bibliography on Phase-Type Distributions	57
References	59
Problems	60
3. Analysis of Single Work Stations	**66**
3.1. Introduction	66
3.2. A Simple Work Station	71
3.3. Distribution of the Number in the System	77
3.4. Work Stations Subject to Failures	81
3.5. Distribution of the Number of Jobs in a Work Station Subject to Failures	87
3.6. The Process-Completion Time	91

x   Contents

3.7. Failure-Repair Policies	93
3.8. The Machine Interference Problem	104
Related Literature	110
References	111
Problems	112

4. Analysis of Flow Lines — 118

4.1. Introduction	118
4.2. Characteristics of Flow Lines	120
4.3. Analysis of Two-Station Flow Lines	122
4.4. Analysis of Flow Lines With More Than Two Stations	141
4.5. Deterministic Processing Times	158
4.6. Bounds on the Output Rate	166
4.7. Design Problems in Flow Lines	168
Related Literature	174
References	176
Problems	180

5. Analysis of Transfer Lines — 185

5.1. Introduction	185
5.2. A Two-Station Line with No Intermediate Buffer: A Regenerative Approach	188
5.3. Transfer Lines with Work-in-Process Buffers	195
5.4. Buffer Clearing Policies in Transfer Lines	197
5.5. Analysis of a Two-Machine Synchronous Line Under BCP1	198
5.6. Operation Under BCP2	207
5.7. Line Behavior and Design Issues	211
5.8. Analysis of Larger Systems	218
5.9. Transfer Lines with No Scrapping	228
Related Bibliography	235
References	238
Problems	240

6. Assembly Systems — 243

6.1. Introduction	243
6.2. An Asynchronous Assembly System	244
6.3. Approximate Analysis of the Two-Station Assembly System	255
6.4. Analysis of Larger Systems	259
6.5. Synchronous Assembly Systems	263
6.6. Relaxing the Assembly Assumption	264
Related Bibliography	267
References	268
Problems	269

7. Pull-Type Manufacturing Systems — **273**

7.1. Introduction — 273
7.2. Production Control Policies — 274
7.3. Analysis of Single-Stage, Single-Product Systems — 284
7.4. A Single-Stage System with Two Products — 300
7.5. Multistage Pull-Type Systems — 309
7.6. The Generalized Kanban Scheme — 323
7.7. Manufacturing Systems with Inventory Procurement — 326
7.8. A Two-Stage, MultiProduct System — 328
7.9. The Look-Back Policy — 334
Related Bibliography — 338
References — 339
Problems — 343
Appendix: A Single-Stage P/I System with PH Delay, Set-up, and Processing Times — 348

Index — **353**

1

Introduction to the Modeling of Manufacturing Systems

A manufacturing system can be viewed as an arrangement of tasks and processes, properly put together, to transform a selected group of raw materials and semifinished products to a set of finished products. The three basic functions of manufacturing are procurement, production and distribution. Clearly, there are other functions in a supporting, role such as planning, design, accounting, personnel management, and marketing, among others. The raw material is procured and converted through the production process to semifinished units, which are sometimes referred to as assemblies or subassemblies, and to finished products to be finally stored in the warehouse to meet the customer demand. Distribution deals with the transfer of the finished products to warehouses at different locations. The flow of materials shown in Figure 1.1 depicts the various stages of manufacturing.

Managing the purchase of raw and semifinished materials and their storage in the facility is usually called procurement. It provides input to the facility, which needs to be well coordinated with its capacity and the throughput. The material flow or the product routing in the facility is a consequence of the process plans of the finished products. Products are processed at production stages and transferred to others according to their process plans. Production stages may consist of equipment, machinery, operators, and the incoming and possibly the outgoing material storage, which comprise the in-process inventory in the facility. Finished products are stored in the finished-product warehouse. The customer demand appears either at the finished-product warehouse or at various other warehouses run by the distribution system. It is quite possible that a customer's demand may not be satisfied immediately and is backlogged or lost.

The production system is called a **flow line** when its stages are arranged in series and all products produced by this system follow the same sequence of processes, as shown in Figure 1.2. A flow line may be designed to be dedicated to a particular product. The processing times at each stage may be constant or variable, and they may differ from one stage to another. In contrast to a flow line, a **job shop** consists of machines placed in some fashion on the shop floor, through which the products flow according to their process plans. The flexibility built into the computer-controlled machining enables the work stations to handle a variety of operations. Thus, the developing manufacturing technologies make it possible to

2   1. Introduction to the Modeling of Manufacturing Systems

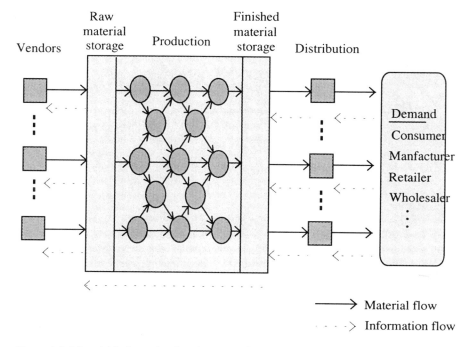

Figure 1.1. Material/information flow in a manufacturing environment.

view shop floors as multiproduct, multistage production lines. In this case, either the whole line switches from one product to another and each time goes through a set-up or a cleaning process, or stages are set up individually according to the arriving batch of products. **Automatic transfer lines** are special cases of flow lines where the system is dedicated to the high-volume production of a particular

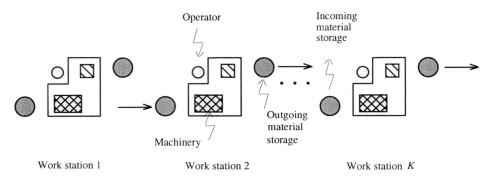

Figure 1.2. An example of a flow line.

product. The stages are usually linked by conveyors and are synchronized in such a way that each stage starts processing simultaneously and completes processing simultaneously. It is possible that transfer lines are dedicated to different products at different points in time.

A higher-level classification of industrial facilities may be **production/ inventory** versus **pure inventory** systems. Though it is hard to visualize a manufacturing facility that does not deal with inventory, the phrase "production/inventory systems" emphasizes the concept of the control of inventories as well as the production activity. There is no production in pure inventory systems, where the main activities are procurement and distribution. Examples include wholesalers, retail stores, and parts stores, among others.

1.1 Typical Decision Problems in Manufacturing Systems

Manufacturing-related decision problems may be put into a standard classification of long-range, intermediate-range, and short-range decisions. The long-range decisions are usually made while designing facilities, shop floors, and warehouses, in the actual design of production lines, in product mixes, and in setting up distribution and customer service policies. For instance, in dedicated flow lines and transfer lines, these decision problems turn out to be work-load design (line balancing), equipment selection, and buffer space allocation problems. The intermediate-range decisions involve setting master production plans and inventory management issues, policies of switching from one product to another, product and customer priorities, procurement plans, maintenance, and work-force plans. Short-term decisions usually deal with daily or weekly production schedules, routing due to failures, and repair management. The guidelines (or objectives) to help make the above decisions include

- maximizing revenues from sales,
- minimizing procurement–production–distribution costs,
- maximizing customer service levels,
- minimizing investment in inventories,
- achieving and maintaining high standards of quality,

among many others, such as maintaining work-force stability and maximizing the returns on investments in assets.

1.2 Performance Evaluation

The impact of a policy decision or a design change on the behavior of a system may be either measured by observation or estimated using a methodology. This is usually called **performance evaluation**. The concept of "performance" is self-explanatory. Humans beings have always measured their performances in almost all of their tasks they carry on to see how well they are doing. This is the only way to find out whether one is doing well or poorly as well as better than or worse than a standard or a competitor. Clearly, the motivation here is to take the course of action that improves the performance. The prerequisite of performance evaluation is to decide on the performance criteria, which may be a single measure or a set of measures. Typical measures in a manufacturing environment are output rates or throughputs (number of units produced per unit time), average inventory levels (inventory levels averaged over a planning horizon), utilization (percentage of time a machine is actually busy), customer service levels (percentage of time customer demand is fully satisfied), average flow times (average time a unit spends in the system), and percentage of down times (percentage of time a machine is under repair), among others. Measuring performance in an existing system is a matter of data collection. Calculation and reporting are routine procedures. However, in the lengthy process of improving the system performance, "what if" questions are frequently asked. For instance, one may consider the possibility of adding a new robot or a product line, or changing an operating policy in an existing system. Then, it is natural to question the impact of these changes on the system's behavior. These situations usually require an analysis of the system. In most cases this analysis is carried out with the help of a mathematical or a computer simulation model of the system.

Performance estimation in systems that are in the design phase is also involved. First of all, these systems may have only a few data points—or none—to adequately estimate the measures of interest. An analysis of the system again becomes necessary in these situations, resulting in a model to estimate its performance. For systems with no data points, contrary to the common belief that not much can be done, there is still a lot to do. Using an appropriate model, it is possible to find a range of values for system parameters that will produce the desired values for the performance measures. As a result, one can make decisions accordingly. In analyzing the model of a system and evaluating its performance when tractable techniques become insufficient, numerical techniques, simulation, or approximations are often used.

1.3 Models of Manufacturing Systems

Models of manufacturing systems can be put into two main categories: optimization and performance models. Optimization models, given a set of criteria and constraints, generate a single or a set of decisions or courses of action. These models are sometimes called generative or prescriptive models, and they rely on the principles of optimization. Performance models, on the other hand, estimate the measures of system performance for a given set of decisions and system parameters. These models are at times called evaluative or normative models, and they rely on techniques of stochastic processes, probability, and simulation. The applications of stochastic processes and their extensions in queueing theory and queueing networks involving exact and approximation methods significantly enriched the modeling effort of manufacturing systems. Also, the recent developments in computer simulation modeling created a tremendous potential for generating test-bed models of manufacturing systems to study a variety of operational policies and their impacts on the system performance.

The loss of productivity in manufacturing systems occurs during the periods where machines are idle as a result of either failures or bottlenecks and excessive accumulation of inventories. The uncertainty in these systems is mainly due to the following factors:

- Processing times may have some variability.
- Failures occur at random points in time.
- Repair times are usually random.
- External impacts of the economy in general are unpredictable.

It is the randomness that makes it difficult to manage the system or to estimate its performance. For this reason, modeling and algorithm development are needed to understand the behavior of these systems and to provide guidelines to improve their performance. Throughout this book we will focus on the models of long-run analysis and use exact as well as approximate probabilistic arguments, queueing theory, renewal theory, and Markov chain techniques to study manufacturing systems.

(Production lines, either flow or transfer lines, are basically series arrangements of work stations. They are quite common in manufacturing facilities in a variety of industries. Because of space limitations and cost considerations, there are target levels (explicit or implicit) for storage between stages in a manufacturing system. (Consequently, the storage limitations give rise to **bottleneck**, or **blocking**, that causes stoppages at work stations due to lack of space or excessive accumulation of in-process inventory in the downstream stages. Similarly, **starvation** may exist in manufacturing systems, causing idleness in stations due to a lack of jobs to process in the upstream stages)

Another type of idleness in manufacturing systems is caused by **failures**. Usually, a failed station is taken into repair as soon as possible, and once it is back up again, it resumes its processing on the same job at the point where it was interrupted. Sometimes, in high-speed transfer lines, for instance, it is likely that the current job is lost upon a failure. The time a job spends in a work station from the time processing starts until the job leaves the station is called **process completion time**. In models of manufacturing systems, sometimes failures are viewed as high-priority jobs and kept in the analysis as vital elements of the model. In others, they are incorporated into the process completion times. There are various other policies of what to do with the interrupted jobs. Failures in one station are usually assumed to occur independently of failures in other stations, thus causing a stoppage only in that particular station. This may not be the case in transfer lines, where a failure on one machine may cause the stoppage of the whole system.

The three issues mentioned above, namely blocking, starvation, and failures, are the main factors causing loss of productivity in production lines and difficulty in the

6   1. Introduction to the Modeling of Manufacturing Systems

analysis of these systems. There have been significant contributions to modeling and analysis of production lines from the literature on queueing networks. The literature consists of analytical, numerical, and simulation models, mostly of the Markovian type. Unfortunately, the analysis of the Markovian model of a production line results in a dimensionality problem because of the size of its state space. For this reason, exact models are limited to rather small systems (see, for instance, Buzacott (1967) and Gershwin and Schick (1983) for transfer lines, Buzacott (1972) and Wijngaard (1979) for flow lines, among many others). Longer lines have been studied through numerous approximations (see Zimmern (1956) and Sevastyanov (1962) for the earlier work and Gershwin (1987), Brandwajn and Jow (1988), Koster and Wijngaard (1989), Altiok and Ranjan (1989), and Gun and Makowski (1989) for the recent work, among others). Approximations gathered mainly around the idea of decomposing the tandem system into subsystems of one or two nodes. These subsystems are analyzed individually within an iterative scheme that relates them to each other. Most of these studies were to compute performance measures such as throughput, average inventory levels, and utilization. In many cases, the long-run probability distribution of the number of jobs in each buffer was found, from which many of the performance measures were directly obtained. Comprehensive reviews and analyses of production lines can be found in Dallery and Gershwin (1992), Buzacott and Shanthikumar (1993), and Gershwin (1994). A recent source on queueing networks with blocking is Perros (1994).

Systems with arbitrary configurations, as shown in Figure 1.3, also attracted attention. In this area, significant advances took place in the theory of queueing networks. Systems with arbitrarily connected work stations having infinite-capacity buffers have been tackled with quite general assumptions (see, for instance, Basket et al. (1975), Kelly (1979) for exact product-form solutions, and Whitt (1983) for approximations). Finite buffer-capacity problems again appear to be quite difficult,

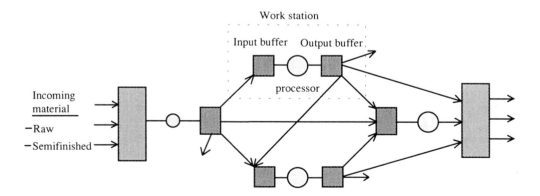

Figure 1.3. An arbitrary configuration of work stations.

yet some recent developments have been highly encouraging (see, for instance, Jun and Perros (1989), Lee and Pollock (1990), and Baynat and Dallery (1993)). These studies propose some type of a decomposition approach to approximate the performance measures of the system. Also, the assembly–disassembly systems have been dealt with by using decomposition approximations (see, for example, Gershwin (1991) and Di Mascolo et al. (1991)). Assembly–disassembly networks are different from classical queueing networks in that a great deal of coordination is needed among the jobs moving through the system. Still, the work done in the area of arbitrary configurations is far behind in satisfying the needs in modeling manufacturing systems. Much needs to be done in gearing the current state-of-the-art toward the manufacturing environment. Models with multiple job-class capability, set-ups, and arbitrary routing and scheduling capabilities with rather general assumptions must be developed. However, these problems are known for their difficulty in systems such as job shops, where scheduling and the dynamic nature of the shop floor make the modeling effort quite difficult.

For a long time, the inventory systems have been modeled as pure inventory control problems with replenishment times being either zero or independent of the outstanding number of orders. This is usually not the case in reality and especially not so in manufacturing environments. The time to put a finished product into the warehouse, that is, the manufacturing lead time, depends on how many orders are placed to be processed in the shop floor. Hence, it becomes critically important to coordinate both production function and inventory control in an effort to model inventory systems within manufacturing environments, as simplified in Figure 1.4. In this direction, the term "production/inventory systems" has evolved and some studies have appeared in the literature. Again, in modeling these systems, queueing theory proved to be quite useful. Gross and Harris (1973) initiated the idea of dependence between replenishment times and the outstanding number of orders. They studied the underlying state-dependent server within the (R, r) continuous-review inventory control policy. Under this policy, the processor stops

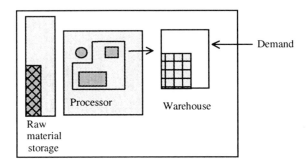

Figure 1.4. A simplified single-stage production/inventory system.

producing when the inventory level reaches the target level R. It starts producing again when the inventory level down-crosses the reorder level r. Other studies followed (see, for instance, Doshi et al. (1978), Gavish and Graves (1980, 1981), Zipkin (1986), Lee and Srinivasan (1987), and Altiok (1989). In all of these studies, a cost-minimizing objective function was considered, and methods were developed to compute the components of the objective function, such as the average inventory and the average back-order levels, and then the optimization problem was tackled.

The coordination of production schedules with external demand gives rise to another classification of manufacturing systems, namely pull versus push systems. A **pull system** involves decisions usually driven by the market. The production effort is focused on keeping sufficient levels of inventory necessary to achieve a certain level of customer service. On the other hand, a **push system** involves decisions driven by the demand forecasts and the resulting production schedule. Thus, in general, in pull systems attention is given to the state of the finished-product warehouse, and in push systems the focus is on how many units to produce according to the production plan. Recently, due to the success of the Kanban concept introduced by the Toyota Corporation after World War II (see Monden (1983)), attention has been given to modeling multistage production systems as pull systems where each stage produces according to the demand generated by the immediate downstream stage. The **Kanban system** is a message-passing mechanism that transfers the demand generated by the downstream stages to the upstream production stages in pull systems. The purpose is to initiate production at each stage when it is necessary and for the necessary quantity, thus giving rise to the concept of just-in-time production. There are different implementations of the Kanban system in various message-passing policies. It is likely that some policies lead to relatively higher inventory accumulation than others. This results in higher customer service levels at the expense of higher investment in inventory. These policies should be chosen according to the company's customer–product priority structure and the cost considerations. The Kanban system was first described by Sugimori et al. (1977) and later analyzed in a simulation model by Kimura and Terada (1981). Recent literature includes Karmarkar and Kekre (1987), Buzacott (1989), and Deleersnyder et al. (1989), which considered Kanban systems to study the problems of batch sizing and inventory level order-triggering. More recently, Di Mascolo et al. (1992) developed queueing network approximations to evaluate the performance of multiproduct pull-type systems under a generalized Kanban scheme.

Various message-passing ideas originating from the Kanban system gave rise to different models of multistage production systems. For instance, an integral control mechanism initiates production at a particular stage by taking into account the total inventory at a group of stages (see de Koster (1988)). Another approach is to divide the production line into segments relative to the displacement of the bottleneck stations (see Lundrigan (1986)). It is also possible to keep the total work-in-process inventory constant by applying the integral control schemes to the entire system (see Spearman et al. (1990)). Finally, Buzacott and Shanthikumar (1993) generalized the message-passing concept to the PAC (Production Authorization Cards) system, which provides a framework for developing coordination and control mechanisms with various desired features.

1.4 Design of Manufacturing Systems

The design of manufacturing systems has been quite a difficult problem. Only some special cases were dealt with to find the cost-optimal values for a set of control parameters through the use of optimization models. Cases where exact approaches were employed usually consisted of single-stage systems because of tractability. When the performance measures have explicit expressions, it is usually not difficult to obtain the optimal or near-optimal values of the system parameters. However, in many cases it is not possible to obtain the performance measures explicitly. In these situations, the algorithms to compute these measures are incorporated into a search procedure that locates the optimal values of parameters in an iterative manner, as depicted in Figure 1.5. This approach has been used to allocate buffers in production lines (see Smith and Daskalaki (1988) and Altiok and Stidham (1983)). There have also been dynamic programming implementations for the buffer allocation problem. For instance, Jafari and Shanthikumar (1989) and Yamashita and Altiok (1991) both used dynamic programming recursions to approximately allocate a finite quantity of buffers to maximize the throughput or to minimize the total buffer space for a desired throughput. In an earlier study, Soyster, et al. (1979) replaced the throughput objective function by a function of its bounds and used a separable programming approach to allocate buffers in transfer lines. In all of these studies, larger buffers are allocated immediately before and after the stages displaying high variability in processing (or completion) times, possibly due to failures.

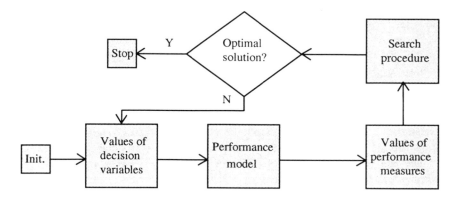

Figure 1.5. An iterative search procedure.

One way of making the design process more efficient is to employ derivative estimation techniques via either analytical methods or simulation. For instance, it is possible to approximate the derivative of the throughput of a transfer line with respect to some system parameters such as the buffer capacities (see Gershwin (1994)) and the repair rates (see Dogan and Altiok (1995)). It is also possible to estimate these derivatives using either finite-difference estimators (see Glynn (1989, 1990) and Glynn et al. (1991)) or likelihood ratio estimators (see Rubinstein (1986) and Glynn (1990)) during a single simulation run. Unfortunately, the simulation approach is not quite applicable to the problems that are discrete in nature, but rather to the work load as well as to the repair-capacity allocation problems that involve continuous decision variables. Also, Ho et al. (1979) proposed a perturbation approach for the derivative estimation again within a single simulation run, this time to allocate a fixed sum of buffer space to maximize the throughput.

In the work-load allocation problem that is formulated to decide on the amount of work to be done at each station in a production line, Hillier and Boling (1966) showed that a bowl-type work-load allocation (in terms of the mean processing times) improves the output rate. Baruh and Altiok (1991) used analytical perturbations in the underlying Markov chain in allocating a fixed total work load to stations in a production line to maximize the throughput. It is common sense to think that the optimization process should be incorporated into the computational effort to obtain performance measures, especially in multistage systems. Future research in optimizing large-scale stochastic systems may develop in that direction to make the whole process of performance analysis and optimization integrated and efficient. A variety of production planning and optimization issues were introduced in Johnson and Montgomery (1974) and Elsayed and Boucher (1985).

Throughout this book, we will introduce probabilistic models for a variety of manufacturing systems. In these models, some of the random variables of interest will be modeled using phase-type distributions. The introduction of phase-type distributions by Neuts (1981), who formalized the earlier work on the phase concept by Erlang (1917) and Cox (1955), has been one of the significant advances in the analysis of stochastic systems in the last decade. Most of the analysis until then used transform techniques i.e., Laplace-Stieltjes transforms and generating functions bearing some computational problems. Phase-type distributions naturally give rise to a Markovian analysis and in many cases make it tractable through the use of recursive or matrix-geometric methods. There has been a considerable effort in fitting these distributions to observed data sets. Thus, the present state-of-the-art in phase-type distributions gives us a chance to deal with difficult problems in manufacturing systems.

Manufacturing systems, in general, are so varied in terms of technology, structure, and practices that there is a tremendous amount of work done in the literature for their analysis. It is not possible to cover every type of a system or every type of an approach in one book. Such a book would be an encyclopedia. Probably, the best approach is to decide first on a perspective, a way of looking at manufacturing systems, and then build around it as much as possible. Here we choose to look at these systems from the views of stochastic modeling, long-run analysis, and queueing theory. We will study in detail the topics and the issues raised in this chapter, emphasizing the methodology and the computational effort whenever possible. We will present exact and approximate models of manufacturing systems and will mainly emphasize the analyses to evaluate their performance. Chapter 2 provides a refresher on probability, Markov chains, and phase-type distributions. In Chapter 3, we start the analysis of production systems at the work–station level. We study single stations that are subject to failures and consider a variety of failure/repair policies. Chapter 4 covers a detailed treatment of flow lines. We present exact and approximate procedures for performance analysis. Chapter 5 extends the material in Chapter 4 to study automatic transfer lines. In Chapter 6, we study assembly-type manufacturing systems with arbitrary configurations. Exact and approximate procedures are presented. Finally, Chapter 7 covers single and multistage pull-type manufacturing systems and a variety of operational policies used to manage these systems.

References

Altiok, T., 1989, (R, r) Production/Inventory Systems, *Operations Research*, Vol. 37, pp. 266–276.

Altiok,T., and R. Ranjan, 1989, Analysis of Production Lines with General Service Times and Finite Buffers: A Two-Node Decomposition Approach, *Engineering Costs and Production Economics*, Vol. 17, pp. 155–165.

Altiok, T., and S. Stidham, Jr., 1983, The Allocation of Interstage Buffer Capacities in Production Lines, *AIIE Transactions,* Vol. 15, no. 4, pp. 292–299.

Baruh, H., and T. Altiok, 1991, Analytical Perturbations in Markov Chains, *European J. Operations Research*, Vol. 51, pp. 210–222.

Basket, F., K. Chandy, R. Muntz, and F. Palacios, 1975, Open, Closed and Mixed Networks of Queues with Different Classes of Customers, *JACM*, Vol. 22, no. 2, pp. 248–260.

Baynat, B., and Y. Dallery, 1993, A Unified View of Product Form Approximation Techniques for General Closed Queueing Networks, to appear in *Performance Evaluation*, Vol. 18, pp. 205–225.

Brandwajn, A., and Y. L. Jow, 1988, An Approximation Method for Tandem Queues with Blocking, *Operations Research*, Vol. 36, pp. 73–83.

Buzacott, J. A., 1967, Automatic Transfer Lines with Buffer Stocks. *Intl. J. of Production Research*, Vol. 5, pp. 183–200.

Buzacott, J. A., 1972, The Effect of Station Breakdowns and Random Processing Times On the Capacity of Flow Lines with In-Process Storage, *AIIE Transactions*, Vol. 4, pp. 308–312.

Buzacott, J. A., 1989, Generalized Kanban/MRP Systems, Technical Report, Dept. of Management Sciences, University of Waterloo.

Buzacott, J. A., and J. G. Shanthikumar, 1993, *Stochastic Models of Manufacturing Systems*, Prentice Hall, Englewood Cliffs, N.J.

Cox, D. R., 1955, A Use of Complex Probabilities in the Theory of Stochastic Processes, *Proc. Cambridge Phil. Society*, Vol. 51, pp. 313–319.

Dallery, Y., and S. B. Gershwin, 1992, Manufacturing Flow Line Systems: A Review of Models and Analytical Results, *Queueing Systems*, Vol. 12, no. 1–2, pp. 3–94.

de Koster, M. B. M., 1988, Approximation of Flow Lines with Integrally Controlled Buffers, *IIE Transactions*, Vol. 20, pp. 374–381.

Deleersnyder, J. L., T. J. Hodgson, H. Muller, and P. O'Grady, 1989, Kanban Controlled Pull Systems: An Analytical Approach, *Management Science*, Vol. 35, pp. 1079–1091.

Di Mascolo, M., R. David, and Y. Dallery, 1991, Modeling and Analysis of Assembly Systems with Unreliable Machines and Finite Buffers, *IIE Transactions*, Vol. 23, no. 4, pp. 315–331.

Di Mascolo, M., Y. Frein, B. Baynat, and Y. Dallery, 1992, Queueing Network Modeling and Analysis of Generalized Kanban Systems, Laboratoire d'Automatique de Grenoble, Saint Martin d'Heres, France.

Dogan, E., and T. Altiok, 1995, Planning Repair Effort in Transfer Lines, IE Tech. Rept. No. 95-111, Rutgers University, N.J.

Doshi, B. T., F. A. Van der Duyn Schouten, and A. J. J. Talman, 1978, A Production-Inventory Model with a Mixture of Backorders and Lost Sales, *Management Science*, Vol. 24, pp. 1078–1086.

Elsayed, E. A. and T. O. Boucher, 1985, *Analysis and Control of Production Systems*, Prentice Hall, Englewood Cliffs, NJ.

Erlang, A. K., 1917, Solution of Some Problems in the Theory of Probabilities of Significance in Automatic Telephone Exchanges, *P.O. Electrical Eng. J.*, Vol. 10, pp. 189–197.

Gavish, B., and S. C. Graves, 1980, A One-Product Production/Inventory Problem Under Continuous Review Policy, *Operations Research*, Vol. 8, pp. 1228–1236.

Gavish, B., and S. C. Graves, 1981, Production/Inventory Systems with a Stochastic Production Rate Under a Continuous Review Policy, *Computers and Operations Research*, Vol. 8, pp. 169–183.

Gershwin, S. B., 1987, An Efficient Decomposition Method for the Approximate Evaluation of Tandem Queues with Finite Storage Space and Blocking. *Operations Research*, Vol. 35, pp. 291–305.

Gershwin, S. B., 1991, Assembly/Disassembly Systems: An Efficient Decomposition Algorithm for Tree-Structured Networks, *IIE Transactions*, Vol 23, no. 4, pp. 302–314.

Gershwin, S. B., 1994, *Manufacturing Systems Engineering*, Prentice Hall, Englewood Cliffs, N.J.

Gershwin, S., and I. C. Schick, 1983, Modeling and Analysis of Two- and Three-Stage Transfer Lines with Unreliable Machines and Finite Buffers. *Operations Research*, Vol. 31, pp. 354–380.

Glynn, P. W., 1989, Optimization of Stochastic Systems via Simulation, *Proc. Winter Simulation Conference*, pp. 90–105.

Glynn, P. W., 1990, Likelihood Ratio Gradient Estimation for Stochastic Systems, *Communications of ACM*, Vol. 33, pp. 75–84.

Glynn, P. W., P. L'Ecuyer, and M. Ades, 1991, Gradient Estimation for Ratios, *Proc. Winter Simulation Conference*, pp. 986–993.

Gross, D., and C. M. Harris, 1973, Continuous Review (s, S) Inventory Models with State-Dependent Leadtimes, *Management Science*, Vol. 19, pp. 567–574.

Gun, L., and A. Makowski, 1989, An Approximation Method for General Tandem Queueing Systems Subject to Blocking, *Queueing Networks with Blocking*, H.G. Perros and T. Altiok, eds., North-Holland, Amsterdam.

Hillier, F., and R. Boling, 1966, The Effect of Some Design Factors on the Efficiency of Production Lines with Variable Operation Times, *J. Industrial Engineering*, Vol. 7, pp. 651–658.

Ho, Y. C., M. A. Eyler, and T. T. Chien, 1979, A Gradient Technique for General Buffer Storage Design in a Production Line, *Intl. J. Production Research*, Vol. 17, pp. 557–580.

Jafari, M., and G. J. Shanthikumar, 1989, Determination of Optimal Buffer Storage Capacities and Optimal Allocation in Multi-Stage Automatic Transfer Lines, *IIE Transactions*, Vol. 21, pp. 130–135.

Johnson, L. A., and D. C. Montgomery, 1974, *Operations Research in Production Planning, Scheduling and Inventory Control*, John Wiley, New York.

Jun, K. P., and H. G. Perros, 1989, Approximate Analysis of Arbitrary Configurations of Queueing Networks With Blocking and Deadlock, *Queueing Networks With Blocking*, H. G. Perros, and T. Altiok, eds., North Holland, Amsterdam.

Karmarkar, U., and S. Kekre, 1987, Batching Policy in Kanban Systems, Grad. School of Management, University of Rochester, NY.

Kelly, F. P., 1979, *Reversibility and Stochastic Networks*, John Wiley, New York.

Kimura, O., and H. Terada, 1981, Design and Analysis of Pull System: A Method of Multi-Stage Control, *Intl. J. Production Control*, Vol. 19, no. 3, pp. 241–253.

Koster, M. B. M., and J. Wijngaard, 1989, On the Use of Continuous vs. Discrete Models in Production Lines with Buffers. *Queueing Networks with Blocking*, H.G. Perros and T. Altiok, eds., Raleigh, North Carolina.

Lee, H., and S. M. Pollock, 1990, Approximate Analysis of Open Acyclic Exponential Queueing Networks with Blocking, *Operations Research*, Vol. 38, pp. 1123–1134.

Lee, H. S., and M. M. Srinivasan, 1987, The Continuous Review (s, S) Policy for Production/Inventory Systems with Poisson Demands and Arbitrary Processing Times, Tech. Rep. No. 87-33, Dept. Ind. Oper. Eng., University of Michigan, Ann Arbor.

Lundrigan, R. 1986, What is the Thing Called OPT? *Production and Inventory Management*, Vol. 27, pp. 2–12.

Monden, Y., 1983, *Toyota Production System*, Industrial Engineering and Management Press, Atlanta, GA.

Neuts, M. F., 1981, *Matrix Geometric Solutions in Stochastic Models - An Algorithmic Approach*. Johns Hopkins University Press, Baltimore, MD.

Perros, H. G., 1994, *Queueing Networks with Blocking*, Oxford Press, New York.

Rubinstein, R. Y., 1986, The Score Function Approach for Sensitivity Analysis of Computer Simulation Models, *Math. Comput. Simul.*, Vol. 28, pp. 351–359.

Sevastyanov, B. A., 1962, Influence of Storage Bin Capacity on the Average Standstill of a Production Line, *Theory of Probability and Its Applications*, Vol. 7, pp. 429–438.

Smith, J. M., and S. Daskalaki, 1988, Buffer Space Allocation in Automated Assembly Lines, *Operations Research*, Vol. 36, pp. 343–359.

Spearman, M. L., D. L. Woodruff, and W. J. Hopp, 1990, CONWIP: A Pull Alternative to Kanban, *Intl. J. Production Research*, Vol. 28, pp. 879–894.

Soyster, A. L., J. W. Schmidt, and M.W. Rohrer, 1979, Allocation of Buffer Capacities for a Class of Fixed Cycle Production Lines, *AIIE Transactions*, Vol. 11, no. 2, pp. 140–146.

Sugimori, Y., K. Kusunoki, F. Cho, and S. Uchikawa, 1977, Toyota Production System and Kanban System-Materialization of Just-In-Time and Respect-for-Human System, *Intl. J. Production System*, Vol. 15, No. 6, pp. 553–564.

Yamashita, H. and T. Altiok, 1991, Buffer Capacity Allocation for a Desired Throughput in Production Lines, IE Tech. Rept. No. 91-112, Rutgers University, N.J.

Whitt, W., 1983, Queueing Network Analyzer, *Bell Technical Journal*, Vol. 62, pp. 2779–2815.

Wijngaard, J., 1979, The Effect of Interstage Buffer Storage on the Output of Two Unreliable Production Units in Series, with Different Production Rates, *AIIE Transactions*, Vol. 11, no. 1, pp. 42–47.

Zimmern, B., 1956, Etudes de la Propogation des Arrets Aleatoires Dans les Chaines de Production, *Revue de Statistique Appliquee*, Vol. 4, pp. 85–104.

Zipkin, P. H., 1986, Models for Design and Control of Stochastic, Multi-Item Batch Production Systems, *Operations Research*, Vol. 34, no. 1, pp. 91–104.

2

Tools of Probability

This chapter provides a refresher on probability and random processes that is essential for a healthy coverage of the remaining chapters of this book. The review will be done to the extent of our interest in the context of manufacturing systems.

2.1 Random Variables and Their Distributions

Usually, a random variable is defined as a function to which every experimental outcome of interest assigns a value. All of the possible values a random variable can take form its **state space**. For instance, if a discrete random variable X represents the status of a machine, the possible set of values X can take may be $S \in$ [Idle, Busy, Down, Blocked], and possibly many more. Clearly, depending on the values they can take, random variables are classified as **discrete** or **continuous**.

If X is a continuous random variable and x is its particular realization, the function $\Pr(X \leq x)$ is known as the **cumulative distribution function** of X and is denoted by $F_X(x)$. Some properties of $F_X(x)$ are as follows:

(i) $0 \leq F_X(x) \leq 1$, $\quad -\infty < x < \infty$;
(ii) $F_X(x_1) \leq F_X(x_2)$, $\quad x_1 \leq x_2$;
(iii) $\lim_{x \to \infty} F_X(x) \to 1$ and $\lim_{x \to -\infty} F_X(x) \to 0$.

Probabilities can be found from $F_X(x)$ by $\Pr(a \leq X \leq b) = F_X(b) - F_X(a)$. The function $f_X(x)$ showing the mass distribution of X is known to be the **density function** of X. It is given by $f_X(x) = dF_X(x)/dx$. Some properties of $f_X(x)$ are

(i) $f_X(x) \geq 0$ for all x,
(ii) $\int_{-\infty}^{\infty} f_X(x)\,dx = 1$,
(iii) $f_X(x)$ is piecewise continuous.

For a discrete random variable X, the probability of its particular realization x is $P(x) = \Pr(X = x)$, known as the **probability distribution function** of X. Clearly, the following hold:

(i) $0 \leq P(x) \leq 1$ for each x in S of X;
(ii) $\sum_{x_i} P(x_i) = 1$,

then, $F_X(x) = \sum_{x_i \leq x} P(x_i)$

2.2 Joint Distributions

Let X and Y be two continuous random variables on a sample space S. Then, $F_{XY}(x, y)$ is known to be the **joint cumulative distribution function** of X and Y, and $f_{XY}(x, y)$ is their **joint density function**. If X and Y are independent, then $F_{XY}(x, y) = F_X(x)F_Y(y)$, and the same is true for their density functions as well. The properties of joint distributions can be summarized as follows:

(i) $0 \leq F_{XY}(x, y) \leq 1$ for $-\infty < x, y < \infty$;
(ii) $F_{XY}(x_1, y_1) \leq F_{XY}(x_2, y_2)$ if $x_1 \leq x_2$ and $y_1 \leq y_2$;
(iii) $F_{XY}(x, \infty) = F_X(x)$ and $F_{XY}(\infty, y) = F_Y(y)$, known as the marginal distributions of X and Y, respectively.

$F_{XY}(x, y)$ can be expressed as

$$F_{XY}(x, y) = \int_{-\infty}^{x} \int_{-\infty}^{y} f_{XY}(t_1, t_2) \, dt_1 \, dt_2.$$

Conversely, $f_{XY}(x, y) = d^2 F_{XY}(x, y)/dx \, dy$. Clearly $f_{XY}(x, y) \geq 0$ for all x and y, and $\int_{-\infty}^{\infty} \int_{-\infty}^{\infty} f_{XY}(x, y) \, dx \, dy = 1$. The marginal density functions are given by

$$f_X(x) = \int_{-\infty}^{\infty} f_{XY}(x, y) \, dy \quad \text{and} \quad f_Y(y) = \int_{-\infty}^{\infty} f_{XY}(x, y) \, dx.$$

The above properties and definitions can also be written for discrete random variables.

2.3 Conditional Distributions

In cases where X and Y are not independent, their joint distribution is obtained with the help of conditional distributions. The distribution of X conditioned on a particular value of Y is $F_{X|Y}(x \mid y) = P(X < x \mid Y = y)$. The conditional and unconditional distributions of X and Y have the following relationship:

$$F_{X|Y}(x \mid y) = \frac{F_{XY}(x, y)}{f_Y(y)}, \qquad (2.1)$$

that is, the conditional distribution function of X given $Y = y$. From the above relationship, we can write $f_{XY}(x, y) = f_{X|Y}(x \mid y) f_Y(y)$, which promotes the idea of first conditioning and then unconditioning to find the unconditional joint density functions. Let us study the following example.

EXAMPLE 2.1. Consider a work station where jobs to be processed arrive randomly. Let us assume that the processing time of a job is random, denoted by X, and has a density function $f_X(x) = \mu e^{-\mu x}$, $x \geq 0$ (exponential distribution). Let the number of jobs that arrive in an interval $[0,t]$ be a random variable denoted by N_t, and assume that N_t has the following distribution function:

$$\Pr(N_t = n) = \frac{(\lambda t)^n e^{-\lambda t}}{n!}, \quad n = 0, 1, 2 \ldots \quad \text{(Poisson distribution)}.$$

Let us find the distribution of the number of new job arrivals during the processing time of a job. To demonstrate the use of conditioning and unconditioning, let us first assume that the processing time is fixed at $X = x$, that is, we condition on a particular value of X. Then, the number of job arrivals during the fixed interval $(0,x)$ is given by

$$\Pr(N_X = n \mid X = x) = \frac{(\lambda x)^n e^{-\lambda x}}{n!}, \quad n = 0, 1, 2 \ldots.$$

However, we know that X is in fact a random variable and has a density function $f_X(x)$. Then, the unconditional distribution function of the number of job arrivals during a processing time is obtained by simply removing the above condition. That is,

$\Pr(N = n$ during a processing time$)$

$$= \int_0^\infty \Pr(N = n \text{ in } [0, x]) f_X(x) \, dx$$

$$= \frac{\mu \lambda^n}{n!} \int_0^\infty x^n e^{-(\lambda + \mu)x} \, dx = \left(\frac{\mu}{\mu + \lambda}\right) \left(\frac{\lambda}{\mu + \lambda}\right)^n, \quad n = 0, 1, 2, \ldots,$$

where $\int_0^\infty x^n e^{-ax} \, dx = n/a^{n+1}$ is utilized. The above work station is known as the M/M/1 queue in queueing theory, and the number of arrivals during a service time in an M/M/1 queue is known to be geometrically distributed, as we have shown above.

2.4 Functions of Random Variables

Quite often a stochastic measure is not a single random variable but a group of random variables or a function of random variables. For example, let X be a function of Y, that is $X = G(Y)$. Then, $F_X(x)$ can be expressed as

$$F_X(x) = \int_{G^{-1}(-\infty)}^{G^{-1}(x)} f_Y(y)\, dy \quad \text{and} \quad f_X(x) = f_Y(G^{-1}(x)) \frac{d}{dx}[G^{-1}(x)]. \tag{2.2}$$

For instance, let $X = aY + b$ with $a > 0$. Then,

$$\Pr(X \leq x) = \Pr(aY + b \leq x) = \Pr\left(Y \leq \frac{x-b}{a}\right)$$

$$= \Pr(Y \leq G^{-1}(x)) \quad \text{and} \quad f_X(x) = f_Y\left(\frac{x-b}{a}\right)\frac{1}{a}.$$

Let us look at some special cases such as the minimum, the maximum, and the sum of random variables. If $X = \min(Y, Z)$, it is not difficult to write $\Pr(X \geq x) = \Pr(Y \geq x, Z \geq x)$. If Y and Z are independent, then

$$\Pr(X \leq x) = 1 - \Pr(Y \geq x)\Pr(Z \geq x).$$

If $X = \max(Y, Z)$, then $\Pr(X \leq x) = \Pr(Y \leq x, Z \leq x)$, and if Y and Z are independent, then $\Pr(X \leq x) = \Pr(Y \leq x)\Pr(Z \leq x)$.

Finally, let $X = Y + Z$, that is, the convolution of Y and Z. We are interested in $\Pr(X \leq x)$ assuming that Y and Z are independent. Let us first condition on Z, that is,

$$\Pr(X \leq x \mid Z = z) = \Pr(Y + Z \leq x \mid Z = z) = \Pr(Y \leq x - z \mid Z = z).$$

After removing the condition, we have

$$F_X(x) = \int_{-\infty}^{\infty} \Pr(Y \leq x - z \mid Z = z) f_Z(z)\, dz. \tag{2.3}$$

In terms of density functions, we can write

$$F_X(x) = \int_{-\infty}^{\infty} \int_{-\infty}^{x-z} f_Y(y) f_Z(z)\, dy\, dz.$$

In case Y and Z are dependent, we have to work with their joint density function.

2.5 Moments of Random Variables

Moments are also helpful in understanding the behavior of random variables. The kth moment around the origin (contrary to $E[(X - b)^k]$ being the kth moment around b) of a random variable X is abbreviated by $E[X^k]$ which is the expectation of a function of the random variable X and is given by

$$E[X^k] = \int_{-\infty}^{\infty} x^k f_X(x)\, dx$$

if X is continuous. Clearly, the first moment gives us the mean of X and the second moment helps us to find the variance of X, that is $V[X] = E[X^2] - E^2[X]$. The mean and variance of a random variable X give us a convenient measure of variability known as the squared coefficient of variation, $Cv_X^2 = V[X]/E^2[X]$. It can be seen that $Cv_X^2 \in [0, \infty)$. Cv_X^2 is close to zero (low variability) if X assumes values close to each other.

The following important relationship exists among the moments of a random variable. Let m_k be the kth moment of a random variable of interest; then

$$m_n^2 \leq m_k m_{2n-k}, \quad k = 1, \ldots, n, \quad n = 1, 2 \ldots, \tag{2.4}$$

which is a special case of Holder's inequality (see Feller (1971), Vol. 2).

Usually, one has to deal with several random variables. Let X and Y be two random variables and S be their sum, then

$$E[S] = E[X] + E[Y],$$
$$V[S] = V[X] + V[Y] + 2\,Cov[X, Y],$$
$$Cov[X, Y] = E[XY] - E[X]E[Y],$$

where $Cov[X, Y]$ is known as the **covariance** of X and Y. $Cov[X, Y] = 0$ if X and Y are independent.

A measure of dependence, which utilizes the covariance, is known as the **correlation coefficient**

$$S = \frac{Cov[X, Y]}{\sqrt{V[X]V[Y]}} \in [-1, 1].$$

Moments of a random variable can also be obtained using its moment-generating function as well as its Laplace-Stieltjes transform. The **moment-generating function** of a continuous random variable X is

$$E[e^{X\theta}] = \int_{-\infty}^{\infty} e^{x\theta} f_X(x)\, dx.$$

Then, the kth moment of X is given by

$$E[X^k] = d^k E[e^{X\theta}]/d\theta^k |_{\theta=0}. \tag{2.5}$$

Some properties of moment-generating functions are:

(i) Two distributions are equivalent if they have the same moment-generating function.
(ii) The moment-generating function of the sum of independent random variables is equal to the product of the individual moment-generating functions.
(iii) $E[e^{X\theta} \mid \theta = 0] = 1$, which is a convenient property to test the moment-generating functions.

The **Laplace-Stieltjes transform (LST)** of a density function is defined as

$$E[e^{-sX}] = \int_{-\infty}^{\infty} e^{-sx} f_X(x)\, dx, \qquad s \geq 0,$$

and

$$E[X^k] = (-1)^k d^k E[e^{-sX}]/ds^k |_{s=0}. \tag{2.6}$$

In some applications, s may be complex with the restriction of $Re(s) > 0$. The above-mentioned properties for the moment-generating functions are also valid for the LSTs.

Let us also briefly review the generating function (z-transform) of a discrete probability distribution. Let $G(z)$ be the generating function of the probability distribution $\{P_n, n = 0, 1, \ldots\}$ of the random variable N. Then, by definition,

$$G(z) = \sum_{n=0}^{\infty} P_n z^n, \qquad |z| \leq 1.$$

where $G(1) = 1$. Moments of N can be found by taking derivatives of $G(z)$, that is,

$$dG(z)/dz|_{z=1} = E[N], \qquad d^2 G(z)/dz^2|_{z=1} = E[N^2] - E[N]. \tag{2.7}$$

The above transform techniques are used not only to obtain the moments of a random variable but also to obtain its probability distribution or density function by inverting the transform functions using methods such as the partial-fraction expansions (see Feller (1968), for instance).

2.5.1 The Continuity Property of Transform Functions

Let $\{X_t\}$ be a sequence of random variables with a density function f_{X_t}. Also, let $\{T_{X_t}(s, t)\}$ be a transform function related to X_t. Then, the sequence of $\{X_t\}$ converges in distribution to X if, for $s > 0$,

$$\lim_{t \to \infty} \{T_{X_t}(s, t)\} \to T_X(s).$$

Also, f_{X_t} converges to f_X. If X is continuous, $T_X(s)$ is the LST, that is, $T_X(s) = E[e^{-sX}]$. In case X is discrete, $T_X(s)$ becomes the generating function, that is, $T_X(s) = G[e^{-s}]$. The key point of the continuity property is that the transform function $T_X(s)$ must be a converging function and has to be continuous at the origin, that is, $T_X(0) = 1$.

2.5.2 The Indicator Function Approach to Obtain Moments

Let us define a discrete random variable $I(A)$ by

$$I(A) = \begin{cases} 1 & \text{if event } A \text{ occurs,} \\ 0 & \text{otherwise,} \end{cases}$$

which is a function indicating the occurrence of event A. We can write $E[I(A)] = Pr(A)$.

Next, for a nonnegative random variable X, let us define $I(x)$ by

$$I(x) = \begin{cases} 1 & \text{if } x < X, \\ 0 & \text{else.} \end{cases}$$

Then,

$$X = \int_0^\infty I(x)\,dx.$$

Consequently, the expected value of X is given by

$$E[X] = E\left[\int_0^\infty I(x)\,dx\right] = \int_0^\infty E[I(x)]\,dx = \int_0^\infty \Pr(X > x)\,dx, \quad (2.8)$$

which is an important identity for the expected value of a continuous, nonnegative random variable. The discrete version of the identity can be written as

$$E[N] = \sum_{k=1}^\infty \Pr(N \geq k).$$

The higher moments of X can also be obtained using the above identity. For the nth moment of X, we have

$$E[X^n] = \int_0^\infty \Pr(X^n > x)\,dx = \int_0^\infty \Pr(X > x^{1/n})\,dx.$$

22   2. Tools of Probability

Using the transformation of $x = y^n$ and $dx = ny^{n-1}\,dy$, we obtain

$$\mathrm{E}[X^n] = \int_0^\infty ny^{n-1}\Pr(X > y)\,dy = n\int_0^\infty y^{n-1}\Pr(X > y)\,dy. \tag{2.9}$$

We refer the reader to Wolff (1989) for further applications of the indicator functions.

2.6 Special Distributions

In this section, we will summarize some of the frequently used distributions in the modeling of stochastic systems. We will refer to most of these distributions throughout this book. Let us first look at the distributions of continuous random variables.

2.6.1 Uniform Distribution

A continuous random variable X having realizations in $a \le x \le b$ is said to have a **uniform distribution** if it has the following density function:

$$f_X(x) = \frac{1}{b-a} \qquad \text{for } a \le x \le b. \tag{2.10}$$

Its cumulative distribution function is $F_X(x) = (x-a)/(b-a)$ for $a \le x \le b$. Its mean and variance are $\mathrm{E}[X] = (a+b)/2$ and $\mathrm{V}[X] = (b-a)^2/12$. Its squared coefficient of variation is $[(b-a)/(b+a)]^2/3$, and its moment-generating function is

$$M^*(\theta) = \frac{e^{b\theta} - e^{a\theta}}{\theta(b-a)}.$$

2.6.2 Exponential Distribution

A continuous nonnegative random variable X is said to be exponentially distributed if it has a density function of

$$f_X(x) = \lambda e^{-\lambda x}, \qquad x \ge 0. \tag{2.11}$$

Its cumulative distribution function is $F_X(x) = 1 - e^{-\lambda x}$, $x \ge 0$. Its mean and variance are $1/\lambda$ and $1/\lambda^2$, respectively, and its squared coefficient of variation is 1. Finally, its moment-generating function is $M^*(\theta) = \lambda/(\lambda - \theta)$.

The **exponential distribution** has a useful property known as the **memoryless property**, which has been extensively used in modeling of stochastic systems. The property says that if the time between two successive occurrences of an event is exponentially distributed, then at any point in time, the time until the next event occurs does not depend on how long it has been since the previous occurrence. That is, let t be the current time and $t + s$ be some future point in time, then

$$\Pr(X \geq t + s \mid X \geq t) = \frac{\Pr(X \geq t + s, X \geq t)}{\Pr(X \geq t)} = \frac{\Pr(X \geq t + s)}{\Pr(X \geq t)}$$

$$= \frac{e^{-\lambda(t+s)}}{e^{-\lambda t}} = e^{-\lambda s} = \Pr(X \geq s),$$

indicating that $\Pr(X \leq s) = 1 - e^{-\lambda s}$, which is the cumulative distribution function of the exponential distribution that is independent of t.

2.6.3 Gamma Distribution

The **gamma distribution** is somewhat a generalization of the exponential distribution. If X is a random variable having a gamma distribution, its density function is

$$f_X(x) = Cx^{b-1}e^{-\lambda x}, \qquad x \geq 0, \qquad (2.12)$$

where $C = \lambda^b/\Gamma(b)$ and $\Gamma(b) = \int_0^\infty x^{b-1}e^{-x}\,dx$ for $b > 0$. Its mean and variance are $E[X] = b/\lambda$ and $V[X] = b/\lambda^2$, respectively. Its squared coefficient of variation is $1/b$, and the moment-generating function is $M^*(\theta) = (\lambda/(\lambda - \theta))^b$.

2.6.4 Normal Distribution

A continuous random variable X is said to be normally distributed if its density function is

$$f_X(x) = \frac{1}{\sigma\sqrt{2\pi}} e^{-(x-\mu)^2/2\sigma^2}, \qquad -\infty < x < \infty, \qquad (2.13)$$

where μ and σ are its mean and standard deviation, respectively, with $\sigma > 0$. Its moment-generating function is

$$M^*(\theta) = e^{\mu\theta + \sigma^2\theta^2/2}.$$

A useful property of the **normal distribution** is given by the central limit theorem, which says the distribution of the sum of independent and identically distributed random variables approaches normal distribution as the number of these random variables approaches infinity. Clearly, the mean and the variance of the sum are the sum of the means and the sum of the variances of the individual random variables, respectively.

2.6.5 Bernoulli Distribution

A discrete random variable N having only two possible values is known to be a **Bernoulli** random variable with the following distribution function:

$$\Pr(N = n) = \begin{cases} p & \text{if } n = 1, \\ 1 - p & \text{if } n = 0. \end{cases} \qquad (2.14)$$

Its mean and variance are $E[N] = p$ and $V[N] = p(1 - p)$, respectively. Its squared coefficient of variation is $(1 - p)/p$, and its moment-generating function is $M^*(\theta) = 1 - p + pe^\theta$, indicating that all of the moments of a Bernoulli random variable are p.

2.6.6 Binomial Distribution

A binomial random variable can be explained as the sum of Bernoulli random variables. N is a **binomially distributed** random variable if $N = X_1 + X_2 + \ldots + X_n$, with X_i being a Bernoulli random variable, resulting in $0 \leq N \leq n$. Then, the distribution function of N is given by

$$\Pr(N = k) = \binom{n}{k} p^k (1 - p)^{n-k}, \qquad k = 0, 1, \ldots, n. \qquad (2.15)$$

Its mean and variance are $E[N] = np$ and $V[N] = np(1 - p)$, respectively, and its squared coefficient of variation is $(1 - p)/np$. Its moment-generating function is

$$M^*(\theta) = (1 - p + pe^\theta)^n,$$

which also implies that N is the sum of n independent Bernoulli random variables.

2.6.7 Geometric Distribution

A geometric random variable can also be viewed as a sequence of Bernoulli random variables. However, in this case the number of Bernoulli random variables is itself a random variable. A **geometric** random variable N is equal to the number of Bernoulli trials until the first event of $X = 1$ (or $X = 0$) is obtained. Its distribution function is

$$\Pr(N = k) = (1 - p)^{k-1} p, \qquad k = 1, 2, \ldots. \qquad (2.16)$$

The mean and the variance of N are $E[N] = 1/p$ and $V[N] = (1 - p)/p^2$, respectively, and the squared coefficient of variation of N is $1 - p$.

Let us look at an important property of the geometric distribution. Suppose we tried k times and did not find the terminating outcome yet. Let us write the probability that the terminating outcome will appear in the $(k+n)$th trial:

$$\Pr(N = k+n \mid N > k) = \frac{\Pr(N = k+n, N > k)}{\Pr(N > k)} = \frac{\Pr(N = k+n)}{\Pr(N > k)}$$

$$= \frac{(1-p)^{k+n-1}p}{(1-p)^k} = (1-p)^{n-1}p,$$

which is itself geometrically distributed with the same parameter p. Moreover, it does not depend on the number of unsuccessful trials k. Thus, the geometric distribution also has the memoryless property. The moment-generating function of the geometric distribution is

$$M^*(\theta) = \frac{pe^\theta}{1-(1-p)e^\theta}.$$

2.6.8 Poisson Distribution

A discrete random variable N is said to have a **Poisson distribution** if its distribution function is

$$\Pr(N = k) = \frac{\lambda^k e^{-\lambda}}{k!}, \qquad k = 0, 1, 2, \ldots. \qquad (2.17)$$

Its mean and variance are $E[N] = V[N] = \lambda$, and consequently its squared coefficient of variation is $1/\lambda$. Its moment-generating function is

$$M^*(\theta) = e^{-\lambda(1-e^\theta)}.$$

One can easily show that if the event occurrences per period have a Poisson distribution, then the time between successive occurrences is exponentially distributed with the same parameter. If X is the time between two successive event occurrences, then the probability of no event occurrence in the interval $[0, t]$ is

$$\Pr(X > t) = \Pr(N = 0 \text{ in } [0, t]) = e^{-\lambda t}.$$

The complementary probability of having at least one occurrence in $[0, t]$ is given by

$$\Pr(X \le t) = 1 - e^{-\lambda t},$$

which is the exponential distribution with mean $1/\lambda$.

2.7 Stochastic Processes

A **stochastic process** can be visualized as a function of a random variable or a set of random variables. Once evaluated at a particular point, the outcome will be the corresponding realization of the underlying random variable. For instance, let X_t be a stochastic process with t representing time. Then, X_t is a collection of random variables each resulting from the evaluation of X_t at different values of t. The parameter set consisting of all possible values for t and the state space consisting of all possible values for X_t for a given t describe the stochastic process. Both domains may take discrete or continuous values.

To completely characterize a stochastic process, one needs the joint distribution function

$$F(X_1, \ldots, X_t) = \Pr(X_1 \leq x_1, \ldots, X_t \leq x_t),$$

for every x_t and t, that characterizes the history of the process. If $X_1, \ldots, X_t$ are independent, then

$$\Pr(X_1 \leq x_1, \ldots, X_t \leq x_t) = \Pr(X_1 \leq x_1) \cdots \Pr(X_t \leq x_t).$$

2.7.1 Markov Chains

The most widely used stochastic processes are the ones possessing the Markovian property. For a discrete-state (discrete-time) stochastic process $\{X_t, t \geq 0\}$,

$$\Pr(X_{t_{n+1}} = i_{n+1} \mid X_{t_n} = i_n, \ldots, X_{t_1} = i_1) = \Pr(X_{t_{n+1}} = i_{n+1} \mid X_{t_n} = i_n),$$

implying that the process moves from one state to another or remains in the current state at every discrete point in time. The next state the process moves to depends only on the current state. Thus, the future of the process only depends on the current state and is not affected by its history. A discrete-state stochastic process possessing the above property is called a Markov process or more often is termed a **Markov chain**. Let $P_{ij} = \Pr(X_{n+1} = j \mid X_n = i)$ be the one-step time-homogeneous transition probability of going from state i to state j. Also, let P be the matrix of one-step transition probabilities. Then, the n-step transition probabilities are obtained by P^n, which has the probabilities of going from state i to state j for any i and j in n steps or in time units. State j is said to be accessible from state i if $P_{ij}^n > 0$ for some $n > 0$. Two states that communicate with each other belong to the same class. There may be several classes in a Markov chain. If all the states communicate, that is, there exists only one class, the Markov chain is said to be irreducible. State i is recurrent if the process will reenter state i in n steps (for any n) with probability 1. Then, state i will be revisited over and over again. $\sum_{n=1}^{\infty} P_{ii}^n = \infty$ if state i is recurrent and $\sum_{n=1}^{\infty} P_{ii}^n < \infty$ if state i is transient. Clearly, recurrence is a class property. If the time until the process returns to state i is finite, then state i is said to be **positive recurrent**. If $P_{ii} = 1$, then state i

is called an absorbing state. There may be several absorbing states in a Markov chain, and eventually the process is absorbed by one of them. State i is **aperiodic** if $P_{ii}^n > 0$ for $n = 1, 2, 3, \ldots$. Positive-recurrent and aperiodic states are also called **ergodic**. After these definitions, we can state the following important theorem.

Theorem. *For an irreducible and ergodic Markov chain, $\lim_{n \to \infty} P_{ij}^n$ exists and is independent of i. Let $\Pi_j = \lim_{n \to \infty} P_{ij}^n$, then*

$$\Pi_j = \sum_{k=0}^{\infty} \Pi_k P_{kj}, \quad \text{with} \sum_{j=0}^{\infty} \Pi_j = 1. \tag{2.18}$$

Π_j is said to be the **stationary probability** of being in state j. The stationary distribution is independent of the initial state that the process starts at time 0. Hence, the stationary distribution of a Markov chain can be obtained by solving (2.18) that has a unique solution.

In continuous time, a Markov chain possesses a similar description: $\{X_t, t \geq 0\}$ is a continuous-time Markov chain if

$$\Pr(X_{t_{n+1}} = j \mid X_{t_1} = i_1, \ldots, X_{t_n} = i_n)$$
$$= \Pr(X_{t_{n+1}} = j \mid X_{t_n} = i_n) = P_{ij}(t_n, t_{n+1}).$$

Let us define the matrix $\mathbf{P}(t) = [P_{ij}(t, t + \Delta t)]$ that is known as the transition probability matrix. As Δt becomes smaller, the probability of having a transition to another state within Δt decreases, while the probability of remaining in the same state increases. That is,

$$\lim_{\Delta t \to 0} P_{ij}(t, t + \Delta t) \to 0,$$
$$\lim_{\Delta t \to 0} P_{ii}(t, t + \Delta t) \to 1.$$

As a result of the above arguments, we can introduce the following transition rates:

$$\lim_{\Delta t \to 0} \frac{P_{ij}(t, t + \Delta t) - 0}{\Delta t} = q_{ij}(t), \quad i \neq j,$$
$$\lim_{\Delta t \to 0} \frac{P_{ii}(t, t + \Delta t) - 1}{\Delta t} = q_{ii}(t), \tag{2.19}$$

which in turn defines the infinitesimal generator matrix $\mathbf{Q}(t)$ as

$$\mathbf{Q}(t) = \lim_{\Delta t \to 0} \frac{\mathbf{P}(t) - \mathbf{I}}{\Delta t}. \tag{2.20}$$

$\mathbf{Q}(t) = [q_{ij}(t)]$ is in essence the transition rate matrix of the Markov chain. Due to the fact that $\sum_j P_{ij}(t, s) = 1$, and, from the definitions of $q_{ij}(t)$, $\sum_j q_{ij}(t) = 0$

is established. This implies that $q_{ii}(t) = -\sum_{j \neq i} q_{ij}(t)$, that is, row sums of $\mathbf{Q}(t)$ are equal to zero.

It is also possible to write $\mathbf{P}(t)$ as

$$\mathbf{P}(t) = \mathbf{Q}(t)\Delta t + \mathbf{I}, \qquad (2.21)$$

which allows us to discretize the continuous-time Markov chain and set its time unit to Δt in case Δt is chosen properly. We will elaborate on this issue in Chapter 4.

Let us define $\Pi_j(t) = \Pr(X_t = j)$ as the probability of being in state j at time t. Then we can write

$$\Pi_i(t + \Delta t) = \Pi_i(t)(1 - \sum_{j \neq i} P_{ij}(t, t + \Delta t)) + \sum_{j \neq i} \Pi_j(t) P_{ji}(t, t + \Delta t),$$

or

$$\Pi_i(t + \Delta t) = \Pi_i(t)(1 - \sum_{j \neq i} q_{ij}(t)\Delta t) + \sum_{j \neq i} \Pi_j(t) q_{ji}(t)\Delta t + o(\Delta t),$$

which can be rewritten as

$$\frac{\Pi_i(t + \Delta t) - \Pi_i(t)}{\Delta t} = -\Pi_i(t) \sum_{j \neq i} q_{ij}(t) + \sum_{j \neq i} \Pi_j(t) q_{ji}(t) + o(\Delta t),$$

$$= \Pi_i(t) q_{ii}(t) + \sum_{j \neq i} q_{ji}(t) \Pi_j(t) + o(\Delta t),$$

$$= \sum_j q_{ji}(t) \Pi_j(t) + o(\Delta t).$$

For $\lim_{\Delta t \to 0}$, we have

$$\lim_{\Delta t \to 0} \frac{\Pi_i(t + \Delta t) - \Pi_i(t)}{\Delta t} = \frac{d\Pi_i(t)}{dt} = \sum_j q_{ji}(t) \Pi_j(t),$$

and in the matrix form,

$$\frac{d\underline{\Pi}(t)}{dt} = \underline{\Pi}(t)\mathbf{Q}(t).$$

which is a matrix differential equation known as the Chapman-Kolmogorov equation. For the time-homogeneous Markov chains, the above equation becomes

$$\frac{d\underline{\Pi}(t)}{dt} = \underline{\Pi}(t)\mathbf{Q}. \qquad (2.22)$$

In steady state, $\Pi_j = \lim_{t \to \infty} \Pi_j(t)$, and the above matrix differential equation becomes $\underline{\Pi}\mathbf{Q} = \underline{0}$ with $\sum_j \Pi_j = 1$. $\underline{\Pi}\mathbf{Q} = \underline{0}$ indicates that, in the long run,

2.7.2 Regenerative Processes

The stochastic process $\{X_t, t \geq 0\}$ is a regenerative process if there exists points in time where the process restarts itself. Let us assume that there exist a time t_1 at which the process continues as if it restarts at time 0. The existence of t_1 implies the existence of similar points in time such as $t_2, t_3, \ldots$ that are called **regeneration points**. The amount of time between any two consecutive regeneration points is called the **regeneration cycle**. That is, the process is going through a set of regenerative cycles. For a regenerative process $\{X_t\}$, let T_c be the random variable representing the cycle length and the indicator function $I_j(u)$ be

$$I_j(u) = \begin{cases} 1 & \text{if } X(u) = j, \\ 0 & \text{if } X(u) \neq j. \end{cases}$$

Then, the proportion of time $\{X_t\}$ spends in state j is equal to

$$\frac{\text{The expected total time spent in state } j}{\text{The expected cycle length}}$$

$$= \lim_{t \to \infty} \Pr\{X(t) = j\} = \Pr(X = j) = \frac{\mathrm{E}[\int_0^t I_j(u)\, du]}{\mathrm{E}[T_c]}. \tag{2.23}$$

For instance, consider the stochastic process with up and down states (i.e., a machine with two states) as shown in Figure 2.1. Assume that every time it starts from scratch, the process regenerates itself at the points of entry to the up state (or the points of entry to the down state). Then, the steady-state probability of being in state up is simply

$$\Pr(\text{up}) = \frac{E[\text{up time}]}{E[\text{up time}] + E[\text{down time}]}.$$

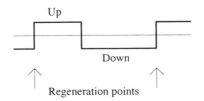

Figure 2.1. The up and down state forming a regeneration cycle.

In a Markov chain, the consecutive entries to any state k form a sequence of regeneration points. Then, the steady-state probability of being in state j is given by

$$\Pi_j = \frac{E[\text{the amount of time in state } j \text{ during a } k\text{–}k \text{ cycle}]}{E[\text{time to return to state } k]}.$$

2.7.3 The Residual Life of a Stochastic Process

Let us observe the stochastic process $\{X(t)\}$ on a random basis without affecting its behavior. The observation points are uniformly distributed over $(0, t)$. The probability of the state of the process at the observation point can be obtained by letting $t \to \infty$. The length of time from the point of observation until the next realization of the stochastic process is called its **residual life**, or at times referred to as its remaining life, denoted by X_r. Usually, one is interested in $\Pr[X_r \leq x]$ or $E[X_r]$. For any $x_r > 0$, let us define R as the residual time at any realization of the process. That is,

$$R = \begin{cases} X & \text{if } X \leq x_r, \\ x_r & \text{if } X > x_r, \end{cases}$$

implying that $R = \min(X, x_r)$. Then, the proportion of time $X_r \leq x_r$ is

$$\Pr(X_r \leq x_r) = \frac{E[R]}{E[X]} = \frac{\int_0^{x_r} \Pr(X > x)\, dx}{E[X]}.$$

The density function of X_r is given by

$$f_{X_r} = \frac{d \Pr(X_r \leq x_r)}{dx_r} = \frac{P(X > x_r)}{E[X]}, \qquad x_r > 0.$$

Using (2.9), we obtain the expected value of X_r,

$$E[X_r] = \int_0^\infty x_r f_{X_r}\, dx_r = \frac{\int_0^{x_r} x_r \Pr(X > x_r)\, dx_r}{E[X]} = \frac{E[X^2]}{2E[X]}, \qquad (2.24)$$

which is frequently used in the analysis of stochastic systems.

Below, we present some examples demonstrating how a continuous-time Markov chain is attained, how it is recursively solved, how generating functions are used, and how the regeneration argument is used in obtaining probabilities and certain performance measures.

EXAMPLE 2.2. In this example, we study the simplest form of the machine–repairman problem that is to be covered in Chapter 4 at greater length. Consider a pool of m machines that are constantly processing jobs and are subject to failures. A machine requests a repair immediately after it fails. There is a single repairman to carry out the repair activity. A failed machine is serviced immediately unless the repairman is busy. Otherwise, a queue is formed in front of the repair facility, as shown in Figure 2.2. A repaired machine is assumed to start processing immediately. The time to failure and the repair time are exponentially distributed with rates δ and γ, respectively. We want to compute the steady-state repairman utilization and the expected number of machines waiting to be repaired.

Let $\{N_t, t \geq 0\}$ be the stochastic process representing the number of down machines at time t. Also, let $n (0 \leq n \leq m)$ be a particular state of this stochastic process, indicating a machine being repaired and $n - 1$ machines waiting to be repaired. The idleness of the repairman is indicated by $n = 0$. Let $P_n(t)$ be the probability of being in state n at time t. If a machine is in operation at time t, the probability that it will fail before time $t + \Delta t$ is $\delta \Delta t$ plus terms which are negligible in the limit $\Delta t \to 0$. The probability that a machine fails in the next Δt time units given that n machines are down is $(m - n)\delta \Delta t$. Also, if a machine is being repaired at time t, the probability that it will start processing before $t + \Delta t$ is $\gamma \Delta t$. Notice that the state of the stochastic process changes when either a failure or a repair occurs. At time t, the time until the next event (either a failure or a repair) occurs is exponentially distributed with a mean of $m\delta^{-1}$ if $N_t = 0$, with a mean of γ^{-1} if $N_t = m$ and with a mean of $[(m - n)\delta + \gamma]^{-1}$ if $N_t = n, 1 \leq n < m$. Clearly, these arguments utilize the memoryless property of the exponential distribution (the first two cases), and the fact that the minimum of a set of exponentially distributed random variables is also exponentially distributed. Thus, the time until the next state change occurs is exponentially distributed, making the next state dependent only on the present state (not the past states due to the memoryless property). Voilà, $\{N_t\}$ becomes a continuous-time, discrete-state Markov chain.

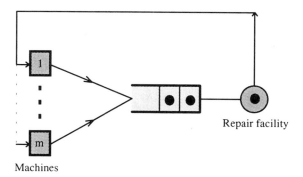

Figure 2.2. The machine–repairman problem.

The time-dependent flow-balance equations of $\{N_t\}$ (given by (2.22)) are

$$\frac{dP_0(t)}{dt} = -m\delta P_0(t) + \gamma P_1(t), \qquad n = 0,$$

$$\frac{dP_m(t)}{dt} = -\gamma P_m(t) + \delta P_{m-1}(t), \qquad n = m,$$

$$\frac{dP_n(t)}{dt} = -[(m-n)\delta + \gamma]P_n(t)$$
$$+ (m-n+1)\delta P_{n-1}(t) + \gamma P_{n+1}(t), \qquad 1 \le n < m.$$

In steady-state, letting[1] $P_n = \lim_{t\to\infty} P_n(t)$, the above set of equations reduces to the following set of difference equations (steady-state flow-balance equations):

$$m\delta P_0 = \gamma P_1,$$

$$\left.\begin{array}{l}[(m-n)\delta + \gamma]P_n = (m-n+1)\delta P_{n-1} + \gamma P_{n+1}, \qquad 1 \le n < m, \\ \gamma P_m = \delta P_{m-1}.\end{array}\right\} \quad 2.25$$

The transition diagram of $\{N_t\}$ governed by (2.25) is given in Figure 2.3.
Notice that one can directly write (2.25) using the argument of "rate in = rate out" for every state in Figure 2.3, and that is what we will do most of the time. From (2.25), it is not difficult to obtain the recursive relationship,

$$(m-n)\delta P_n = \gamma P_{n+1}, \qquad n < m. \tag{2.26}$$

Another way to obtain (2.26), which at times is referred to as the local balance equation, is to equate the rates of the flows from both ends into a "cut" such as the one shown in Figure 2.3. Successive substitutions in (2.25) for $n = m-1, \ldots, 0$

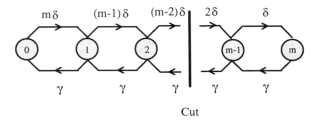

Figure 2.3. The transition diagram of $\{N_t\}$ in Example 2.2.

---

[1] It is quite common to use the notation P_n instead of Π_n as the steady-state probability of being in state n.

result in

$$P_{m-k} = \frac{1}{k!} \left(\frac{\gamma}{\delta}\right)^k P_m, \quad k = 0, 1, \ldots, m. \quad (2.27)$$

Finally, P_m is obtained through normalization, that is

$$P_m = \left[1 + \frac{1}{1!}\left(\frac{\gamma}{\delta}\right)^1 + \cdots + \frac{1}{m!}\left(\frac{\gamma}{\delta}\right)^m\right]^{-1}. \quad (2.28)$$

The repairman utilization is $1 - P_0$. The expected number of machines waiting to be repaired is given by

$$\overline{N}_q = \sum_{k=1}^{m} k P_k - (1 - P_0), \quad (2.29)$$

where the first set of terms represents the number of down machines and $1 - P_0$ is the expected number of machines being repaired.

An easier way of computing $\overline{N}_q$ would be to use the steady-state relationship of "the effective input rate = the effective output rate" at the repair facility. In the long run, the average number of machines entering the repair facility is equal to the average number of machines being repaired per unit time. That is,

$$\left[m - \overline{N}_q - (1 - P_0)\right]\delta = \gamma(1 - P_0),$$

resulting in

$$\overline{N}_q = m - \frac{\gamma + \delta}{\delta}(1 - P_0). \quad (2.30)$$

The steady-state flow-balance equations will be heavily utilized throughout this book. For a detailed analysis of Markov chains, we refer the reader to Bhat (1984), Çinlar (1975), and Wolff (1989), among many other books in the fields of applied probability and stochastic processes.

EXAMPLE 2.3. Consider a single processor that processes jobs on a first-come, first-serve (FCFS) basis. Jobs arrive according to a Poisson process with rate λ. The processing time is exponentially distributed with mean μ^{-1}. The processor experiences failures on a random basis while processing jobs as well as while being idle. The time until a failure occurs is exponentially distributed with mean δ^{-1}. A failure causes the transfer of all the jobs waiting or being processed in the system to another processor. For practical purposes, we assume that all the jobs in the system are lost upon a failure and the processor starts going through a repair process. The repair time is exponentially distributed with mean γ^{-1}. The arrival process is turned off during the repair time. We would like to obtain measures such as the steady-state mean number of jobs in the system, the actual server utilization, and the probability that an arriving job is processed in this station, among possible others.

Although one may develop direct methods to compute the above measures, to demonstrate some of the techniques we have introduced in this chapter, we will first focus on the steady-state probability distribution of the number of jobs in the system. Let us consider the stochastic process $\{N_t, t \geq 0\}$ representing the number of jobs as well as the status of the processor. $\{N_t\}$ draws realizations from $\{d, 0, 1, 2, \ldots\}$, where d simply represents the down state. $N_t = 0, 1, 2, \ldots$ represents the number of jobs in the system and the processor's status. Notice that every state has an access to state d, which in turn has a direct access to state 0. $\{N_t\}$ forms a continuous-time, discrete-state Markov chain that possesses a unique steady-state probability distribution. Let P_n, $n = d, 0, 1, \ldots,$, be the steady-state distribution of $\{N_t\}$. The transition diagram of the above Markov chain is shown in Figure 2.4.

We can directly write the steady-state flow-balance equations as follows:

$$\gamma P_d = \delta \sum_{n=0}^{\infty} P_n, \tag{2.31}$$

$$(\lambda + \delta) P_0 = \gamma P_d + \mu P_1, \tag{2.32}$$

$$(\lambda + \delta + \mu) P_n = \lambda P_{n-1} + \mu P_{n+1}, \quad n > 0. \tag{2.33}$$

Equation (2.31) can be rewritten as

$$\gamma P_d = \delta(1 - P_d),$$

from which we obtain $P_d = \delta/(\delta + \gamma)$. P_d can also be obtained intuitively by simply observing the up and down cycles that the system is going through. From (2.32) we can write

$$P_1 = \frac{[(\lambda + \delta) P_0 - \gamma \delta/(\delta + \gamma)]}{\mu}, \tag{2.34}$$

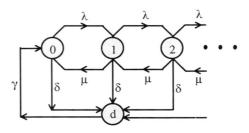

Figure 2.4. Transition diagram of $\{N_t\}$ in Example 2.3.

and from (2.33) we have

$$P_n = \frac{[(\lambda + \delta + \mu)P_{n-1} - \lambda P_{n-2}]}{\mu}, \qquad n = 2, 3, \ldots. \tag{2.35}$$

Notice that $\{N_t\}$ is always stable, since state 0 can be visited from every state, indicating that there will always be a nonzero P_0 no matter what the values of the system parameters are. This argument also indicates that if we know P_0, it is possible to compute P_1, P_2, and so forth. Unfortunately, $\sum_{n=0}^{\infty} P_n + P_d = 1$ is not helping us to find P_0, since expressing $P_n, n = 2, 3, \ldots$, in terms of P_0 is not feasible (for now), especially for large n.

Let us divert our attention to obtaining P_0 using the properties of the generating function of P_n. Let us define

$$G(z) = \sum_{n=1}^{\infty} P_n z^n + P_0 + P_d = \overline{G}(z) + P_0 + P_d, \tag{2.36}$$

and use (2.33) to find an explicit expression for it. $\overline{G}(z)$ can be obtained from (2.33) as follows:

$$(\lambda + \delta + \mu) \sum_{n=1}^{\infty} P_n z^n = \lambda \sum_{n=1}^{\infty} P_{n-1} z^n + \mu \sum_{n=1}^{\infty} P_{n+1} z^n,$$

which can be rewritten as

$$(\lambda + \delta + \mu)\overline{G}(z) = \lambda \sum_{k=0}^{\infty} P_k z^{k+1} + \mu \sum_{j=2}^{\infty} P_j z^{j-1}$$

$$= \lambda z [\overline{G}(z) + P_0] + \frac{\mu}{z}[\overline{G}(z) - P_1 z].$$

Hence,

$$\overline{G}(z) = \frac{\lambda z P_0 - \mu P_1}{(\lambda + \delta + \mu) - \lambda z - (\mu/z)}.$$

Since P_1 is expressed as a function of P_0 and P_d using (2.32), we arrive at

$$\overline{G}(z) = \frac{P_0(\lambda z - \lambda - \delta) + \gamma P_d}{\lambda + \delta + \mu - \lambda z - (\mu/z)}. \tag{2.37}$$

The usual way to obtain an expression for P_0 is to use $G(1) = 1$. Unfortunately, P_0 vanishes at $z = 1$ in our case. For this reason, we will use the continuity property of generating functions, explained in Section 2.5.1. The property states that the generating function (the sum of an infinite number of terms) of a converging sequence always converges. In our case, the sequence is a probability distribution and it converges. Therefore, $G(z)$ converges and exists. This is guaranteed when the numerator has exactly the same number of zeroes (roots) as the denominator, and moreover the numerator has the same zeroes as the denominator. Clearly, the denominator of $G(z)$ is the denominator of (2.37) denoted by $d(z)$. The uniqueness of P_0 already suggests that there exists a single root of $d(z)$, say z_0, at which the numerator of $G(z)$ will also have a zero. We have

$$d(z) = (\lambda + \delta + \mu) - \lambda z - \frac{\mu}{z}. \tag{2.38}$$

The zero of (2.38) is at

$$z_0 = \frac{(\lambda + \mu + \delta) - \sqrt{(\lambda + \mu + \delta)^2 - 4\mu\lambda}}{2\lambda} < 1. \tag{2.39}$$

Then, P_0 can be obtained by using the fact that the numerator of $G(z)$ is also zero at z_0, that is,

$$P_0(\lambda z_0 - \lambda - \delta) + \gamma P_d = 0. \tag{2.40}$$

Once P_0 is found, P_n, $n = 1, 2, \ldots$, can be found from (2.34) and (2.35). We have come a long way to compute P_0. However, the approach is useful in showing the fine technicalities of the generating functions.

Now we have gone through the above analysis, let us focus on the following argument, which helps us to find P_0 in a rather simpler manner. Observe that (2.33) remains the same for $n \geq 1$, which suggests the following **multiplicative (or geometric) solution**:

$$P_n = q^{n-1} P_1, \qquad q < 1, \qquad n \geq 1. \tag{2.41}$$

Then, for any $n > 1$, (2.33) yields

$$\mu q^2 - (\lambda + \delta + \mu)q + \lambda = 0,$$

providing the following value for q:

$$q = \frac{(\lambda + \delta + \mu) - \sqrt{(\lambda + \delta + \mu)^2 - 4\lambda\mu}}{2\mu}.$$

Notice the similarity between the expressions for q and z_0. Finally (2.31) and (2.33) together with (2.41) yield the following for P_0 and P_1:

$$P_0 = \frac{[\lambda + \delta + \mu(1-q)]P_u}{\lambda/(1-q) + \lambda + \delta + (1-q)\mu}$$

and

$$P_1 = \frac{\lambda P_u}{\lambda/(1-q) + \lambda + \delta + (1-q)\mu},$$

where $P_u = 1 - P_d$.

In this type of an approach, one has be concerned with the legitimacy of P_0 and P_1. In our case, it can be easily observed that both P_0 and P_1 are legitimate probabilities.

Now let us continue with the performance measures of our interest. The processor utilization is the proportion of time that the processor is actually busy, that is $1 - P_0 - P_d$. The long-run average number of jobs in the system can be found from the derivative of $G(z)$ and is

$$G(1) = \overline{N} = \frac{P_0 \mu - (1 - P_d)(\mu + \lambda)}{\delta}. \tag{2.42}$$

The important relationship of the effective input rate is equal to the effective output rate, for the processor also provides the same equation in the form of

$$\lambda(1 - P_d) - \delta \overline{N} = \mu(1 - P_o - P_d).$$

The rationale behind the above relation is that jobs arrive during up times and depart while the processor is actually busy. Failures arrive (by definition, during up times) according to a Poisson distribution and clear the system. Since Poisson arrivals see time averages (PASTA property[1]), each failure is expected to clear as many jobs as the average number of jobs in the system at any time. Consequently, the average number of jobs lost per unit time due to failures is given by $\delta \overline{N}$.

---

[1] The Poisson Arrivals See Time Averages (PASTA) argument states that the long-run fraction of customer arrivals who see the system at a particular state is the same as the long-run fraction of time the system is in that state. The proof is given by Wolff (1982). It can be verified by using the following simple relationship (Wolff (1982)) from regenerative processes. Consider the regenerative process $\{X_t\}$ with random variables T and N representing the length of a cycle and the number of arrivals during a cycle. Also, let B be a particular state of the process, with T_B being the amount of time the process spends in state B during a cycle, and N_B, the total number of arrivals in state B during a cycle. Then, Wolff shows that

$$E[N_B] = \lambda E[T_B],$$

which can also be written for the cycle itself that

$$E[N] = \lambda E[T],$$

resulting in

$$\frac{E[N_B]}{E[N]} = \frac{E[T_B]}{E[T]},$$

where the left-hand side is the probability that an arrival sees the system in state B, whereas the right-hand side is the arbitrary-time probability that the system is in state B.

38   2. Tools of Probability

The probability that an arriving job departs from the system before a failure occurs is also of interest. First of all, the probability that a job completes its processing given that it sees n jobs in the system at the time of its arrival is $\Pr(Y \geq S_{n+1})$, where Y represents the time to breakdown and S_{n+1} is the sum of the processing times of a total of $(n + 1)$ jobs, which includes the arriving job, the job in service, and $(n - 1)$ jobs waiting in the queue. That is, the probability that an arriving job completes processing depends on how many jobs it sees at the time of its arrival. Again, due to the PASTA property of the Poisson arrivals, we can write the following:

$$\Pr(\text{an arriving job completes processing}) = \sum_{n=0}^{\infty} P_n \Pr(Y \geq S_{n+1})$$

$$= P_0 \Pr(Y > S_1) + P_1 \sum_{n=1}^{\infty} q^{n-1} \left(\frac{\mu}{\mu+s}\right)^{n+1}$$

$$= \frac{P_0 \mu}{\mu + \delta} + \frac{P_1 \mu^2}{[\mu(1-q) + \delta](\mu + \delta)}.$$

As we will explain in the next section, S_{n+1} is an Erlang random variable that is the sum of $n + 1$ exponentially distributed random variables each with rate μ. The analysis of Example 2.3 for more general assumptions can be found in Altiok (1989).

EXAMPLE 2.4. [Wolff (1989)]

Consider an inventory problem where demand arrives according to a Poisson distribution with rate λ. Every time the inventory level drops to zero, a batch of Q units is ordered. The time until the batch arrives is known as the lead time and is exponentially distributed with mean $1/\mu$. The customer demand that arrives during the lead time is lost. Let us compute the average inventory-on-hand, the loss probability, and some related measures using the regeneration argument.

Let $N(t)$ represent the inventory level at time t. As shown in Figure 2.5, $N(t)$ regenerates itself every time it reaches Q, since demand arrivals are independent of each other. There have to be Q units of demand arrivals for the inventory level to drop to zero. Similarly, the points where $N(t)$ drops to zero also constitute a set of regeneration points. Let T be the cycle time, then

$$E[T] = \frac{Q}{\lambda} + \frac{1}{\mu} = \frac{Q\mu + \lambda}{\lambda \mu}.$$

Clearly, the average amount of time $N(t)$ spends in levels $1, \ldots, Q$ is the mean demand interarrival time, and therefore

$$\lim_{t \to \infty} \Pr(N(t) = 1) = \cdots = \Pr(N(t) = Q) = \frac{1}{\lambda E[T]} = \frac{\mu}{Q\mu + \lambda}.$$

2.7. Stochastic Processes

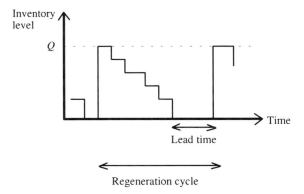

Figure 2.5. The regenerative behavior of $N(t)$.

Also, the average amount of time $N(t)$ remains in level 0 is $1/\mu$, resulting in

$$\lim_{t \to \infty} \Pr(N(t) = 0) = \frac{\lambda}{Q\mu + \lambda},$$

which can also be verified by the sum of the probabilities being equal to 1. The average on-hand inventory is given by

$$\overline{N} = \frac{Q(Q+1)\mu}{2(Q\mu + \lambda)}.$$

The loss probability, that is, the probability of not satisfying the demand upon customer arrival, is simply $\Pr(N = 0)$, due to the PASTA property of Poisson arrivals. Notice that the above results are also valid for any lead time distribution. The distribution of the lost demand is geometrically distributed, which can be shown using similar arguments to the ones used in Example 2.1.

Let us consider a cost optimization in the above scenario. Assume that a fixed cost K is incurred every time an order of size Q is placed. Also, a holding cost h is charged to carry a unit of inventory from one period to another. Then, the total expected cost per unit time is given by

$$\mathrm{E}[TC] = \frac{K}{\mathrm{E}[T]} + h\overline{N}.$$

The cost-minimizing value of Q can be found by taking the derivative of $\mathrm{E}[TC]$. It is the nonnegative root of the quadratic equation

$$Q^2 + \frac{2\lambda}{\mu} Q + \frac{\lambda}{\mu} - \frac{2\lambda K}{h} = 0.$$

2.8 Phase-Type Distributions

Phase-type distributions have been extensively used in modeling stochastic systems such as production systems, maintenance systems, storage systems, and computer and communication systems, among many others. This is partially due to the fact that their structure gives rise to a Markovian state description and partially due to their potential for algorithmic analysis, which is usually carried out using matrix algebra. These distributions will be frequently used in modeling production systems throughout this book. In this section, we summarize their formalism, their use, and their properties to the extent of our interest within the present context.

A phase-type distribution can be considered as the distribution of the time until absorption in a Markov chain with a single absorbing state. Specifically, consider a Markov chain with states $\{1, 2, \ldots, k, k+1\}$, with $k+1$ being the absorbing state. Let the transition-rate matrix be

$$Q = \begin{bmatrix} \mathbf{T} & \mathbf{T}^o \\ \underline{0} & 0 \end{bmatrix},$$

where $\mathbf{T}$ is a $k \times k$ matrix with $\mathbf{T}_{ij} \geq 0$ for $i \neq j$, $\mathbf{T}_{ii} < 0$ for $i = 1, \ldots, k$, and $\mathbf{T}\underline{e} + \underline{\mathbf{T}}^o = \underline{0}$, where $\underline{e}$ is a column vector of ones. Let the initial probability vector of the Markov chain be $(\underline{\alpha}, 0)^T$ with $\underline{\alpha}^T \underline{e} = 1$. Then, the distribution of the random variable representing the time until absorption in the above Markov chain is said to be of phase-type with an $(\underline{\alpha}, \mathbf{T})$ representation. One may interpret $\mathbf{T}$ as the matrix of the rates of transition among the phases and $\underline{\mathbf{T}}^o$ as the vector of the rates of transition from the transient states $[1, \ldots, k]$ to the absorbing state $k+1$. Let us also define the order of a phase-type distribution as the number of transient states in its absorbing Markov chain.

The following probabilistic construction is given in [Neuts (1981), p. 48]. State $k+1$ is an instantaneous state upon which an entry restarts the process from state (or phase) i with probability α_i. This concept is convenient to model demand or job arrival processes and processing times in production systems. Indefinite repetitions of the above process result in a new Markov chain with states $\{1, \ldots, k\}$ and with an infinitesimal generator

$$\mathbf{Q}^* = \mathbf{T} + \tilde{\mathbf{T}}^o \mathbf{A}^o$$

where $\tilde{\mathbf{T}}^o$ is a $k \times k$ matrix with identical columns of $\underline{\mathbf{T}}^o$, and $\mathbf{A}^o$ is a diagonal matrix defined as $\mathbf{A}^o = \{a_{ii}^o = \alpha_i, a_{ij_{i \neq j}}^o = 0\}$. We are assuming that $\alpha_{k+1} = 0$

without loss of generality. The representation $(\underline{\alpha}, \mathbf{T})$ is said to be irreducible if and only if $\mathbf{Q}^*$ is an irreducible matrix. We will restrict our attention to irreducible representations throughout this book.

EXAMPLE 2.5. Consider the phase-type distribution shown in Figure 2.6. It has three phases, with the third phase being the absorbing phase (or state). Let us assume that it is the distribution of the job interarrival time at a work station. As soon as a job arrives, the next job to arrive starts in phase 1 with probability α_1 or in phase 2 with probability α_2. It may go from one phase to another, and it may do so more than once. It eventually arrives when it visits the third phase. That is, the time from the start until the third phase is visited represents the interarrival time.

The parameters of the above phase-type distribution are as follows:

$$\underline{\alpha}^{\mathbf{T}} = (\alpha_1, \alpha_2), \quad \mathbf{T} = \begin{bmatrix} -\mu_1 & a_1\mu_1 \\ a_2\mu_2 & -\mu_2 \end{bmatrix} \quad \text{and} \quad \underline{\mathbf{T}}^o = \begin{bmatrix} (1-a_1)\mu_1 \\ (1-a_2)\mu_2 \end{bmatrix}.$$

The absorbing phase is an instantaneous state, and the process immediately restarts from the beginning when it is visited. Thus, states 1 and 2 give rise to a two-state Markov chain with the following transition-rate matrix:

$$\mathbf{Q}^* = \begin{bmatrix} (1-a_1)\mu_1\alpha_1 - \mu_1 & (1-a_1)\mu_1\alpha_2 + a_1\mu_1 \\ (1-a_2)\mu_2\alpha_1 + a_2\mu_2 & (1-a_2)\mu_2\alpha_2 - \mu_2 \end{bmatrix}.$$

Basically, it is the state of the arriving job, which is either the first phase or the second at any point in time, that gives rise to a Markov chain.

Some operations on phase-type distributions lead again to phase-type distributions. Among many, we will mention two that will be quite useful in the following chapters:

(1) The convolution of two phase-type distributions is also a phase-type distribution (i.e., Erlang distribution).
(2) A finite mixture of phase-type distributions is also a phase-type distribution (i.e., hyperexponential distribution).

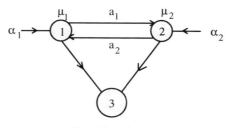

Absorbing phase

Figure 2.6. A two-phase phase-type distribution (Example 2.5).

Let X be a phase-type random variable with an $(\alpha, \mathbf{T})$ representation. Then, the density function of X is

$$f_X(x) = \underline{\alpha}^T e^{(\mathbf{T}x)} \underline{\mathbf{T}}^o, \qquad x \geq 0,$$

the LST of its density function is

$$F_X^*(s) = \underline{\alpha}^T (s\mathbf{I} - \mathbf{T})^{-1} \underline{\mathbf{T}}^o, \qquad \text{Re}(s) \geq 0;$$

and its central moments are given by

$$E[X^n] = (-1)^n n! (\underline{\alpha}^T \mathbf{T}^{-n} \underline{e}), \qquad n \geq 1.$$

The LST and the moment formulas are quite useful in dealing with phase-type distributions or in fitting them to data.

2.8.1 Mixtures of Generalized Erlang Distributions

Numerous phase-type distributions with special structures are used in modeling stochastic systems. For instance, mixtures of generalized Erlang distributions (MGEs), which are often called Coxian distributions, have been used in the analysis of manufacturing, computer, and communication systems. In principle, an MGE distribution is a mixture of the convolutions of exponential distributions, as shown in Figure 2.7. In case the processing time is modeled using an MGE distribution, one can visualize a job going through the phases of the distribution, spending a random amount of time in each phase, and leaving the system from one of the phases. The holding time in each phase is exponentially distributed with rate μ_i in phase i. The conditional probability that a job visits phase $i + 1$ given that phase i is completed is abbreviated by a_i. It is usually assumed that $a_0 = 1$. The MGE

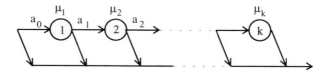

Figure 2.7. Graphical representation of the MGE distribution

2.8. Phase-Type Distributions

distribution has the following $(\underline{\alpha}, \mathbf{T})$ representation:

$$\mathbf{T} = \begin{bmatrix} -\mu_1 & \mu_1 a_1 & & & \\ & -\mu_2 & \mu_2 a_2 & & \\ & & -\mu_3 & \ddots & \\ & & & \ddots & \\ & & & -\mu_{k-1} & \mu_{k-1} a_{k-1} \\ & & & & -\mu_k \end{bmatrix}, \quad \underline{\alpha}^T = (1, 0, \ldots, 0), \tag{2.43}$$

and

$$\underline{\mathbf{T}}^o = [\mu_1(1 - a_1), \mu_2(1 - a_2), \ldots, \mu_k]^T.$$

The domain of the squared coefficient of variation of the MGE distribution is $(0, \infty)$. This makes it a proper candidate to represent data or approximate other distributions using the method of moments, for instance. We will look into this matter in Section 2.8.3.

In the frequently encountered case of $k = 2$, the density function of X is given by

$$f_X(x) = C_1 \mu_1 e^{-\mu_1 x} + C_2 \mu_2 e^{-\mu_2 x}, \quad x \geq 0, \tag{2.44}$$

where

$$C_1 = \frac{\mu_1(1 - a_1) - \mu_2}{\mu_1 - \mu_2} \quad \text{and} \quad C_2 = 1 - C_1, \quad \text{with } \mu_1 \neq \mu_2.$$

Its LST is

$$F_X^*(s) = \frac{s\mu_1(1 - a_1) + \mu_1 \mu_2}{s^2 + s(\mu_1 + \mu_2) + \mu_1 \mu_2}, \tag{2.45}$$

and its mean and Cv^2 are as follows:

$$E[X] = \frac{1}{\mu_1} + \frac{a_1}{\mu_2} \quad \text{and} \quad \text{Cv}_X^2 = 1 - \frac{2a_1 \mu_1 [\mu_2 - \mu_1(1 - a_1)]}{(\mu_2 + a_1 \mu_1)^2}. \tag{2.46}$$

Notice that $\text{Cv}_X^2 \in [0.5, \infty)$ for $k = 2$.

EXAMPLE 2.6. Consider the two-stage MGE distribution with $\mu_1 = 2$, $\mu_2 = 3$, and $a_1 = 0.2$. The $(\underline{\alpha}, \mathbf{T})$ representation is

$$\underline{\alpha}^T = (1, 0) \quad \text{and} \quad \mathbf{T} = \begin{bmatrix} -2 & 0.4 \\ 0 & -3 \end{bmatrix}.$$

Its density function is given by

$$f_X(x) = 2.8e^{-2x} - 1.2e^{-3x}, \qquad x \geq 0.$$

Its central moments are

$$E[X^n] = (-1^n)n!(1, 0) \begin{bmatrix} -2 & 0.4 \\ 0 & -3 \end{bmatrix}^{-n} \begin{bmatrix} 1 \\ 1 \end{bmatrix},$$

which can also be obtained from the LST of its density function. Its mean and Cv^2 are

$$E[X] = 0.5667 \quad \text{and} \quad Cv_X^2 = 0.9031.$$

An important special case of the MGE distribution is the Erlang distribution with the graphical representation shown in Figure 2.8. The Erlang distribution has the following $(\underline{\alpha}, \mathbf{T})$ representation:

$$\mathbf{T} = \begin{bmatrix} -\mu & \mu & & & \\ & -\mu & \mu & & \\ & & -\mu & \ddots & \\ & & & \ddots & \mu \\ & & & & -\mu \end{bmatrix}, \qquad \underline{\alpha}^{\mathbf{T}} = (1, 0, \ldots, 0), \qquad (2.47)$$

with $\underline{\mathbf{T}}^o = (0, \ldots, 0, \mu)^{\mathbf{T}}$.

The density function of the Erlang distribution is given by

$$f_X(x) = \frac{\mu(\mu x)^{k-1} e^{-\mu x}}{(k-1)!}, \qquad x \geq 0. \qquad (2.48)$$

Since a k-phase Erlang random variable is the sum of k exponential random variables, the LST of its density function is clearly the kth power of the LST of a single exponential random variable with rate μ. The mean and the Cv_X^2 are given by

$$E[X] = \frac{k}{\mu} \quad \text{and} \quad Cv_X^2 = \frac{1}{k}.$$

Clearly, as k increases and μ is scaled, the Erlang random variable approaches a deterministic random variable equal to $E(X)$.

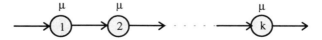

Figure 2.8. Graphical representation of the Erlang distribution.

2.8. Phase-Type Distributions

EXAMPLE 2.7. A three-stage Erlang distribution with phase rate $\mu = 2$ has a density function

$$f_X(x) = 4x^2 e^{-2x}, \qquad x \geq 0,$$

with mean $E[X] = 1.5$, and $Cv_X^2 = 1/3$.

A special case of the MGE distribution—and a relaxed version of the Erlang distribution—is the case where $a_i = 1$ for all i in Figure 2.7, and it is referred to as the **generalized Erlang distribution**. Its Cv^2 can take any value in $(0, 1]$. For $k = 2$, its density function and its LST can be obtained from (2.44) and (2.45).

Another widely used special case of the MGE distribution, and again a relaxed version of the Erlang distribution, is the case where $\mu_i = \mu$ for all i, $a_1 < 1$, and $a_i = 1$ for all $i \neq 1$. In literature, it is also referred to as the generalized Erlang distribution. Its graphical representation is given in Figure 2.9. Clearly, it is a mixture of a k-phase Erlang distribution and an exponential distribution. Its density function and the corresponding LST can be obtained using the ones of the Erlang and the exponential distributions. The domain of its Cv^2 is $(0,1]$.

EXAMPLE 2.8. Let us consider the above two generalized Erlang distributions, each having two phases. When $a_1 = 1$, for $\mu_1 = 2$ and $\mu_2 = 3$ the density function can be obtained using (2.44):

$$f_X(x) = 6e^{-2x} - 6e^{-3x}, \qquad x \geq 0,$$

with mean $E[X] = 5/6$ and $Cv_X^2 = 0.52$.

When $a_1 \neq 1$ and $\mu_1 = \mu_2 = \mu$, for $\mu = 2$ and $a_1 = 0.2$ the density function cannot be obtained from (2.44) since C_i approaches infinity when $\mu_1 = \mu_2$. We can derive it using an equivalent mixture of the exponential and the Erlang distributions, as shown in Figure 2.10. It is given by

$$f_X(x) = 1.6e^{-2x} + 0.8xe^{-2x}, \qquad x \geq 0.$$

Finally, let us mention the **generalized exponential distribution** obtained by letting $k = 1$ and $a_0 < 1$ in Figure 2.7, which implies that some mass is located at $X = 0$ with probability $F_X(0) = 1 - a_0$. Its density function is

$$f_X(x) = (1 - a_o)u_x(0) + a_o\mu_1 e^{-\mu_1 x}, \qquad x \geq 0, \qquad (2.49)$$

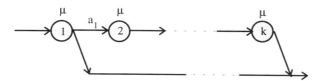

Figure 2.9. Graphical representation of the generalized Erlang distribution.

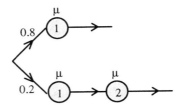

Figure 2.10. A mixture of the exponential and the Erlang distribution.

where $u_x(0) = 1$ for $x = 0$. Its LST and Cv^2 are given by

$$F_X^*(s) = \frac{a_o \mu_1}{\mu_1 + s} \quad \text{and} \quad \text{Cv}_X^2 = \frac{2 - a_o}{a_o} \in [1, \infty).$$

A k-phase **hyperexponential distribution** is obtained by letting $a_i = 0$ and $\alpha_i > 0$, for all i in Figure 2.7. Notice that a_0 and α_1 both represent the probability of entering phase 1. Practically speaking, a hyperexponential random variable is a proper mixture of exponential random variables. Its graphical representation is given in Figure 2.11. In the case of modeling the repair time, the hyperexponential distribution implies that, on the average, $100\alpha_i$ percent of the repairs use phase i as the re‚air time. After the choice of the branch is made, the time spent is exponentially distributed.

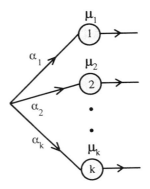

Figure 2.11. Graphical representation of the hyperexponential distribution.

2.8. Phase-Type Distributions

The $(\underline{\alpha}, \mathbf{T})$ representation of the hyperexponential distribution of order k is given by

$$\underline{\alpha}^T = (\alpha_1, \alpha_2, \ldots, \alpha_k), \qquad \mathbf{T} = \begin{bmatrix} -\mu_1 & & & \\ & -\mu_2 & & \\ & & \ddots & \\ & & & -\mu_k \end{bmatrix}.$$

The density function and the corresponding LST of a k-phase hyperexponential distribution are

$$f_X(x) = \sum_{i=1}^{k} \alpha_i \mu_i e^{-\mu_i x}, \quad x \geq 0, \quad \text{and} \quad F_X^*(s) = \sum_{i=1}^{k} \alpha_i \frac{\mu_i}{\mu_i + s}. \qquad (2.50)$$

The nth moment of the hyperexponential random variable is given by

$$E[X^n] = \sum_{j=1}^{k} \frac{n! \alpha_i}{\mu_i^n}, \qquad (2.51)$$

and its Cv^2 takes values in $[1, \infty)$.

EXAMPLE 2.9. Consider the two-phase hyperexponential distribution with $\mu_1 = 2$, $\mu_2 = 3$, and $\underline{\alpha}^T = (0.8, 0.2)$. The $\mathbf{T}$ matrix in its $(\underline{\alpha}, \mathbf{T})$ representation is given by

$$\mathbf{T} = \begin{bmatrix} -2 & 0 \\ 0 & -3 \end{bmatrix},$$

and its moments are given by

$$E[X^n] = (-1)^n n! (0.8, 0.2) \begin{bmatrix} -2 & 0 \\ 0 & -3 \end{bmatrix}^{-n} \begin{bmatrix} 1 \\ 1 \end{bmatrix}.$$

Its density function is given by

$$f_X(x) = 1.6 e^{-2x} + 0.6 e^{-3x}, \qquad x \geq 0,$$

with mean $E[X] = 0.4667$ and $Cv_X^2 = 1.0405$.

2.8.2 Relationship Between Phase Structure and Variability

In dealing with phase-type distributions, it is possible to completely ignore the pictorial representations and work only with the density functions. However, phase

48   2. Tools of Probability

structures are helpful in bringing intuitive explanations to the concept of variability and the equivalence relations among phase-type distributions. In fact, there exists a strong relationship between their phase structure and their squared coefficient of variation. On the other hand, a phase-type distribution may have infinitely many structural representations, each with its own order and an (α, **T**) representation. Then, its order will be the minimum of the orders of all of its representations. The one with the minimum order is called the minimal representation. Let us demonstrate the multiplicity of the representations in the simplest phase-type distribution, that is, the exponential distribution.

EXAMPLE 2.10. An exponential distribution with rate μ has a phase-type representation with an order of k provided that the transition rate from each phase to phase $k + 1$ (the absorbing phase) is μ. Without loss of generality, let us consider the phase-type distribution given in Figure 2.12, where the transition rate out of each phase to the absorbing phase is μ. Let us show that it is equivalent to an exponential distribution with rate μ.

The LST of the phase-type distribution in Figure 2.12 can be written as

$$F^*(s) = \alpha_1 F_1^*(s) + \alpha_2 F_2^*(s).$$

$F_1^*(s)$ can be written explicitly as

$$F_1^*(s) = \frac{\mu_1}{\mu_1 + s}\left(\frac{\mu}{\mu_1}\right)$$

$$+ \sum_{n=0}^{\infty}\left[\left(\frac{\mu_1}{\mu_1 + s}\right)^{n+1}\left(\frac{\mu_2}{\mu_2 + s}\right)^{n+1}\left(1 - \frac{\mu}{\mu_1}\right)^{n+1}\left(1 - \frac{\mu}{\mu_2}\right)^{n}\right]\frac{\mu}{\mu_2}$$

$$+ \sum_{n=0}^{\infty}\left[\left(\frac{\mu_1}{\mu_1 + s}\right)^{n+1}\left(\frac{\mu_2}{\mu_2 + s}\right)^{n+1}\left(1 - \frac{\mu}{\mu_1}\right)^{n+1}\left(1 - \frac{\mu}{\mu_2}\right)^{n+1}\right]\frac{\mu}{\mu_1}$$

$$= \left(\frac{\mu_1}{\mu_1 + s}\right)\frac{\mu}{\mu_1} + \frac{\mu}{\mu_2}G(s) + \frac{\mu_1}{\mu_1 + s}\left(1 - \frac{\mu}{\mu_2}\right)\frac{\mu}{\mu_1}G(s), \quad (2.52)$$

where

$$G(s) = \frac{\left(\frac{\mu_1}{\mu_1+s}\right)\left(\frac{\mu_2}{\mu_2+s}\right)\left(1 - \frac{\mu}{\mu_1}\right)}{1 - \left(\frac{\mu_1}{\mu_1+s}\right)\left(\frac{\mu_2}{\mu_2+s}\right)\left(1 - \frac{\mu}{\mu_1}\right)\left(1 - \frac{\mu}{\mu_2}\right)}.$$

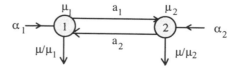

Figure 2.12. A phase-type representation of the exponential distribution.

2.8. Phase-Type Distributions

It is not difficult to see that $F_1^*(0) = 1$. Some algebra will simplify (2.52) and lead to

$$F_1^*(s) = \frac{\mu(s + \mu_1 + \mu_2 - \mu)}{s^2 + (\mu_1 + \mu_2)s + \mu(\mu_1 + \mu_2 - \mu)}, \quad (2.53)$$

which is nothing but $F_1^*(s) = \mu/(\mu + s)$.

On the other hand, $F_1^*(s)$ and $F_2^*(s)$ are the same except μ_1 and μ_2 are interchanged in $F_2^*(s)$, and therefore $F_2^*(s) = \mu/(\mu + s)$, resulting in $F^*(s) = \mu/(\mu + s)$ due to $\alpha_1 + \alpha_2 = 1$. That is, $F^*(s)$ is the LST of the density function of the exponential distribution with rate μ. Thus, we conclude that the exponential distribution has a k-phase phase-type representation with $k = 1, 2, \ldots$ and therefore has an order of 1. In the trivial case, the hyperexponential distribution with phase rates equal to μ is in fact an exponential distribution with rate μ.

The relationship between phase structures and variability is demonstrated in Figure 2.13. For a given order k ($k = 3$ in Figure 2.13) and a fixed mean, the least-variable phase structure is the one of the Erlang distribution. Any deviation from the Erlangian structure, no matter how slight it is, increases variability. For instance, the generalized Erlang distribution obtained through a minor modification of the

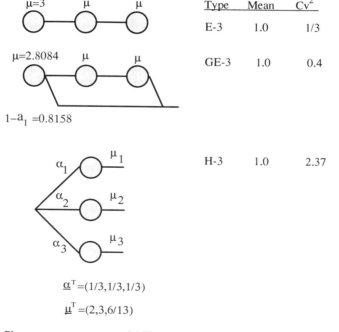

Figure 2.13. Phase structure versus variability.

Erlang distribution has a $Cv^2 = 0.4$. Contrary to the series structure, if the phases are arranged in parallel forming a hyperexponential distribution, the variability further increases.

The squared coefficient of variations of different phase-type distributions are put in perspective in Figure 2.14. While the Cv^2 of the Erlang and the exponential distributions can assume only specific values, the Cv^2 of the generalized Erlang distribution can assume any value in $(0,1]$, and the Cv^2 of the hyperexponential distribution can take any value greater than 1. Since these distributions are special cases of the MGE distribution, it is also added to the figure with the full coverage of its Cv^2.

For the moment, let us focus on the MGE-2 distribution with the density function given by (2.44). The capability of its density function to accommodate any $Cv^2 \geq 0.5$ is depicted in Figure 2.15, where the mean is kept fixed at 10.

The Cv^2 of the MGE-2 distribution is given by

$$Cv_X^2 = 1 - \frac{2a_1\mu_1[\mu_2 - \mu_1(1 - a_1)]}{(\mu_2 + a_1\mu_1)^2}.$$

Note that

$$Cv_X^2 \begin{cases} < 1 & \text{if } \mu_1(1 - a_1) < \mu_2, \\ = 1 & \text{if } \mu_1(1 - a_1) = \mu_2, \\ > 1 & \text{if } \mu_1(1 - a_1) > \mu_2. \end{cases} \quad \text{(necessary conditions)}$$

Thus, it is possible to specify the parameters of the MGE-2 distribution such that its Cv^2 falls into a desired range. At this stage, the values of the parameters for a specific Cv^2 can be obtained by trial and error. We will later introduce special MGE-2 distributions with parameters expressed as functions of their Cv^2.

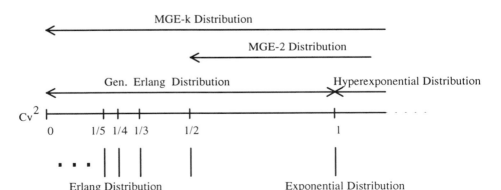

Figure 2.14. The variability of phase-type distributions.

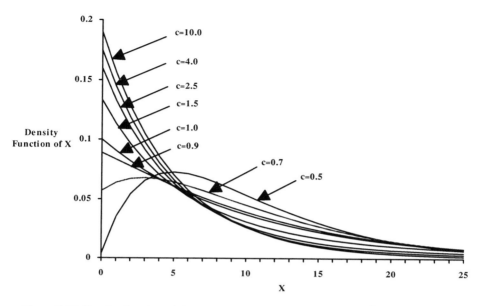

Figure 2.15. Density function of the MGE-2 distribution with varying Cv^2.

Also, notice that it is possible to find exact equivalence between MGE-k and hyperexponential distributions with k phases by matching the coefficients in their LSTs. This can also be achieved by employing a graphical approach and exploiting the minimal MGE representation of the exponential distribution (see Problem 2.16).

In the special case of the exponential distribution, it is not sufficient to have $\mu_1 > \mu_2$. It is necessary to have $\mu_1(1 - a_1) = \mu_2$. Also, though we have indicated earlier that the order of the exponential distribution is 1, for future reference, let us refer to the phase structure (with $k = 2$) shown in Figure 2.16 as the minimal MGE representation of the exponential distribution.

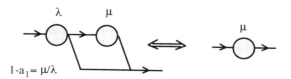

Figure 2.16. The minimal MGE representation of the exponential distribution ($\lambda > \mu$).

2.8.3 Moment-Matching Approximations

In this section, we introduce the issue of modeling data or other distributions by the MGE distributions using the method of moments. The LST of any distribution function can be approximated arbitrarily closely by a rational function. Therefore, in principle, phase-type distributions can be used to model any distribution. The main problem remains to be the lack of a method to determine the structure and the parameters of the approximating phase-type distribution. Furthermore, even if an approximate or an exact phase representation is obtained, if the number of phases is large, it may be difficult and sometimes impossible to deal with the underlying Markov process due to the size of its state space. Here we will introduce simple and convenient phase structures to model random phenomena and use the method of moments to obtain the values of their parameters. Given a phase structure Ω with its parameter set $\underline{\omega}$ to model random phenomena Φ, let

$$m_i^{\Omega} = f_i(\underline{\omega}), \qquad i = 1, 2, \ldots, \tag{2.54}$$

and let m_i be the ith central moment of Φ. We assume that m_i is known. Then, in case there are k parameters to be quantified in $\underline{\omega}$, the method of moments proceeds with

$$m_i^{\Omega} = m_i, \qquad i = 1, 2, \ldots, k, \tag{2.55}$$

that is, a set of k equations and k unknowns. In principle, k may assume any integer value provided that $f_i(\underline{\omega})$ is known for each i and (2.55) is solvable. However, there is every reason to simplify (2.54) to reduce k, not only to be able to solve (2.55) but also to end up with a smaller number of phases in Ω. This will assure a smaller state space in the Markovian analysis of the system of interest.

For the purpose of modeling a random variable using phase-type distributions, we divide the domain of the squared coefficient of variation into two main regions, $\text{Cv}^2 > 1$ and $\text{Cv}^2 < 1$, and introduce a phase structure for each region. Also, in the remainder of this chapter, we will replace Cv^2 by c for ease of notation.

Case I: $c > 1$

As shown in Figures 2.14 and 2.15, the MGE-2 is the simplest distribution that is capable of modeling random phenomena with $c > 1$. Thus, it is a good candidate to be the choice for Ω with the set of unknown parameters $\underline{\omega}^T = (\mu_1, \mu_2, a)$ to be determined by solving (2.55). The LST of the probability density function of a two-stage MGE distribution (with $a_0 = 1$ and a_1 replaced by a) is given by

$$F_X^*(s) = \frac{s\mu_1(1-a) + \mu_1\mu_2}{s^2 + s(\mu_1 + \mu_2) + \mu_1\mu_2}.$$

2.8. Phase-Type (PH) Distributions   53

The first three moments, m_i^Ω, $i = 1, 2, 3$, are obtained by taking the derivatives of $F_X^*(s)$:

$$m_1^\Omega = \frac{1}{\mu_1} + \frac{a}{\mu_2}, \tag{2.56}$$

$$m_2^\Omega = \frac{2(1-a)}{\mu_1^2} + \frac{[2a\mu_1\mu_2 - 2a(\mu_1 + \mu_2)^2]}{\mu_1^2 \mu_2^2}, \tag{2.57}$$

$$m_3^\Omega = \frac{6(1-a)}{\mu_1^3} - \frac{[12a\mu_1\mu_2(\mu_1 + \mu_2) - 6a(\mu_1 + \mu_2)^3]}{\mu_1^3 \mu_2^3}. \tag{2.58}$$

Assuming that m_is are known, and using (2.55), the parameters μ_1, μ_2, and a can be obtained as follows: Using (2.56),

$$a = \frac{\mu_2}{\mu_1}(m_1\mu_1 - 1), \tag{2.59}$$

and inserting (2.59) into (2.57) and (2.58), we obtain the following two equations:

$$2m_1(\mu_1 + \mu_2) - m_2\mu_1\mu_2 = 2, \tag{2.60}$$

$$6m_1(\mu_1 + \mu_2)^2 - 6m_1\mu_1\mu_2 - 6(\mu_1 + \mu_2) - m_3\mu_1^2\mu_2^2 = 0. \tag{2.61}$$

The substitution of $X = \mu_1 + \mu_2$ and $Y = \mu_1\mu_2$ yields

$$Y = \frac{6m_1 - 3m_2/m_1}{(6m_2^2/4m_1) - m_3}, \tag{2.62}$$

$$X = \frac{1}{m_1} + \frac{m_2 Y}{2m_1}. \tag{2.63}$$

The next task is to find the values of μ_1 and μ_2 for which the sum and the product are known. That is,

$$\mu_1 = \frac{X + \sqrt{X^2 - 4Y}}{2} \quad \text{and} \quad \mu_2 = X - \mu_1. \tag{2.64}$$

Finally, a can be found from (2.59). The choice of the positive root in (2.64) stems from the fact that μ_1 must be greater than μ_2 for c to be greater than 1. The necessary condition for the legitimacy of (2.64) is

$$\frac{m_3}{m_1^3} > \frac{3}{2}(c+1)^2. \tag{2.65}$$

That is, the resulting MGE-2 distribution is legitimate (positive rates and a positive branching probability) and unique if (2.65) is satisfied.

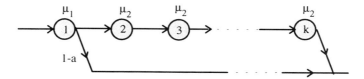

Figure 2.17. MGE-k distribution with $c > 1$.

In case (2.65) is not satisfied, one can still use a three-moment approximation at the expense of adding more phases to the MGE distribution. For instance, the MGE-k distribution shown in Figure 2.17 can be used for a three-moment approximation for any $c > 1$. The procedure to obtain its parameters is as follows: Find the smallest k (the number of phases) such that

$$k \geq \left(\frac{m_3 m_1}{m_2^2} - 1\right)^{-1},$$

that is, k is the smallest integer satisfying the above inequality. Next, solve for α, which is the smaller root of

$$\alpha^2[4m_1^2(k+1) - 3m_2 k] + \alpha[km_3 - (4+k)m_1 m_2] + (k+1)m_2^2 - km_1 m_3 = 0.$$

Then,

$$\mu_1 = \frac{1}{\alpha}, \quad \mu_2 = \frac{k(m_1 - \alpha)}{m_2 - 2m_1\alpha} \quad \text{and} \quad a = \frac{k(m_1 - \alpha)^2}{(k-1)(m_2 - 2m_1\alpha)}. \tag{2.66}$$

In case (2.65) is not satisfied, another approach would be to choose a third moment that is closest to m_3 and that satisfies (2.65). Thus, m_3 can be found by

$$m_3 = \frac{3m_2^2}{2m_1} + \varepsilon m_1^3.$$

Note that ε should be large enough to prevent Y from going to infinity.

It is also possible to resort to two-moment approximations in case either the third moment is not available or again (2.65) is not satisfied. For instance, given the mean m_1 and the squared coefficient of variation c of Φ, the simplest Ω would be again an MGE-2 distribution with

$$\mu_1 = \frac{2}{m_1}, \quad \mu_2 = \frac{1}{m_1 c}, \quad \text{and} \quad a = \frac{0.5}{c}, \tag{2.67}$$

which also gives us the chance of finding the parameters of the MGE-2 distribution for a specific value of $c > 1$.

2.8. Phase-Type Distributions

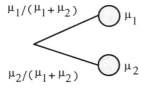

Figure 2.18. The weighted hyperexponential distribution.

For a two-moment approximation, it is also possible to find a hyperexponential distribution with two phases and a weighted mixing distribution for the branching probabilities in terms of the phase rates μ_1 and μ_2, as shown in Figure 2.18. From the expressions of the first two moments it is not difficult to obtain

$$\mu_1 + \mu_2 = \frac{2}{m_1}, \quad \mu_1 \mu_2 = \frac{2}{m_2}, \qquad (2.68)$$

which always has a solution for $c \geq 1$.

Case II: $c < 1$

The empirical studies in the literature suggest the adequacy of the first two moments when $c < 1$. This is due to the lesser significance of the third moment, which captures the skewness of the distribution, in the case of low variability. The MGE distribution that can be used to model random phenomena with $c < 1$ is in the form of the generalized Erlang distribution shown in Figure 2.19. Given m_1 and c, the number of phases, k, should be such that

$$\frac{1}{k} \leq c \leq \frac{1}{k-1}, \qquad (2.69)$$

and the parameters μ and a are respectively given by

$$1 - a = \frac{2kc + k - 2 - \sqrt{k^2 + 4 - 4kc}}{2(c+1)(k-1)}, \qquad (2.70)$$

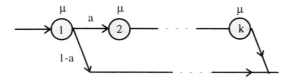

Figure 2.19. k-phase generalized Erlang distribution.

56   2. Tools of Probability

and

$$\mu = \frac{1 + (k-1)a}{m_1}. \tag{2.71}$$

For $k = 2$, a simpler MGE-2 distribution has the following parameters:

$$\mu_1 = \frac{1}{m_1 c}, \qquad \mu_2 = \frac{2}{m_1}, \qquad a = 2(1-c), \tag{2.72}$$

provided that $0.5 \le c < 1$. Both of the above MGE distributions are convenient for the purpose of finding the parameters for a desired squared coefficient of variation. It is difficult, however, to assert that one is a better fit than the other.

EXAMPLE 2.11. Phase-type approximations have the following implication. In general, distributions and data sets (in simulation or in analytical studies) can be represented by phase-type distributions with the same first two or three moments. Then, it is natural to ask the question of how close the CDFs or the density functions of the original and approximating distributions are.

Let us consider a hyperexponential distribution with four phases and with parameters

$$\underline{\alpha} = (0.2, 0.1, 0.3, 0.4) \quad \text{and} \quad \underline{\mu} = (0.05, 0.02, 0.06, 0.0667).$$

Its squared coefficient of variation is 1.517. Let us approximate it using an MGE distribution with a lesser number of phases. The two-phase MGE distribution that has the same first three moments has parameters

$$\mu_1^{(3)} = 0.0607, \qquad \mu_2^{(3)} = 0.02037, \qquad a^{(3)} = 0.07236,$$

obtained using (2.64). The two-moment MGE representation of the above hyperexponential distribution has two phases with parameters

$$\mu_1^{(2)} = 0.1, \qquad \mu_2^{(2)} = 0.03297, \qquad a^{(2)} = 0.32967,$$

obtained using (2.67). A comparison between the exact and the approximate CDFs is given in Figure 2.20. The three-moment approximation and the exact CDFs coincide in the figure. However, both of the approximate distributions seem to be acceptable, while the three-moment approximation produces slightly better results.

Let us look at another case where a normally distributed random variable is approximated by an Erlang random variable. The mean and the standard deviation of the normal distribution are $\mu_n = 100$ and $\sigma_n = 50$, respectively. The corresponding Erlang random variable (shown in Figure 2.7) has $k = 4$ phases and a phase rate $\mu_e = 0.04$. The two distributions have the same first two moments. Their density functions are given in Figure 2.21. In this example, the density functions are

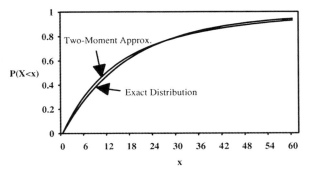

Figure 2.20. Comparison of CDFs ($c = 1.517$).

not that close to each other. This is expected in a way, since the Erlang distribution converges to a normal distribution as k tends to infinity (as opposed to $k = 4$ in this case) due to the central limit theorem. This example indicates that one may not always obtain a perfect match in these approximations. However, it may still be possible to replace the normal distribution by its Erlang counterpart and carry out a Markovian analysis in the system of interest. This enables us to study analytically those problems that are difficult to analyze otherwise. A precautionary note here is that there is always an error in this type of approximations, and it is important to invest every effort to minimize it.

Related Bibliography on Phase-Type Distributions

The phase concept was first introduced by Erlang (1917) in the form of Erlang distribution, which has been widely used in queueing modeling of stochastic systems. Later, Cox (1955) showed that any distribution having a rational Laplace-Stieltjes transform can be represented by a set of exponential phases. Since the LST of

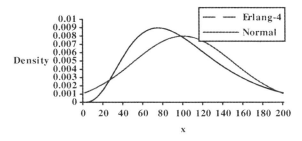

Figure 2.21. Comparison of the density functions ($c = 0.25$).

any density function can be approximated arbitrarily closely by a rational function (see, for example, Newman and Reddy (1979) for e^x and related functions), distributions of Cox type can in principle represent any distribution approximately, if not exactly. Phase-type distributions fall into a proper subset of those introduced by Cox (1955). Their definitions, formalism, properties, and their use in queueing theory and its applications can be found in Neuts (1981), Assaf and Landberg (1985), and O'Cinneide (1990), among others. Neuts developed matrix-geometric methods to obtain the steady-state probabilities in queues with phase-type arrival and/or service processes. The MGE and the hyperexponential distributions were heavily utilized by Kleinrock (1975).

The relationship between phase structure and variability, and consequently the equivalence relations, were studied by Marie (1978), Cumani (1982), Iversen and Nielsen (1986), and Commault and Chemla (1992). The MGE equivalent of an exponential distribution was mentioned by Marie (1978), and the MGE equivalent of a hyperexponential distribution was proposed by Iversen and Nielsen (1986). The generalized exponential distribution was used by Kouvatsos (1988), who developed algorithms based on the concept of maximum entropy to study queueing networks.

Various moment-matching approximations have appeared in the literature. The two-moment characterization of the generalized Erlang distribution ($Cv^2 < 1$) given by (2.70) to (2.71) were developed by Sauer and Chandy (1975). Bux and Herzog (1977) developed a mathematical programming approach to restrict the CDF of an MGE distribution with uniform rates to have the same first two-moments and to be within a certain tolerance with a given distribution (shape constraints). The main disadvantage of such an approach is that once the phases are restricted to have a uniform rate, the number of phases increases rapidly. The two-moment approximations (2.67) and (2.72) were developed by Marie (1978). Altiok (1985) matched the three moments of a given distribution with $Cv^2 > 1$ by the ones of a two-stage MGE distribution and obtained the existence condition (2.65) with expressions for the parameters of the approximating distribution in terms of the given moments (2.56) to (2.58). Altiok's two-stage MGE distribution coincides with Whitt's (1982) two-stage hyperexponential distribution. Whitt (1982) developed existence conditions and expressions (for two- and three-moment matching) for the parameters of the approximating distributions that are hyperexponential for $Cv^2 > 1$ and shifted exponential for $Cv^2 < 1$. Pierrat (1987) was able to eliminate condition (2.65) at the expense of adding more phases. Pierrat introduced the three-moment representation given by (2.66) for distributions with $Cv^2 > 1$. Johnson and Taaffe (1988) suggested a three-moment approximation using a mixture of two Erlang distributions of common order to approximate distributions with any squared coefficient of variation. Botta et al. (1987) and Botta and Harris (1986) suggested generalized hyperexponential distributions that are linear combinations of exponential distributions with mixing parameters (positive or negative) adding up to unity. Botta and Harris (1986) showed that generalized hyperexponential distributions are dense in all CDFs and have a unique representation, which is a property not shared by phase-type distributions. However, there are some inherent implementation difficulties in these distributions. For instance, the number of exponential phases to be included in the hyperexponential distribution is not known. The maximum-likelihood estimation method with the use of numerical procedures seems to be the only tool to estimate the parameters of the approximating distribution.

There have been empirical studies on the impact of moment approximations on the performance measures in queueing systems. Bere (1981) and Whitt (1984) both showed that the third moment can have a significant impact on measures such as the average levels and the steady-state distributions when the Cv^2 of the random variable approximated by a phase-type distribution is greater than 1.0. So, one may arrive at the conclusion that if the third moment of the distribution being approximated is known, one shall put every effort to use it in the approximation. The inclusion of a larger number of moments necessitates the solution of the nonlinear moment equations, as in Johnson (1990).

References

Altiok, T., 1985, On the Phase-Type Approximations of General Distributions. *IIE Transactions*, Vol. 17, pp. 110–116.

Altiok, T., 1989, Queueing Modeling of a Single Processor with Failures. *Performance Evaluation*, Vol. 9, pp. 93–102.

Assaf, D., and N. A. Landberg, 1985, Presentation of Phase-Type Distributions as Proper Mixtures. *J. Applied Probability*, Vol. 22, pp. 247–250.

Bere, B., 1981, Influence du Moment d'Ordre 3 sur les Files D'Attente à Lois de Services Generales. M.S. thesis, U.E.R. Mathematiques d'Informatique, Université de Rennes, France.

Bhat, U. N., 1984, *Elements of Applied Stochastic Processes*. John Wiley, New York.

Botta, R. F., and C. M. Harris, 1986, Approximation with Generalized Hyperexponential Distributions: Weak Convergence Results. *Queueing Systems*, Vol. 2, pp. 169–190.

Botta, R. F., C. M. Harris, and W. G. Marchal, 1987, Characterization of Generalized Hyperexponential Distribution Functions. *Comm. Stat. Stochastic Models*, Vol. 3, pp. 115–148.

Bux, W., and U. Herzog, 1977, The Phase Concept: Approximation of Measured Data and Performance Analysis. *Computer Performance* (K M. Chandy and M. Reiser, eds.), North-Holland.

Çinlar, E., 1975, *Introduction to Stochastic Processes*. Prentice Hall, Englewood Cliffs, NJ.

Commault, C., and J. P. Chemla, 1992, *On Dual and Minimal Phase-Type Representations*. Lab. d'Automatique de Grenoble, ENSIEG, Saint Martin d'Heres, France.

Cox, D. R., 1955, A Use of Complex Probabilities in the Theory of Stochastic Processes. *Proc. Cambridge Philosophical Society*, Vol. 51, pp. 313–319.

Cumani, A., 1982, On the Canonical Representation of Homogeneous Markov Processes Modeling Failure-Time Distributions. *Microelectronics and Reliability*, Vol. 22, pp. 583–602.

Erlang, A. K., 1917, Solution of Some Problems in the Theory of Probabilities of Significance in Automatic Telephone Exchanges. *P.O. Electrical Engineering J.*, Vol. 10, pp. 189–197.

Feller, W., 1968, *An Introduction to Probability Theory and Its Applications, Vol. I.* John Wiley, New York.

Feller, W., 1971, *An Introduction to Probability Theory and Its Applications, Vol. II.* John Wiley, New York.

Heyman, D. P., and M. J. Sobel, 1982, *Stochastic Models in Operations Research, Vol. I.* McGraw-Hill, New York.

Iversen, V. B., and B. F. Nielsen, 1986, Some Properties of Coxian Distributions with Applications. *Modeling Techniques and Tools for Performance Analysis '85* (Abu El Ata, ed.), North Holland.

Johnson, M. A., 1990, *Selecting Parameters of Phase-Type Distributions: Combining Nonlinear Programming Heuristics and Erlang Distributions.* Dept. of Mechanical and Industrial Engineering, University of Illinois–Urbana/Champaign.

Johnson, M. A., and M. R. Taaffe, 1988, *Matching Moments with a Class of Phase-Type Distributions: Mixtures of Erlang Distributions of Common Order.* School of IE Tech. Report No. 88-10, Purdue University, Indiana.

Kleinrock, L., 1975, *Queueing Systems, Vol. 1: Theory.* John Wiley, New York.

Kouvatsos, D. D., 1988, A Maximum Entropy Analysis of the G/G/1 Queue at Equilibrium. *J. Oper. Res. Soc.*, Vol. 39, no. 2.

Marie, R., 1978, Modelisation par Reseaux de Files d'Attente. Ph.D. thesis, l'Université de Rennes, Rennes, France.

Neuts, M. F., 1981, *Matrix-Geometric Solutions in Stochastic Models—An Algorithmic Approach.* John Hopkins University Press, Baltimore, MD.

Newman, D. J., and A. R. Reddy, 1979, Rational Approximation to e^x and to Related Functions. *J. Approximation Theory*, Vol. 25, pp. 21–30.

O'Cinneide, C. A., 1990, Characterization of Phase-Type Distributions. *Stochastic Models*, Vol. 6, no. 1, pp. 1–57.

Pierrat, J., 1987, *Modelisation de Systèmes à Evenements Discrets Sujets à des Pannes.* L'Institut National Polytechnique de Grenoble, France.

Sauer, C. H., and K. M. Chandy, 1975, *Approximate Analysis of Central Server Models.* IBM T. J. Watson Research Center, Yorktown Heights, NY.

Whitt, W., 1982, Approximating a Point Process by a Renewal Process, I: Two Basic Methods. *Operations Research*, Vol. 30, pp. 125–147.

Whitt, W., 1984, On Approximations for Queues III: Mixtures of Exponential Distributions. *B.S.T.J.*, Vol. 63, pp. 162–175.

Wolff, R. W., 1982, Poisson Arrivals See Time Averages. *Operations Research*, Vol. 30, pp. 223–231.

Wolff, R. W., 1989, *Stochastic Modeling and the Theory of Queues.* Prentice Hall, Englewood Clifffs, NJ.

Problems

2.1. Let X be a random variable with the following density function:

$$f_X(x) = b(2x^2 - 1)^2, \quad 1 \leq x \leq 3.$$

(a) Find the value of b for $f_X(x)$ to be a density function.
(b) Find $E[X]$, $V[X]$, and Cv_X^2.

2.2. Let X and Y be two random variables with the following joint density function:

$$f_{XY}(x, y) = 2a\left(y - \frac{x}{4}\right)^b, \quad 0 \le x \le y \le 2.$$

(a) Find the relationship between a and b that makes $f_{XY}(x, y)$ legitimate.
(b) Find $P(Y - X \ge 0.1)$ for $a = 1$ and the corresponding b.
(c) Find $P(X > .5 \mid Y < 1)$ for the values of a and b in (b).
(d) Are X and Y independent?

2.3. Let X be a random variable with a density function

$$f_X(x) = \frac{1}{2}(1 + x)e^{-x}, \quad x \ge 0.$$

Find $E(X)$ and Cv_X^2 utilizing the Laplace-Stieltjes transforms.

2.4. Let X and Y be two random variables with the following joint density function:

$$f_{XY}(x, y) = xe^{-x(1+y)}, \quad x \ge 0, \quad y \ge 0.$$

(a) Find the marginal density functions $f_X(x)$ and $f_Y(y)$.
(b) Find $P(Y > X)$.
(c) Find $P(X + Y < 2)$.
(d) Are X and Y independent?

2.5. Consider the following density function of random variables X and Y:

$$f_{X,Y}(x, y) = \lambda^2 e^{-\lambda y}, \quad 0 \le x \le y.$$

Obtain $f_X(x)$, $f_Y(y)$, $f_{X|Y}$, and $f_{Y|X}$. What is $E[X \mid Y < 1]$?

2.6. Let X and Y be two exponentially distributed random variables both with parameter μ.

(a) Find the density of $Z = X + Y$.
(b) Find the density function of $Z = X - Y$.
(c) Find the density function of $Z = X/Y$.

2.7. Let X and Y be two exponentially distributed random variables with parameters μ_1 and μ_2, respectively.

(a) Find $E[Z]$, where $Z = \min(X, Y)$.
(b) Find $E[Z]$, where $Z = \max(X, Y)$.
(c) Repeat (a) for Y positive and fixed.

2. Tools of Probability

2.8. Let X be a random variable that is uniformly distributed in (2,6).

   (a) Find $E[X^2]$.
   (b) Find $E[X \mid X < 5]$.

2.9. Derive the generating function (z-transform) of the geometric distribution $P_n = q(1-q)^{n-1}$, $n = 1, 2, \ldots$. Obtain its mean and variance using its generating function.

2.10. Let K_1 and K_2 be two Poisson random variables with parameters λ_1 and λ_2, respectively. Show that $K_1 + K_2$ is also a Poisson random variable with rate $\lambda_1 + \lambda_2$.

2.11. Consider a machine that produces a unit of a product at every T time units. The machine is subject to failures where the time until a failure occurs is exponentially distributed with rate δ. Suppose the machine is just repaired. Find the probability distribution of the number of units produced until the next failure occurs. What is the average number of jobs to be finished until the next failure occurs?

2.12. Consider the machine–repairman problem presented in Example 2.2.

   (a) Extend the problem to the case with r repairmen. Write down its steady-state balance equations. Develop a recursive method for the solution, if possible. For $r = 3$, $m = 5$, $\delta = 1$, and $\gamma = 2$, solve for the steady-state probabilities of the number of down machines and obtain the repairman utilization and the average number of down machines.
   (b) Extend the problem with $r = 1$ to the case with a state-dependent repair rate. In particular, let $\gamma_n = n\gamma$ be the repair rate when n machines are down. Obtain the average number of down machines and the repairman utilization for $\gamma = 2$. Hence, one can compare (a) and (b).

2.13. Let us consider an inventory system where the demand arrivals follow a Poisson distribution with rate λ. Demand is satisfied from the available stock or is back-ordered. Inventory level is controlled by using an (r, Q) policy, which states that an order of Q units is placed as soon as the inventory level drops to the reorder level r. The lead time, that is, the time until the order comes back, is exponentially distributed with rate μ. A new order is placed, at the point of an order arrival and not earlier, if the on-hand inventory plus the order quantity is still below the level r. Hence, the outstanding number of orders cannot exceed 1. Cases with $r \leq Q$ guarantee a single outstanding order.

   (a) Find the steady-state average inventory level in the system at any point in time. You may have to use the continuity property of the generating functions.
   (b) What is the probability that a demand is back-ordered?
   (c) Repeat (a) and (b) for $r = 2$, $Q = 3$, $\lambda = 1$, and $\mu = 1$.

2.14. (Continuation of 2.13)

   (a) Solve (a) and (b) of 2.13 with the data provided in (c) for the case where the unsatisfied demand is immediately lost. Show the recursive nature of the steady-state balance equations, and use it to solve for the steady-state probabilities.

(b) Extend (a) to the case where there may be multiple outstanding orders. Assume that the lead time for orders are iid exponential random variables with rate μ. In addition, what is the average number of outstanding orders?

2.15. Consider a two-phase Erlang distribution.

(a) What is the impact of placing a feedback loop from phase 1 to phase 1 as shown below on its mean and its Cv^2?

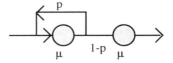

(b) Repeat (a) for a feedback loop from phase 2 to phase 1 as shown below.

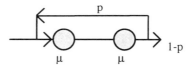

2.16. Consider a three-phase MGE distribution with parameters $\underline{\mu} = (5, 4, 3)$ and $\underline{a} = (0.16, 0.125)$.

(a) Show its $(\underline{\alpha}, \mathbf{T})$ representation.
(b) Find its mean, variance, and the Cv^2 using the $(\underline{\alpha}, \mathbf{T})$ representation.
(c) Obtain the parameters of the equivalent hyperexponential distribution using the minimal representation of the exponential distribution.

Hint: In the case of $k = 2$, a graphical equivalence argument would work as follows:

(a)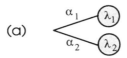

Replace the 2nd phase by its minimal MGE representation

(b)

λ_2 / λ_1

Convert the entire representation to MGE-2

(c)

$\alpha_1 + \alpha_2 \lambda_2 / \lambda_1$

64   2. Tools of Probability

where $\lambda_1 > \lambda_2$. Since we know the parameters of the MGE-2 distribution, namely (μ_1, μ_2, a), we obtain the parameters of the hyperexponential equivalent by

$$\lambda_1 = \mu_1, \qquad \lambda_2 = \mu_2, \qquad \alpha_2 = \frac{a\lambda_1}{\lambda_1 - \lambda_2}.$$

2.17. Develop expressions for the parameters of the hyperexponential distribution that has the same density function as the two-phase MGE distribution with parameters μ_1, μ_2, and a. Extend the problem to the three-phase MGE distribution.

2.18. This question amounts to obtaining a phase-type distribution for a given Cv^2. Given two moments m_1 and m_2, develop moment-matching expressions for a two-phase hyperexponential distribution using balanced means, that is, $p_1/\lambda_1 = p_2/\lambda_2$. Notice that weighted branching probabilities would give the same result.

2.19. Consider the following phase-type distribution with a feedback loop.

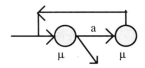

(a) Obtain expressions for its mean and Cv^2.
(b) What is the range of its Cv^2?
(c) Obtain expressions for μ and a as a function of the first two moments m_1 and m_2.

2.20. Consider the hyperexponential distribution with parameters $\underline{\lambda} = (1, 5, 25)$ and $\underline{p} = (0.40, 0.20, 0.40)$.

(a) Find all its possible two- and three-moment approximate MGE-2 representations.
(b) Compare the CDFs of the hyperexponential distribution and that of its approximate representations. Which approximation do you recommend to model the hyperexponential distribution?

2.21. Show that condition (2.65) is necessary when $Cv^2 > 1$. You may refer to Altiok (1985).

2.22. A discrete MGE distribution is a proper mixture of the convolutions of independent geometric distributions. Consider the following two-phase discrete MGE distribution with parameters p_1, p_2, and a.

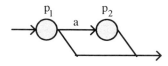

(a) Develop expressions for its mean and Cv^2.

(b) What is the range of its Cv^2?

(c) Develop moment-matching expressions for p_1, p_2, and a using the first two and three moments.

2.23. An alternate approach to model random variables with $Cv^2 < 1$ is to use the shifted exponential distribution with a density function

$$f_X(x) = \lambda e^{-\lambda(x-r)}, \qquad x \geq r.$$

(a) Find the mean and the Cv^2 of the shifted exponential distribution.

(b) Find expressions for λ and r in terms of the first two moments.

The disadvantage of using the shifted exponential distribution in modeling random phenomena is that its domain is bounded below by r. It is recommended only if the nature of the random phenomenon to be modeled shows such a behavior.

2.24. Again let us consider the machine–repairman problem studied in Example 2.2, this time with an MGE-2 repair time distribution with parameters $\mu_1 = 2$, $\mu_2 = 3$ and $a = 0.2$. Let us assume there are four machines each, with a failure rate $\delta = 2$. Calculate the average number of down machines and the repairman utilization.

2.25. Carry out the analysis of the inventory problem in Example 2.4 with an Erlang-2 lead time distribution with phase rate μ.

3

Analysis of Single Work Stations

3.1 Introduction

We start the analysis of manufacturing systems by first explaining the way we perceive them. Then we will focus on some simple cases of work stations with and without failures and demonstrate how these systems can be analyzed. A work station can be thought of as a work place in the shop floor as depicted in Figure 3.1, where jobs arrive to be processed or machined. They are kept in the buffer zone (work-in-process (WIP) inventory) until their process begins and are transferred to other areas in the facility after their process is completed. The machines may experience random failures. Our approach in studying work stations in manufacturing environments is to perceive them as queueing systems. The flow of incoming jobs, either one by one or in batches, forms the arrival stream, which can be identified by the statistical characteristics of the time between consecutive arrivals. Service times are usually the processing times of jobs in the work station. Processing times may be augmented by random failures and repairs. Thus, a work station may very well be looked at as a queueing system with single or multiple servers, with finite or infinite waiting room (input buffer), and with a service policy depending on the operating characteristics of the work place. Queueing theory is a fairly well-developed field, and a considerable amount of research is being done on a variety of queueing systems and queueing networks. We demonstrate in this and other chapters that with some level of abstraction, queueing theory and queueing networks can be very useful in the analysis and design of manufacturing systems. In fact, queueing modeling of numerous systems have already been studied and can be found in Gross and Harris (1974), Cooper (1990), and Kleinrock (1975), among various others. Next we give a short introduction on the formulation and the analysis of queueing systems.

A waiting line develops in every service-providing facility whenever it cannot immediately cope with the customer's requests for service. A queueing system is characterized by its customer (job or demand) arrival process, its service process, the number of servers, the service discipline, and its queue capacity. The arrival process is defined by the time between consecutive arrivals, referred to as the **interarrival time**. For example, let A_i be the interarrival time between the $(i-1)$st

3.1. Introduction   67

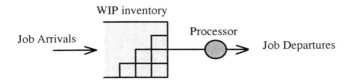

Figure 3.1. A simple work station.

and the (i)th jobs. It is usually assumed that the A_is are i.i.d. (independent, identically distributed) random variables. The arrival rate λ is the number of job arrivals per unit time, that is, $\lambda = 1/E(A)$. The service time is the time the server devotes to a particular job. Let X_i be the service time of the ith job. Again, it is usually assumed that the X_is are i.i.d. random variables with a mean service time $E(X)$ and a service rate $\mu = 1/E(X)$. In a multiserver facility, the servers are assumed to be identical, having the same service-time distribution. The servers are selected equally likely in case more than one server is available for service. The order in which jobs are taken into service defines the service discipline. FCFS, LCFS, SIRO (service in random order), and PR (priority service) are typical service disciplines. The waiting space is where jobs accumulate while the station is busy. The capacity of this buffer area is usually finite. However, the area may also be so large that its capacity is considered infinite for practical purposes.

The standard notation to represent a queueing system is to separate by slashes the letters and the numbers representing the arrival and the service processes, the number of servers, and the queue capacity. In this notation, M stands for exponential (Markovian in particular), D for deterministic, and G for general interarrival or service-time distributions. Then, M/M/1 represents exponential interarrival and service times and a single server; M/D/1/N represents exponential interarrival times, deterministic service times, a single server, and a limited queue capacity of N. Capacity N implies one position for the server (or one for each server in the multiserver case) and $N - 1$ waiting positions. An arrival who finds no available position in the queue is usually assumed to be lost.

The purpose of studying a queueing system is to understand its behavior and to estimate its performance as reflected by certain measures. Usually these measures are the long-run average number of jobs in the queue and/or in the system, the average waiting time (or delay) per job, the average server utilization, and the output rate, among others. One way of computing these measures is to first compute the steady-state probability distribution of the number of jobs in the system at any point in time. That is, if P_n denotes the steady-state probability of having n jobs in the system at any point in time, then the average number of jobs in the system is given by

$$\overline{N}_S = \sum_{n=0}^{\infty} n P_n, \qquad (3.1)$$

or by the finite sum in the limited-capacity case. In a single-server system, the average number of jobs in service is $1 - P_0$, and therefore, the average number of jobs waiting in the queue becomes

$$\overline{N}_q = \overline{N}_S - (1 - P_0). \tag{3.2}$$

In general, $1 - P_0$ is the server utilization for any single-server queueing system with a limited or an unlimited queue capacity. For the unlimited capacity systems, the utilization can be expressed as the ratio of the input rate and the service rate, that is, $\rho = \lambda/\mu$, and consequently P_0 is obtained from $\rho = 1 - P_o$. The server is said to be exhausted if $\rho \geq 1$. In this case, the server is not able to clear the queue, and consequently the number of jobs in the system eventually approaches infinity. In queueing theory terminology, the system is said to be unstable when $\rho \geq 1$. For stability, we must have $\rho < 1$, or equivalently $P_o > 0$, making the long-run measures exist. Clearly, a finite-capacity system rejecting jobs in excess of its capacity is always stable.

The fact that the server utilization is $\rho = \lambda/\mu$ can be explained more rigorously using the following regeneration argument. Let us first assume that the interarrival and service times are independent and independent among themselves. Any stable queueing system goes through cycles of idle (empty) and busy periods. When the system enters the empty-and-idle state (queue is empty and the server is idle), the next event will definitely be a job arrival, and the time until this event takes place is independent of the history of the work station. Thus, the number of jobs in the system, which is usually the stochastic process of our interest, probabilistically regenerates itself at these points in time, as shown in Figure 3.2. The period between any two successive visits to the empty-and-idle state becomes a regeneration cycle. Let T_B be the length of the busy period and T be the cycle time. In a system with no capacity restrictions, all the jobs that arrive during a cycle are served during the busy period in the same cycle. That is,

$$\lambda E[T] = \mu E[T_B],$$

resulting in

$$\frac{\lambda}{\mu} = \frac{E[T_B]}{E[T]}, \tag{3.3}$$

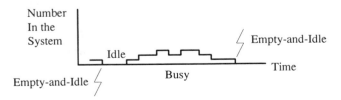

Figure 3.2. Cycles of busy and idle periods.

3.1. Introduction

which is nothing but the percentage of time the server is busy, that is $\rho = \lambda/\mu$.

The length of the busy period and of the regeneration cycle are also important measures of interest. Since the idle time is the time until the first job arrives during the idle period, we can write the following:

$$\text{Pr(Server is idle)} = 1 - \rho = \frac{1/\lambda}{E[T]},$$

which in turn gives us

$$E[T] = \frac{1/\lambda}{1 - \rho}. \qquad (3.4)$$

Then, using (3.3) and (3.4), we obtain

$$E[T_B] = \frac{E[X]}{1 - \rho}. \qquad (3.5)$$

Let us elaborate on $E[T_B]$. The job that starts the busy period arrives at an empty system. However, new jobs (offsprings of the first job) may arrive during its service time, and every job may have its own offsprings. Thus, $E[T_B]$ can be looked at as the time to clear the family of jobs generated by the first job that has the mean service time of $1/\mu$ or $E[X]$. We will recall this argument in the next section when we introduce work stations processing many types of jobs according to a priority structure or the SPT (shortest processing time first) rule.

The output rate of a queueing system can be expressed using the general definition of utilization. The server is producing with rate μ when it is busy, resulting in the average output rate

$$\bar{o} = \mu(1 - P_0), \qquad (3.6)$$

which reduces to λ for systems with an infinite-capacity queue due to conservation of the flow of jobs.

Before we talk about the average delay per job, let us introduce the following important relationship known as Little's formula.[1]

$$\overline{N} = \lambda \overline{W}, \qquad (3.7)$$

---

[1] Ross (1980) gives an intuitive explanation for Little's formula by superimposing a reward structure on the queueing system. Assume that every entry pays the system a dollar per unit time during the period it stays in the system. That is, the system earns j dollars per unit time when j customers are present. Then the average reward per unit time becomes the average number of units in the system. Also, the average amount paid by each customer becomes the average system time. Since the customers arrive at a rate λ per unit time, the relationship (3.7) holds.

70   3. Analysis of Single Work Stations

where $\overline{N}$ is the average number of jobs and $\overline{W}$ is the average waiting time per job in the system. Little's formula is valid for any storage system where units arrive, spend some time in the system, and leave. The three parameters in (3.7) are to be consistent with each other. That is, we can rewrite (3.7) as $\overline{N}_S = \lambda \overline{W}_S$ for the system as well as $\overline{N}_q = \lambda \overline{W}_q$ for the queue. Note that $\overline{W}_S$ is also the mean flow time of a job in the system. If the queue has a limited capacity, then λ has to be the effective arrival rate, that is, the average rate of arrivals that enter the system (remember, some may be lost).

A widely used relationship in finite-capacity queueing systems is that the long-run average effective arrival rate into the system is equal to the average departure rate from the system. That is, for a finite capacity N,

$$\lambda(1 - P_N) = \mu(1 - P_o), \tag{3.8}$$

which we have already introduced in Example 2.2. Equation (3.8) can also be obtained by applying the above regeneration argument to systems with finite-queue capacities. If T_N is defined as the total time with N jobs in the system during a cycle, then the jobs entering the system per unit time is given by

$$\lambda E[T] - \lambda E[T_N] = \mu E[T_B].$$

Dividing both sides by $E[T]$, we obtain (3.8).

Queueing systems with exponential interarrival and service times are usually called **Markovian queues**, i.e., (M/M/1). In a Markovian queue, the number of servers and the queue capacity may have any integer value. The common approach to studying these systems focuses on the stochastic process $\{N_t, t \geq 0\}$ representing the number of jobs in the system at any time t. $\{N_t\}$ gives rise to a continuous-time, discrete-state Markov chain, and its steady-state probabilities can be obtained in a manner similar to what we have shown in the examples in Chapter 2. For the finite-capacity systems, usually a normalization has to be carried out at the end to make sure that $\sum P_n = 1$ is satisfied.

Queueing systems with nonexponential arrival and/or service processes are called **non-Markovian queues**. Examples of non-Markovian queues include M/D/1, M/GI/1 (GI meaning general, independent), GI/M/k, GI/GI/1, and so forth. Methods to obtain the steady-state distribution of the number in the system at any time in a non-Markovian queue are more involved. It no longer suffices to observe the number in the system at arbitrary times, but rather at specific points in time. For instance, in an M/G/1 queue, $\{Q_n, n = 1, 2, \ldots\}$ representing the number left in the system by the nth departing job is a continuous-time, discrete-state Markov chain. So, the system is observed at departure points. We can write

$$Q_n = \begin{cases} Q_{n-1} - 1 + V_n & \text{if } Q_{n-1} > 0, \\ V_n & \text{if } Q_{n-1} = 0, \end{cases}$$

where V_n is the number of job arrivals during the service of the nth job. Clearly, the transition-rate matrix or the steady-state balance equations of $\{Q_n\}$ involve the probability distribution of V_n. These equations can again be solved for $P(Q_n)$ recursively. Notice, however, that the $P(Q_n)$ is the steady-state probability distribution of the number of units in the system as seen at departure points. Consequently, the measures obtained from this distribution will have values as seen by a departing job. A similar approach focuses on the number seen by an arriving job in a GI/M/1 queue resulting in the arrival-point probability distribution. This is called the imbedded Markov chain method and can be found in standard queueing theory books, such as Kleinrock (1975) and Gross and Harris (1985).

In general, performance measures are based on arbitrary-time observations. Therefore, one needs to find ways to obtain the arbitrary-time probability distribution in non-Markovian queues. In a GI/G/1 queue, it is shown that the departure-point probabilities are identical to the arrival-point probabilities (see, for example, Cooper (1990)). Furthermore, due to the PASTA property of the Poisson arrivals, as explained in Chapter 2, the arrival-point, the departure-point, and the arbitrary-time probability distributions of the number of jobs in an M/G/1 queue are all identical to each other. However, this is valid only for the infinite-capacity systems. For the finite-capacity case, the arbitrary-time probability distribution still needs to be obtained from the departure-point probabilities. We will explain some of these concepts as we use them in the remainder of the book. This introduction to queueing theory is sufficient for our purposes. We refer the reader to the books in queueing theory for more details.

In the following sections, we first consider a simple work station and introduce a queueing-theoretic and a Markovian approach consisting of some of the well-known arguments in queueing theory with our way of looking at manufacturing systems. Then, we extend this approach to study work stations subject to failures.

3.2 A Simple Work Station

Consider a work station in a manufacturing facility, which may be processing several types of jobs, each having a particular processing time. Jobs may arrive from several other work stations. The aggregation of these jobs through some level of abstraction into a single flow of jobs with an aggregate arrival process and a single aggregate service process is a common practice. Since large numbers of independent arrival streams tend to follow a Poisson distribution (see Çinlar (1975), p. 88), one may assume that the aggregate flow of jobs has a Poisson distribution with rate λ. Let the random variable X represent the processing time with an arbitrary density function. Under these assumptions, our work station in Figure 3.1 can be modeled as an M/G/1 queueing system for which numerous well-established theoretical results are available in the references mentioned earlier. Here, we will briefly introduce some basic arguments to obtain the long-run average number of jobs waiting in the buffer, and the average delay that a job experiences before its processing starts in the above work station.

72   3. Analysis of Single Work Stations

Let us first focus on the mean delay of a job until its processing starts. The delay of an arriving job is zero if the station is idle at the time of arrival. Otherwise, there is a positive delay consisting of two parts. First, it has to wait until the job that is on the machine at the time of arrival is processed. Then, it has to wait until all the jobs that have arrived earlier are cleared. Let $\overline{W}$ and $\overline{N}$ be the long-run expected delay of a job and the mean number of waiting jobs in the buffer at the point of a job arrival, respectively. Since Poisson arrivals see time averages (the PASTA property), a Poisson job arrival is expected to see as many as the average number of jobs in the buffer at an arbitrary point in time. That is, $\overline{W}$ and $\overline{N}$ are also the long-run average measures observed at any point in time. Again, due to the PASTA argument, the probability that an arriving job finds the processor busy is the same as the processor utilization, that is, $\rho = \lambda E[X]$. Thus, the expected delay of a job in the buffer can be expressed as

$$\overline{W} = \rho \left\{ E[X_r] + \frac{\overline{N}}{\rho} E[X] \right\}. \tag{3.9}$$

where X_r represents the remaining processing time of the current job at the time of the arrival. The second term inside the brackets represents the work to be done ahead of an arriving job, after the current job is completed. Notice that the terms inside brackets represent the delay conditioned on the job arriving at a busy system. The product by ρ gives us the unconditional delay.

Using the concept of residual time and the fact that the arrival stream is Poisson, we can write

$$E[X_r] = \frac{E[X^2]}{2E[X]}, \tag{3.10}$$

which is introduced in Section 2.7.3. Then, using (3.7) and (3.10), (3.9) is reduced to

$$\overline{W} = \frac{\rho E[X^2]}{2E[X]} + \lambda \overline{W} E[X],$$

resulting in the average delay given by

$$\overline{W} = \frac{\rho(\text{Cv}_X^2 + 1)E[X]}{2(1 - \rho)}. \tag{3.11}$$

Recall from Chapter 2 that Cv_X^2 is the squared coefficient of variation of the processing time. Finally, again using $\overline{N} = \lambda \overline{W}$, we obtain

$$\overline{N} = \frac{\rho^2(\text{Cv}_X^2 + 1)}{2(1 - \rho)}, \tag{3.12}$$

which is the well-known Pollaczek-Khintchine formula for the long-run average number of jobs waiting in an M/G/1 queueing system. In some cases, one may want to include the job on the machine in the average work-in-process inventory in the station, which then becomes $\overline{N}_S = \overline{N} + \rho$. Consequently, the mean flow time of a job in the system becomes $\overline{W} + E[X]$. Notice that (3.11) yields the average waiting time of a job as observed at the points of arrival. However, due to the PASTA property, (3.11) automatically becomes an arbitrary-time relationship. Hence, the Poisson assumption becomes the key to obtain the simple expressions given by (3.11) and (3.12).

In the case of exponentially distributed processing times, that is, $Cv_X^2 = 1$, $\overline{N}$ is given by

$$\overline{N} = \frac{\rho^2}{1-\rho},$$

which is the average number in the queue in an M/M/1 queueing system.

In the case of deterministic processing times, that is, $Cv_X^2 = 0$, (3.12) becomes

$$\overline{N} = \frac{\rho^2}{2(1-\rho)},$$

which is the minimum possible average number of jobs in the buffer in any M/G/1 work station. Clearly, $\overline{N}$ increases as the processing-time variability increases. For instance, it is doubled in the case of exponential processing times ($Cv_X^2 = 1$). Also, a higher number of jobs accumulates in systems with higher utilization. In fact, the impact of ρ on the system behavior is much more significant than that of Cv_X^2 as illustrated in the following example.

EXAMPLE 3.1. Consider an M/G/1 work station with a job arrival rate of $\lambda = 1$ per unit time and a processing time of Erlang type with $E[X] = 0.80$ and $Cv_X^2 = 0.5$. The resulting processor utilization is 0.80, and the output rate of the system is equal to λ, that is, one unit per unit time. The average number of units waiting in the buffer is $\overline{N} = 2.4$, which is also the average delay per job in the buffer. Figure 3.3 shows the behavior of $\overline{N}$ with respect to Cv_X^2 and ρ in this particular example. As expected, $\overline{N}$ approaches infinity as ρ tends to 1. Again, $\overline{N}$ approaches infinity as does the Cv_X^2. One can generalize that higher processing-time variability results in a larger accumulation in the input buffer of any work station. However, as seen from Figure 3.3, the impact of ρ on $\overline{N}$ is more dramatic than that of Cv_X^2.

As mentioned earlier, the analysis of a work station with **arbitrary interarrival and processing times (G/G/1)** is quite complicated and relies on complex-variable techniques or random walk models. These can be found in sources such as Wolff (1989), Kleinrock (1975), and Gross and Harris (1974). Next we introduce an approximation due to Kraemer and Langenbach-Belz (1976) for the average number of jobs waiting in the buffer in a work station with arbitrary interarrival and

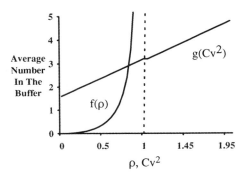

Figure 3.3. Behavior of $\overline{N}$ with respect to ρ and Cv_X^2.

processing times (GI/G/1 system):

$$\overline{N} = \frac{\rho^2 (Cv_A^2 + Cv_X^2)}{2(1 - \rho)} g, \qquad (3.13)$$

where $g \equiv g(Cv_A^2, Cv_X^2, \rho)$ is given by

$$g(Cv_A^2, Cv_X^2, \rho) = \begin{cases} \exp\left(\dfrac{-2(1-\rho)}{3\rho} \dfrac{(1 - Cv_A^2)^2}{(Cv_A^2 + Cv_X^2)} \right), & Cv_A^2 < 1, \\ \exp\left(\dfrac{(1 - \rho)(1 - Cv_A^2)}{(Cv_A^2 + 4 Cv_X^2)} \right), & Cv_A^2 > 1. \end{cases}$$

Work stations with **more than one machine to process the jobs** are referred to as multiserver systems in the queueing theory terminology. The machines are usually assumed to be identical. Each available machine (server) is equally likely to be chosen to process an arriving job. In the case of multiple machines with arbitrary interarrival and processing times (GI/G/m system), the average number of jobs waiting in the buffer can be approximated by

$$\overline{N} \cong \left(\frac{Cv_A^2 + Cv_X^2}{2} \right) \overline{N}^{M/M/m}, \qquad (3.14)$$

where $\overline{N}^{M/M/m}$ is the average number of jobs in the buffer in an M/M/m work station and is given by

$$\overline{N}^{M/M/m} = \left[\frac{\rho (m\rho)^m}{m!(1 - \rho)^2} \right] P_o, \qquad (3.15)$$

3.2. A Simple Work Station

P_o is the steady-state probability that the M/M/m work station is empty. The probabilities, $\{P_n\}$, are obtained by solving the steady-state flow-balance equations:

$$P_0 = \left[\sum_{n=0}^{m-1} \frac{(m\rho)^n P_0}{n!} + \frac{(m\rho)^m}{m!(1-\rho)}\right]^{-1}, \quad P_n = \begin{cases} \dfrac{(m\rho)^n P_0}{n!}, & 1 \leq n \leq m, \\ \dfrac{(m\rho)^n P_0}{m^{n-m} m!}, & n \geq m. \end{cases}$$

Here, the stability condition is $\rho = \lambda/m\mu < 1$.

The analysis of **work stations processing different types of products** is quite difficult. There are issues of set-ups, priorities, and switching between the products according to some criteria. The broader problem is more difficult and will be dealt with in Chapter 7 within the context of pull-type production systems. A rather simplified version of this problem is studied in queueing theory under priority queues. Let us assume a single-machine work station processing N types of products that arrive according to Poisson processes with rate λ_i for product type i. The processing time of product i is arbitrarily distributed with mean $E[X_i]$. Furthermore, products are processed according to a priority structure. Products of type 1 have the highest priority, while products of type N have the lowest. When the current job is completed, the job with the highest priority among those waiting in the buffer is chosen as the next to be processed. Finally, the priority structure is nonpreemptive, implying that the processing of a job cannot be interrupted by another, regardless of the priorities. The system is stable if $\sum_{i=1}^{N} \rho_i < 1$ where $\rho_i = \lambda_i E[X_i]$.

Let us focus on the delay of each type of jobs in the buffer. First of all, an arriving job, regardless of its priority, sees $\overline{N}_i$ of class i, $i = 1, N$, in the buffer at the time of its arrival. Its delay in a busy system consists of the remaining processing time of the current job, the time to clear all the higher-priority jobs that are present at the time of its arrival, and the new arrivals of higher priority during its delay in the buffer. The current job can be a type i job with probability ρ_i. Therefore, the expected remaining processing time of the current job is given by

$$E[X_r^c] = \sum_{i=1}^{N} \rho_i \frac{E[X_i^2]}{2E[X_i]}.$$

Then, a type 1 job, upon arrival, first waits until the current job is completed. It then waits for all the type 1 jobs that have arrived earlier. Other priority classes do not interfere with this job, since it has the highest priority. Thus, the delay of a type 1 job is given by

$$\overline{W}_1 = \overline{N}_1 E[X_1] + E[X_r^c],$$

and using Little's formula, we obtain the average number of type 1 jobs in the buffer:

$$\overline{N}_1 = \frac{\lambda_1 E[X_r^c]}{1 - \rho_1}. \tag{3.16}$$

A type 2 job first waits for $\overline{N}_1 E[X_1] + \overline{N}_2 E[X_2] + E[X_r^c]$ time units and for the time to clear the type 1 jobs arriving during its delay. Then, using the expected length of the busy period (3.5) we presented earlier in this section, we can write

$$\overline{W}_2 = \frac{\overline{N}_1 E[X_1] + \overline{N}_2 E[X_2] + E[X_r^c]}{1 - \rho_1},$$

resulting in

$$\overline{N}_2 = \frac{E[X_r^c]}{(1 - \rho^{[1]})(1 - \rho^{[2]})},$$

where $\rho^{[k]} = \sum_{j=1}^{k} \rho_j$. In a similar manner, we obtain $\overline{N}_j$ for job j,

$$\overline{N}_j = \frac{E[X_r^c]}{(1 - \rho^{[j-1]})(1 - \rho^{[j]})}. \tag{3.17}$$

It is also possible to process the jobs according to **the shortest processing time (SPT) rule**. That is, upon arrival, each job is assigned a processing time and placed in a position in the queue such that every job in front of it has a shorter processing time. Assuming a Poisson arrival process, let c be the processing time of an arriving job. The expected delay of this job includes the processing times of the jobs with $X < c$ in front of it upon arrival and of the jobs with $X < c$ that arrive after its arrival. The first part of its delay is $\rho E[X_r] + \overline{N}_c E[X|X < c]$. Using the same argument for the expected length of the busy period from earlier in this section, we can write

$$\overline{W}_c = \frac{\rho E[X_r] + \overline{N}_c E[X|X < c]}{1 - \rho_c}, \tag{3.18}$$

where $\overline{N}_c$ is the average number of jobs waiting in the buffer with a processing time less than c,

$$E[X|X < c] = \int_0^c x f_X(x) dx / P_c, \quad \text{with } P_c = \int_0^c f_X(x) dx,$$

and

$$\rho_c = \lambda P_c E[X|X < c] = \lambda \int_0^c x f_X(x) dx.$$

3.3. Distribution of the Number in the System   77

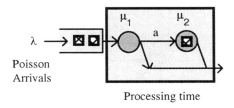

Figure 3.4. An M/G/1 work station with MGE-2 processing times.

Since the proportion of the arrivals that have $X < c$ is P_c, we have $\overline{N}_c = \lambda P_c \overline{W}_c$, resulting in

$$\overline{N}_c = \frac{\rho^2 (\mathrm{Cv}_X^2 + 1)}{2(1 - \rho_c)^2}, \tag{3.19}$$

which is the average number of jobs in the buffer with a processing time less than c. The average number of jobs in the buffer regardless of the processing times necessitates lifting the condition of $X < c$ in (3.19), involving a cumbersome integration.

There exist many other scenarios as to how a work station operates. For instance, it is possible to use policies of starting production only after a certain quantity of jobs arrive, processing in batches, and other policies. Furthermore, jobs may also arrive in batches. Some of these cases will be covered in the context of pull-type production systems in Chapter 7, especially work stations with set-ups and threshold policies that are used to start and stop production.

3.3 Distribution of the Number in the System

One is often interested in the long-run distribution of the number of jobs in the system at any point in time. In fact, the arbitrary-time performance measures in systems with finite capacities can only be obtained by first finding this particular distribution. Let us focus on a specific M/G/1 work station and derive its steady-state distribution of the number of units in the system. Assume that the processing times are phase-type and in particular of MGE-2 type. We begin by assuming unlimited WIP capacity. As shown in Figure 3.4, each unit starts its process in phase 1 and proceeds to phase 2 with probability a. The process is terminated after phase 1 with probability $1 - a$ or after phase 2 with probability 1. The mean processing time is $\mathrm{E}[X] = 1/\mu_1 + a/\mu_2$, and the station utilization is $\rho = \lambda \mathrm{E}[X]$. For stability, $\rho < 1$ must be satisfied. Let $\{N_t, J_t, t \geq 0\}$ be a stochastic process with N_t representing the number of units in the system, and J_t the phase that the processor is in at time t. Then, $\{N_t, J_t\}$ constitutes a continuous-time Markov chain with a unique steady-state distribution if $\rho < 1$. If (n, j) is a particular state

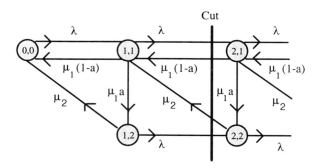

Figure 3.5. Transition diagram of $\{N_t, J_t\}$.

and $P(n, j)$ is its steady-state probability, the state transitions of $\{N_t, J_t\}$ occur according to the transition diagram given in Figure 3.5.

The steady-state flow-balance equations of $\{N_t, J_t\}$ are as follows:

$$\lambda P(0, 0) = \mu_1(1 - a)P(1, 1) + \mu_2 P(1, 2), \tag{3.20}$$

$$(\lambda + \mu_1)P(n, 1) = \lambda P(n - 1, 1) + \mu_1(1 - a)P(n + 1, 1)$$
$$+ \mu_2 P(n + 1, 2), \quad n \geq 1, \tag{3.21}$$

$$(\lambda + \mu_2)P(n, 2) = \lambda P(n - 1, 2) + \mu_1 a P(n, 1), \tag{3.22}$$

where $P(0, 1) = P(0, 0) = 1 - \rho$ and $P(0, 2) = 0$.

Consider for the moment the cut in Figure 3.5. In the long-run, the rate of transitions into the cut from the right must be equal to the one from the left. That is,

$$\lambda P(1, 1) + \lambda P(1, 2) = \mu_1(1 - a)P(2, 1) + \mu_2 P(2, 2),$$

or in general

$$\lambda P(n - 1, 1) + \lambda P(n - 1, 2) = \mu_1(1 - a)P(n, 1) + \mu_2 P(n, 2). \tag{3.23}$$

Let us next rewrite (3.22) and (3.23) in a form that pertains to a recursive solution. For $n = 1$,

$$\mu_1(1 - a)P(1, 1) + \mu_2 P(1, 2) = \lambda(1 - \rho),$$
$$\mu_1 a P(1, 1) - (\lambda + \mu_2)P(1, 2) = 0, \tag{3.24}$$

and for $n > 1$,

$$\mu_1(1 - a)P(n, 1) + \mu_2 P(n, 2) = \lambda P(n - 1, 1) + \lambda P(n - 1, 2),$$
$$\mu_1 a P(n, 1) - (\lambda + \mu_2)P(n, 2) = -\lambda P(n - 1, 2). \tag{3.25}$$

3.3. Distribution of the Number in the System

Thus, (3.24) and (3.25) naturally lead to a recursive solution where two equations are solved to obtain $P(n, 1)$ and $P(n, 2)$ for each $n = 1, 2, \ldots$. The explicit equations for the recursive solution are given below:

$$P(0, 0) = (1 - \rho);$$

for $n = 1$,

$$P(1, 1) = (1 - \rho)c_1(\lambda + \mu_2), \qquad P(1, 2) = \mu_1 a c_1 (1 - \rho);$$

for $n > 1$,

$$P(n, 1) = c_1(\lambda + \mu_2)\left[P(n-1, 1) + c_2 P(n-1, 2)\right],$$
$$P(n, 2) = \frac{1}{(\lambda + \mu_2)}\left[a\mu_1 P(n, 1) + \lambda P(n-1, 2)\right]; \qquad (3.26)$$

where

$$c_1 = \frac{1}{\mu_1(1-a) + \mu_1\mu_2\lambda^{-1}} \quad \text{and} \quad c_2 = \frac{\lambda + 2\mu_2}{\lambda + \mu_2}.$$

The marginal probabilities of the WIP inventory level are given by $P_n = P(n, 1) + P(n, 2)$ for $n > 0$.

It is also possible to convert the above procedure to obtain P_n to a **matrix-geometric procedure**. We can rewrite (3.24) and (3.25) as matrix equations. Let us first define the probability vector $\tilde{\underline{P}}(n)$ by

$$\tilde{\underline{P}}(n) = \begin{bmatrix} P(n, 1) \\ P(n, 2) \end{bmatrix}.$$

Then, from (3.24) we have

$$\begin{bmatrix} \mu_1(1-a) & \mu_2 \\ \mu_1 a & -(\lambda + \mu_2) \end{bmatrix} \tilde{\underline{P}}(1) = \lambda \begin{bmatrix} 1-\rho \\ 0 \end{bmatrix},$$

that is,

$$\tilde{\underline{P}}(1) = \underline{R}(1 - \rho),$$

where

$$\underline{R} = \begin{bmatrix} c_1(\lambda + \mu_2) \\ c_1\mu_1 a \end{bmatrix}.$$

3. Analysis of Single Work Stations

On the other hand, (3.25) can be rewritten as

$$\begin{bmatrix} \mu_1(1-a) & \mu_2 \\ \mu_1 a & -(\lambda + \mu_2) \end{bmatrix} \underline{\tilde{P}}(n) = \lambda \begin{bmatrix} 1 & 1 \\ 0 & -1 \end{bmatrix} \underline{\tilde{P}}(n-1),$$

resulting in the following matrix-recursive relation:

$$\underline{\tilde{P}}(n) = \mathbf{Q}\underline{\tilde{P}}(n-1), \qquad n > 1, \qquad (3.27)$$

where

$$\mathbf{Q} = \begin{bmatrix} c_1(\lambda + \mu_2) & \lambda c_1 \\ \mu_1 a c_1 & \mu_1 c_1 \end{bmatrix}.$$

From the above recursion, it is not difficult to obtain the following explicit matrix-geometric expression (since it involves the geometric term $\mathbf{Q}^{n-1}$):

$$\underline{\tilde{P}}(n) = (1-\rho)\mathbf{Q}^{n-1}\mathbf{R}, \qquad n > 0. \qquad (3.28)$$

Clearly, $P_o = 1 - \rho$ and $P_n = \underline{e}^T \underline{\tilde{P}}(n)$, where $\underline{e}^T$ is the vector of ones.

In the finite-capacity case, the above approach remains pretty much the same, except that we have a finite number of equations and a finite number of probabilities to compute. That is, for a buffer capacity of N, in addition to the set of flow-balance equations (3.20) to (3.22) for $n < N$, we have the following boundary conditions for $n = N$:

$$\mu_1 P(N, 1) = \lambda P(N-1, 1),$$
$$\mu_2 P(N, 2) = \lambda P(N-1, 2) + \mu_1 a P(N, 1),$$

resulting in

$$\underline{\tilde{P}}(N) = \frac{\lambda}{\mu_1 \mu_2} \begin{bmatrix} \mu_2 & 0 \\ a\mu_1 & \mu_1 \end{bmatrix} \underline{\tilde{P}}(N-1). \qquad (3.29)$$

In this case, one way to obtain the steady-state probabilities is to start with $P(0, 0) = 1$ and obtain $P_1, \ldots P_n$ using (3.28) and (3.29), and normalize all the probabilities.

EXAMPLE 3.2. Let us reconsider the work station studied in Example 3.1 to obtain the long-run probability distribution of the number of jobs in the system. The parameters of the station are $\lambda = 1.0$, $E[X] = 0.80$ and $Cv_X^2 = 0.5$. The processing-time distribution is Erlang type with two phases. The parameters of the matrix-geometric algorithm are

$$c = \frac{\lambda}{\mu_1 \mu_2} = 0.16, \qquad \mathbf{Q} = \begin{bmatrix} 0.56 & 0.16 \\ 0.40 & 0.40 \end{bmatrix}, \quad \text{and} \quad \underline{\mathbf{R}} = \begin{bmatrix} 0.56 \\ 0.40 \end{bmatrix}.$$

The long-run probability of having n jobs in the work station for various values of n are given in Table 3.1. The recursive and the matrix-geometric approaches provide exactly the same numerical results.

TABLE 3.1. Steady-state probabilities in Example 3.2

n	P_n
0	0.2
1	0.19199999427795
2	0.15231999546051
3	0.11550719655462
4	0.08651570942163
5	0.06457392959555
50	1.1641042416049D-07
100	4.8241576678165D-14
200	8.2847753555917D-27
300	1.4227873013049D-39
500	4.1962221057978D-65
700	1.2375904638055D-90
1000	6.2683708842392D-129

As demonstrated in Table 3.1, both of the above approaches are stable in the sense that they produce legitimate values for the probabilities no matter how large n is; that is, both methods are numerically stable. Empirical evidence also shows that the procedure is stable no matter what the values of the work station parameters are. Verification and testing of the results can be easily done by assuming exponential processing time and implementing the algorithm using the MGE-2 representation of the exponential distribution. The results of the algorithm must coincide with the distribution of the number of jobs in an M/M/1 queueing system. The algorithm can be easily extended to include processing times with more than two phases.

In the case of MGE interarrival times and exponentially distributed processing times, the underlying Markov chain is defined by $\{N_t, I_t\}$ where I_t is defined as the phase of the arrival process that the arriving job is in at time t, and N_t is again the number of jobs in the system at time t. The solution of the flow-balance equations for the steady-state probabilities of the number of jobs in the system is quite similar to the case of the M/MGE/1 work station. If the processing time is also MGE-distributed, the state description becomes $\{N_t, I_t, J_t\}$, where the phase that the job being processed is in at time t, (that is, J_t), is also added to the state of the system. In this case, the solution approach for the steady-state probabilities remains the same in principle, yet it becomes more involved. We will study this case within the concept of two-node production lines in the following chapter.

3.4 Work Stations Subject to Failures

Failures are an important source of idleness and randomness in manufacturing systems. The failure process is characterized by the statistical properties of the

82   3. Analysis of Single Work Stations

time between consecutive failures or by the counting process that keeps track of the number of failures within a given period of time. In the analysis of a single work station, it is generally assumed that upon a failure the work station experiences a stoppage and goes through a repair process with the assumption of an unlimited supply of repair crew. The repair time is clearly a loss of productive time. Usually, in a manufacturing environment, failures occur while the machines are actually processing jobs. These failures are called **operation-dependent**. On the other hand, the new breed of highly computerized machines may fail homogeneously in time regardless of the machine status. These failures are called **operation-independent**. Upon a failure, the unit being processed on the machine may be discarded, may go through the same process from the beginning, may receive a new independent processing time, or may continue being processed from the point of interruption. In the analysis of systems subject to failures, one approach is to treat failures and repairs as active elements in the problem. Another approach is to incorporate them into the time a unit spends on the machine and thereby eliminate the concept of failures from the analysis entirely. We will introduce both approaches in this chapter.

Consider a work station where jobs arrive from a Poisson process with rate λ per unit time. The processing time X has an arbitrary distribution, and jobs are processed based on a FIFO manner. Failures also follow a Poisson process with rate δ per unit time. The repair process starts immediately after a failure occurs. The repair time Y has also an arbitrary distribution with mean γ^{-1}. Let us assume that the process resumes on the same job after the repair is finished. We assume that the work station has an infinite-capacity buffer and is stable. We are interested in obtaining the long-run measures such as the average number of jobs in the system, the mean flow time of a job, the actual station utilization, and the probability that the system is down.

Due to failures and repairs, the work station goes through the cycles of up and down periods. Let P_d and P_u be the long-run probabilities of the system being down (in repair) and being up (idle or actually processing), respectively. Also, due to the random nature of the job arrivals, the work station exhibits another cyclic behavior, that is, it also goes through the cycles of idle and busy periods. Every time the system is cleared of jobs, that is, the work station enters the empty-and-idle state, a cycle of random length T starts and lasts until the system is cleared again. Due to our assumption that the job arrivals occur independently of each other (Poisson arrivals), once the system is cleared of jobs, the time until the processor is busy again is independent of what happened before. In other words, the system probabilistically regenerates itself every time it visits the empty-and-idle state. The existence of a sequence of regeneration points depends on the independence assumptions built into the system and the values of the system parameters. The Poisson assumption for the job arrivals and the failures automatically gives rise to the regeneration argument, which simplifies the analysis of the work station. Similar to the argument that resulted in (3.3), regardless of the failure/repair activity, the jobs that arrive during a regeneration cycle are processed during the actual busy periods of the station within the same cycle. Therefore, the actual station utilization remains to be $\rho = \lambda E[X]$, as if the station does not experience any failures. The busy period denoted by B starts when the station starts processing a job for the first time in a given cycle. It includes the possible repairs and ends when the system enters the next empty-and-idle state. If T_i is the ith up time during the busy period B and S is the total productive time during a cycle, then

$$S = T_1 + T_2 + T_3 \quad \text{and} \quad B = S + Y_1 + Y_2$$

in the case depicted in Figure 3.6.

The flow time of a job depends on the status of the work station at its arrival time. If the job arrives at an idle station, its flow time will simply be the busy period. If it arrives at any one of the up times (T_is), it first waits for the remaining processing time of the current job on the machine. Then it waits for the processing

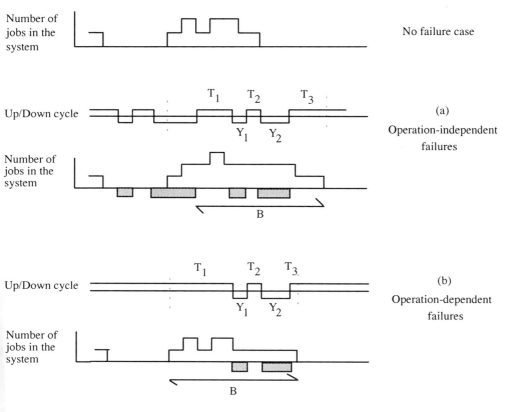

Figure 3.6. Sample path of the processor's status and the number of jobs in the system under different failure mechanisms.

84   3. Analysis of Single Work Stations

times of the jobs arrived earlier. It further waits for the possible repair activities that may be carried out during the processing of these jobs. If it arrives during a repair time, it additionally waits for the remaining repair time. Let X_r and Y_r represent the remaining processing time and the remaining repair time at the time of a job arrival, respectively. Since the failure process is independent of the processing time, the expected time a job spends on the processor is given by $E[X]\{1 + \delta E[Y]\}$. Notice that this period includes all the possible repairs that the station may go through while processing a unit. Similarly, the expected remaining time of the current job, which a newly arriving unit has to wait for, given that it arrives during processing and not repair, becomes $E[X_r]\{1 + \delta E[Y]\}$. Let $P(B)$ be the steady-state probability that the processor is in period B. Due to the PASTA property, it is also the probability that a job arrives during B rather than outside of B. Also, P_d is the probability that an arriving job finds the station down. In the light of this discussion, (3.9) can be expanded to express the average delay of a job (excluding its processing time) at a station subject to failures as follows:

$$\overline{W} = \left\{ P(B)E[X_r] + \overline{N}E[X] \right\} \{1 + \delta E[Y]\} + P_d E[Y_r], \tag{3.30}$$

where $E[X_r]$ and $E[Y_r]$ can again be obtained using (3.10), the concept of residual time, and the fact that the arrival process is Poisson. Also, $P(B)$ can be expressed as the proportion of the average busy period to the average cycle time:

$$P(B) = \frac{E[B]}{E[T]} = \frac{E[S] + \delta E[S]E[Y]}{E[T]} = \frac{E[S]\{1 + \delta E[Y]\}}{E[T]}$$
$$= \rho \{1 + \delta E[Y]\}. \tag{3.31}$$

Thus, (3.30) coupled with $\overline{N} = \lambda \overline{W}$ gives us

$$\overline{N} = \frac{P(B)^2(\mathrm{Cv}_X^2 + 1) + 2\lambda P_d E[Y_r]}{2[1 - P(B)]}, \tag{3.32}$$

and $\overline{W}$ can be obtained accordingly. The mean flow time of a job is again $\overline{W} + E[X]$. Special cases are given below:

(1) The work station may not experience failures; then

$$\overline{N} = \frac{\rho^2 \left(\mathrm{Cv}_X^2 + 1\right)}{2(1 - \rho)},$$

which agrees with (3.12).

(2) Both the processing and the repair times may be exponentially distributed; then

$$\overline{N} = \frac{P(B)^2 + \lambda P_d E[Y]}{1 - P(B)}. \tag{3.33}$$

(3) The processing time may be deterministic; then

$$\overline{N} = \frac{P(B)^2 + 2\lambda P_\mathrm{d} E[Y_r]}{2[1 - P(B)]}. \tag{3.34}$$

Again we observe that $\overline{N}$ of the case with the deterministic processing times is smaller than that of the cases with nonzero processing-time variability.

In the above expressions for $\overline{N}$, the steady-state probability that the system is down, that is, P_d, is yet to be determined. The value of P_d depends on the type of failures. In the case of **operation-independent failures**, stations may fail homogeneously in time, meaning that they may fail even if they are idle. This is not uncommon in computer-controlled machinery, where some of the failures are due to the problems in computer software and/or hardware. Clearly, the system goes through cycles of up and down times, and the up/down process is completely independent of the state of the system. The sample-path behavior of the operation-independent failures and the associated up/down process are shown in Figure 3.6(a). Once the system is repaired, the time until the next failure occurs is again exponentially distributed with a mean of δ^{-1}. Then, the expected cycle time of the up/down process becomes $\delta^{-1} + \gamma^{-1}$, and consequently

$$P_\mathrm{u} = \frac{\gamma}{\gamma + \delta} \quad \text{and} \quad P_\mathrm{d} = \frac{\delta}{\gamma + \delta}. \tag{3.35}$$

In the case of **operation-dependent failures**, the station experiences failures only when a job is being processed, as depicted in Figure 3.6(b). Again, we assume that the repair activity immediately takes over after the repair is completed. In this case, P_d and P_u have to be reevaluated, since (3.35) is no longer valid. However, the system again goes through the up/down cycles but only when there are jobs in the system, that is, only during period B. Consequently, we can write

$$P_\mathrm{d} = \frac{\delta}{\delta + \gamma} P(B), \tag{3.36}$$

which can also be shown by the following probabilistic argument:

$$\Pr(\text{the station is down} \mid \text{there are jobs to process}) = \frac{\delta E[S]E[Y]}{E[B]},$$

By removing the condition, we obtain

$$\Pr(\text{the station is down and there are jobs to process})$$
$$= P_\mathrm{d} = \frac{\delta E[S]E[Y]E[B]}{E[B]E[T]} = \delta \rho E[Y],$$

which is equivalent to (3.36). P_u is given by $1 - P_\mathrm{d}$.

86   3. Analysis of Single Work Stations

Let us emphasize the fact that the expressions for $\overline{N}$ and ρ do not depend on whether or not the failures are operation-dependent, as long as the repair activity takes over immediately after the failure, and the process resumes from the point of interruption after the repair is finished. However, the value of $\overline{N}$ depends on the failure type simply because P_d does so. The equivalence in the expressions should not be surprising. The main difference between the two types of failures is that an arriving job may find the processor down and the station empty in the case with operation-independent failures, while this is not possible in the operation-dependent case. However, as far as the arriving job is concerned, this down time is not any different than the ones in period B. Therefore, the expressions for the expected delay are the same. Clearly, the P_d of the case with operation-dependent failures is smaller than the one of the independent-failure case. Therefore, the average delay per job is shorter and the average job accumulation is smaller with operation-dependent failures.

The output rate in both of the above cases remains to be λ since the waiting space (buffer) has an infinite capacity and the system is assumed to be stable. The stability condition for both of the above cases can be expressed as $P(B) = \rho[1 + \delta E[Y]] < 1$, implying $\rho < P_u$. Let us observe the above results in the following example.

EXAMPLE 3.3. Consider the following work station subject to failures. The job arrival process is Poisson with rate $\lambda = 3.0$. The first and second moments, respectively, of the processing time are $E[X] = 0.20$ and $E[X^2] = 0.06$, with $Cv_X^2 = 0.50$. The first and second moments, respectively, of the repair time are $E[Y] = 2.50$ and $E[Y^2] = 35$, with $Cv_Y^2 = 4.60$. The time until a failure occurs is exponentially distributed with rate $\delta = 0.15$. The repair activity immediately follows a failure. The process resumes after the repair is completed. Let us first assume that the failures are operation-independent, and using (3.35) we obtain

$$P_u = 0.7273, \qquad P_d = 0.2727, \qquad \rho = 0.6,$$

indicating that the stability condition is satisfied. From (3.31), $P(B) = 0.8250$. Also,

$$E[X_r] = 0.15 \quad \text{and} \quad E[Y_r] = 7.$$

Then, using (3.32), we obtain the average number of jobs waiting to be processed, that is, $\overline{N} = 35.641$ and the average delay per job in the buffer is $\overline{W} = 11.880$.

In the case of operation-dependent failures, we have

$$P_u = 0.775, \qquad P_d = 0.225 \qquad \text{with } P(B) = 0.8250.$$

Again using (3.32), we obtain $\overline{N} = 29.917$ and $\overline{W} = 9.972$. Thus, it is clear that operation-independent failures tend to cause longer delays and larger accumulation of jobs in the buffer.

3.5 Distribution of the Number of Jobs in a Work Station Subject to Failures

Let us consider a work station experiencing **operation-dependent** failures where the remaining processing of the interrupted job resumes right after the repair is completed. Assume that the job arrival process is Poisson with rate λ and the processing times are exponentially distributed with rate μ. The time to failure is also exponentially distributed with rate δ. The repair-time distribution is MGE-2 type with phase rates γ_1 and γ_2 and the branching probability a from the first phase to the second. From the previous analysis, it is clear that the system is stable if $\rho[1 + \delta E[Y]] < 1$ or $\rho < P_u$, while $\rho = \lambda/\mu$ and $P_d = \rho \delta E[Y]$.

Similar to the case with no failures, let us again observe the stochastic process $\{N_t, J_t, t > 0\}$, where N_t is the number of jobs in the system and J_t is the status of the machine at the work station at time t. J_t is u if the machine is busy, d_1 if it is in the first phase, and d_2 if it is in the second phase of the repair process. $\{N_t, J_t\}$ is a continuous-time Markov chain with the transition diagram given in Figure 3.7. Notice the difference and the connection between this case and the one given in Example 2.3, observing Figures 2.4 and 3.7. Let (n, j) be a particular state and $P(n, j)$ its steady-state probability. The empty-and-idle state is represented by "state 0". The steady-state probability that there are n jobs in the system is given by

$$P_n = P(n, u) + P(n, d_1) + P(n, d_2), \tag{3.37}$$

with $P_o = 1 - \rho - P_d$.

Once again, we use the argument that for any such cut that is shown in Figure 3.7, the long-run number of entries into the cut in either direction remains the same. That is,

$$P(n, u) = \rho P_{n-1}, \quad n > 0. \tag{3.38}$$

The repair-state probabilities can be immediately obtained from the flow-balance equations for phases 1 and 2 of the repair process. For states (n, d_1) and (n, d_2), we have

$$\begin{bmatrix} \gamma_1 + \lambda & 0 \\ -\gamma_1 a & \gamma_2 + \lambda \end{bmatrix} \begin{bmatrix} P(n, d_1) \\ P(n, d_2) \end{bmatrix} = \lambda \begin{bmatrix} P(n-1, d_1) \\ P(n-1, d_2) \end{bmatrix} + \delta \begin{bmatrix} P(n, u) \\ 0 \end{bmatrix}, \tag{3.39}$$

where $n > 0$, resulting in

$$\underline{P}(n, d) = \begin{bmatrix} \dfrac{1}{\gamma_1 + \lambda} & 0 \\ \dfrac{\gamma_1 a}{(\gamma_1 + \lambda)(\gamma_2 + \lambda)} & \dfrac{1}{\gamma_2 + \lambda} \end{bmatrix} \left\{ \lambda \underline{P}(n-1, d) + \delta \begin{bmatrix} P(n, u) \\ 0 \end{bmatrix} \right\}. \tag{3.40}$$

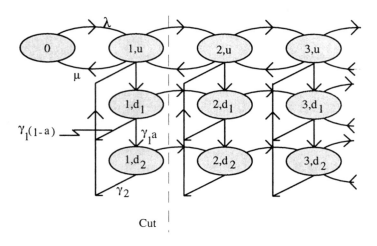

Figure 3.7. State transitions in a work station subject to operation-dependent failures.

where $n > 0$, with $\underline{P}(n, d)$ being the vector of the probabilities $P(n, d_1)$ and $P(n, d_2)$, and $\underline{P}(0, d)$ the vector of zeroes. Hence, since $P_d = \delta \rho E[Y]$, $P_o = 1 - \rho - P_d$, and $P(1, u)$ is obtained from the flow-balance equation of "state 0". Consequently, P_n can be obtained by using (3.37) and (3.40). This simple procedure can be readily extended to an MGE-k repair-time distribution for any k. Furthermore, the processing-time distribution can be MGE-k type also. The main approach would remain the same, involving vectors of the busy-period probabilities as well as the repair-state probability vectors.

The arrival process may be turned off during the repair time. In this case, (3.3) can be rewritten so that the probability that the processor is actually busy becomes

$$\rho_a = \sum P(n, u) = \frac{\lambda(1 - P_d)}{\mu} = \rho(1 - P_d),$$

and consequently $P_o = (1 - \rho)(1 - P_d)$, still with $\rho = \lambda/\mu$.

Furthermore, any repair-state probability $P(n, d)$ can be expressed as a function of only the up-time probability $P(n, u)$, leading to a relation between the probabilities of the cases with and without failures, that is,

$$P_n = cP'_n, \qquad n > 0,$$

where P'_n is the probability of having n jobs at the station that do not experience failures ($\delta = 0$), and c is a constant. Then, $\sum P_n = c \sum P'_n = c\rho = 1 - P_o$,

3.5. Distribution of the Number of Jobs to Failures

resulting in

$$c = \frac{1 - P_o}{\rho} = 1 - P_d - \frac{P_d}{\rho}.$$

$P_n = cP'_n$ allows us to compute the probabilities of the system with failures using the probabilities of the system without failures, or vice versa.

In the above case, the expression for P_d is not the same as (3.36) anymore. However, it can be obtained using

$$P_d = \sum P(n, d) = g_1[P(n, u)] = g_2(P_d),$$

where g_1 and g_2 are some functions of $P(n, u)$ and P_d resulting from the flow-balance equations, thus providing an expression for P_d.

For example, when the repair time is exponentially distributed, the flow-balance for state (n, d) gives us

$$P(n, d) = \frac{\delta}{\gamma} P(n, u).$$

Then,

$$\sum P(n, d) = \frac{\delta}{\gamma} \sum P(n, u) = \frac{\delta}{\gamma} \rho(1 - P_d) = P_d,$$

resulting in

$$P_d = \frac{\delta \rho}{\gamma + \delta \rho} \quad \text{and} \quad c = \frac{\delta + \gamma}{\delta \rho + \gamma}.$$

The probabilities $P(n, u)$ and $P(n, d)$ can again be found using the flow-balance equations:

$$P(n, u) = \rho^n (1 - \rho)(1 - P_d),$$

and

$$P(n, d) = \frac{\delta}{\gamma} P(n, u), \quad n > 0.$$

In case the repair time is an MGE-2 random variable, the down-time probabilities are given by

$$\underline{P}(n, d) = \delta P(n, u) \begin{bmatrix} 1/\gamma_1 \\ a/\gamma_2 \end{bmatrix}, \quad n > 0.$$

3. Analysis of Single Work Stations

Op.-Dep. Failures and Arrival Process is "On"

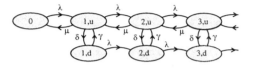

$\rho = \lambda/\mu \quad P_d = \rho\delta/\gamma$
$P_0 = 1 - \rho - P_d$
Pr(Actually busy) = ρ

Op.-Dep. Failures and Arrival Process is "Off"

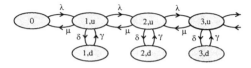

$P_d = \Sigma_n P(n,d) = \Sigma_n \frac{\delta}{\gamma} P(n,u) = \rho(1-P_d)\frac{\delta}{\gamma}$
$P_0 = 1 - \rho(1-P_d) - P_d = (1-\rho)(1-P_d)$
Pr(Actually busy) = $\lambda(1-P_d)/\mu = \rho(1-P_d)$

Op.-Indep. Failures and Arrival Process is "On"

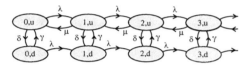

$P(0,u) = 1 - \rho - P_d \quad P(0,d) = P_0 P_d$
Pr(Actually busy) = ρ, $\quad P_d = \delta/(\delta+\gamma)$
$P_0 = 1 - \rho - P_d + P_0 P_d = 1 - \rho/P_u$

Op.-Indep. Failures and Arrival Process is "Off"

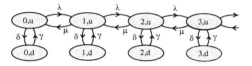

Pr(Actually busy) = $\rho(1-P_d) \quad P_d = \delta/(\delta+\gamma)$
$P(0,u) = 1 - $ Pr(Actuallybusy) $- P_d = 1 - \rho(1-P_d)$
$P(0,d) = \frac{\delta}{\gamma}P(0,u) \quad P_0 = P(0,u) + P(0,d) = 1 - \rho$

Figure 3.8. A structural summary of different failure scenarios

The basic structure of the underlying Markov chains in systems with operation-dependent failures with the job arrival processes "on" and "off" during down times are depicted in Figure 3.8.

If the failures are **operation-independent**, the only minor addition to the case where the arrival process is "on" during down times is that

$$P(0, u) = 1 - \rho - P_d \quad \text{and} \quad P(0, d) = P_o P_d, \quad \text{with } P_d = \frac{\delta}{\delta + \gamma},$$

resulting in $P_0 = 1 - (\rho/P_u)$, with $\rho = \lambda/\mu$. It can be observed from Figure 3.8—and it is also intuitive—that the expression for the probability of having n jobs in a work station with operation-independent failures is exactly the same as its counterpart in the case with operation-dependent failures. The sole difference between the two cases again comes from the down-time probability.

When the job arrival process is turned off, the probabilities $\{P_n\}$ of the station with operation-independent failures remain the same as those in the case where the same work station does not experience any failures. This is expected because the state of the system (the WIP level) does not change during the repair times. One can also show this result using the flow-balance arguments. All of the above-mentioned cases are summarized in Figure 3.8.

The finite-capacity scenario requires normalization of the probabilities after the boundary equations are solved.

3.6 The Process-Completion Time

There is a variety of failure-repair policies that can be implemented in a work station. For instance, after the repair is completed, processing may resume at the point of interruption, or it may restart from scratch and the new processing time may be either the same as the old one or different. The interrupted job may also be scrapped. This choice depends upon the process and how expensive the raw material is. Furthermore, it is possible to remove all the jobs from the work station upon a failure and turn the arrival process off during the repair time, which is studied in Example 2.3. Clearly, the values of the performance measures of our interest will be different for different policies. A common random variable of importance in all these cases is the time a job spends on the processor, referred to as the **process completion time** and denoted by C. The process completion time is equal to the processing time if no failures occur. However, in the case where the process resumes after the repair is completed, the process completion time is longer than the processing time due to possible repairs. In case the current job is scrapped upon a failure, it becomes the time until either the process is completed or a failure is encountered. For any given failure-repair policy, once the process completion time is characterized, it may be viewed as the processing time on an imaginary processor that does not experience failures. The performance measures of the original work station and that of the imaginary one are expected to be the same provided that other details (e.g., failures when the machine is idle) are taken care of. It is more likely that the analysis of the imaginary machine is simpler under a variety of failure-repair policies due to the potential phase-type approximation of the process completion time.

In the case of **process resumption after repair**, the expected time a job spends on the processor is

$$E[C] = E[X]\{1 + \delta E[Y]\}.$$

Accordingly, the utilization of the imaginary processor (percentage of the time that the imaginary processor has a job on it) is given by $P(B) = \lambda E[C]$, which coincides with (3.31). Hence, in the case of **operation-dependent failures**, the expected delay of an arriving job as a function of the process completion time is given by

$$\overline{W} = P(B)E[C_r] + \overline{N}E[C], \tag{3.41}$$

where $E[C_r]$ is the expected remaining process completion time at the point of a job arrival. $\overline{W}$ and $\overline{N}$ are obtained using (3.41), and $\overline{N} = \lambda\overline{W}$. Notice that (3.41) is nothing more than (3.30) and even (3.9), where X is replaced by C. This can be shown by providing $E[C_r]$ with a proper definition. C_r depends on the processor's status at the point of a job arrival. For any job arriving during period B, the expected remaining completion time is $E[X_r]\{1 + \delta E[Y]\}$. In addition, if the processor is down at the time of the arrival, $E[Y_r]$ becomes part of the remaining completion time. The probability of this event is simply the probability that a job arrives during a down period given that it arrives during B, and from (3.36), it is $P_d/P(B) = \delta/(\delta + \gamma)$. Thus, the expected remaining process completion time is given by

$$E[C_r] = E[X_r]\{1 + \delta E[Y]\} + \frac{E[Y_r]\delta}{\delta + \gamma}. \tag{3.42}$$

With the above definitions of C and C_r, (3.41) becomes equivalent to (3.30). Thus, we have regrouped some of the terms in (3.30) to be able to write (3.41). The potential benefit in doing that is to directly use (3.41) in those cases where we have information on the completion time.

In the case of **operation-independent failures**, one can still follow the process completion-time approach, yet with a slight revision. Remember that in this case the station may fail even when it is idle, and therefore it is not possible to fully incorporate all the failures into the process completion time. On the other hand, it is still possible to use the completion-time approach in which the processor may still fail but only when it is idle. The down-time probability that the imaginary processor may fail, but only when it is idle, is given by

$$P_d^o = \frac{\delta[1 - P(B)]}{\delta + \gamma}, \tag{3.43}$$

where $P(B) = \lambda E[C]$. Then, for the case of operation-independent failures with the policy of process resumption after repair, the delay of an arriving job is given by

$$\overline{W} = P(B)E[C_r] + \overline{N}E[C] + P_d^o E[Y_r], \tag{3.44}$$

where $E[C_r]$ is still given by (3.42). Again, (3.44) is equivalent to (3.30). We can generalize from here that for any failure-repair scenario, the expected delay of an arriving job can be obtained using the information about the process completion time and the down time. The expressions (3.41) and (3.44) are valid for any failure-repair policy. This result is convenient to compute the measures of performance when the moments of the process completion time are known. It is also possible to model the completion time by a phase-type distribution and use a matrix-recursive or a matrix-geometric algorithm to compute the steady-state probabilities of the number of jobs in the system.

3.7 Failure-Repair Policies

Having the concept of process completion time established, let us continue its analysis under different failure-repair policies. We will obtain expressions for the moments of the process completion time under different scenarios so that expressions (3.41) and (3.43) can be used to obtain certain measures of a work station. Next we will consider the following policies: process resumes after repair; process restarts from the beginning with the same processing time; only one failure may take place during the processing time with a fixed probability; and finally, the interrupted job is scrapped.

3.7.1 Process Resumes After Repair

Let us redefine the random variables of interest again just to refresh the concept. Let X be the processing time, Z be the time to failure, and Y be the repair time, with density functions $f_X(x)$, $f_Z(z)$, and $f_Y(y)$, respectively. Also, let δ and γ be the failure and the repair rates, respectively. Finally, let C be the completion time, with $f_C(c)$ being its density function. It is assumed that, after the repair is completed, the process resumes on the same job from the point of interruption.

While processing a job, the work station goes through a sequence of up times (U_i represents the ith up time) and down times (Y_i represents the ith down time). The end of an up time is either a failure or the completion of the process. It is clear that the down times are in fact the repair times. The behavior of a busy work station constitutes an alternating stochastic process with up/down states; a graphical representation of this is given in Figure 3.9. The stopping rule for this stochastic process is that the sum of the up times is equal to the processing time. Let K represent the number of failures to occur during the processing time of a job. If k failures occur, the process is completed at the end of the $(k+1)$st up time, and consequently we can write

$$X = U_1 + U_2 + \cdots + U_{k+1}.$$

Let us condition on $K = k$ failures and a fixed processing time x. Then the conditional LST of the density function of the process completion time becomes

$$E[e^{-sC} \mid K = k, X = x] = E[e^{-s(U_1 + \cdots + U_{k+1} + Y_1 + \cdots + Y_k)} \mid k = k, X = x].$$

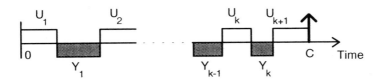

Figure 3.9. Stochastic behavior of a busy work station.

94   3. Analysis of Single Work Stations

Since $\{Y_i, i = 1, \cdots, k\}$ are i.i.d. random variables, independent of the processing time, we can write

$$\mathrm{E}[e^{-sC} \mid K = k, X = x] = e^{-sx}\{\mathrm{E}[e^{-sY}]\}^k. \tag{3.45}$$

If the LST of the process completion time and the repair times are denoted by $F_C^*(s)$ and $F_Y^*(s)$, respectively, then (3.45) becomes

$$\mathrm{E}[e^{-sC} \mid K = k, X = x] = e^{-sx}\{F_Y^*(s)\}^k. \tag{3.46}$$

Now let us remove the conditions, starting with the one on the number of failures:

$$\mathrm{E}[e^{-sC} \mid X = x\} = e^{-sx} \sum_{k=0}^{\infty} \frac{[F_Y^*(s)\delta x]^k}{k!} e^{-\delta x} = e^{[-sx-\delta x+\delta x F_Y^*(s)]}, \tag{3.47}$$

which would be the unconditional LST of the density function of C if the processing time were fixed at x. Removing the condition on X for the general case, we obtain

$$\mathrm{E}[e^{-sC}] = \int_0^{\infty} e^{-sx-\delta x+x\delta F_Y^*(s)} f_X(x)dx,$$

which in turn can be rewritten as

$$F_C^*(s) = F_X^*[s + \delta - \delta F_Y^*(s)]. \tag{3.48}$$

Equation (3.48) expresses the LST of the process completion time as the LST of the processing time evaluated at $s + \delta - \delta F_Y^*(s)$. For instance, in the case of Erlang processing time with n phases with μ being the rate at each phase, (3.48) becomes

$$F_C^*(s) = \left[\frac{\mu}{s + \delta + \mu - \delta F_Y^*(s)}\right]^n. \tag{3.49}$$

Considering the complexity of (3.48), moments of the process completion time become important since one can use them in an MGE approximation to obtain a simplified view of the work station. The first three moments of the process completion time (m_i, $i = 1, 2, 3$) in the case of Erlang-n processing times and general repair times are obtained using

$$m_k = (-1)^k \frac{d^k}{ds^k} F_C^*(s)|_{s=0}, \quad k = 1, 2, 3,$$

and are as follows:

$$m_1 = \frac{n}{\mu}(1 + \delta\overline{y^1}),$$

3.7. Failure-Repair Policies

$$m_2 = \left(1 + \frac{1}{n}\right) m_1^2 + \frac{n\gamma \overline{y^2}}{\mu}, \tag{3.50}$$

$$m_3 = \overline{y^3} \frac{\delta n}{\mu} + \frac{m_1 \overline{\delta y^2}}{\mu}(2n+1) + \left(1 + \frac{2}{n}\right) m_1 m_2.$$

The squared coefficient of variation of C is

$$\operatorname{Cv}_C^2 = \frac{1}{n} + \frac{n\delta \overline{y^2}}{\mu m_1^2}, \tag{3.51}$$

where $\overline{y^k}$ is the kth moment of the repair time. Notice that $\operatorname{Cv}_C^2 > 1$ if $n = 1$.

If the processing time is deterministic, say $X = x$, then (3.50) becomes

$$\begin{aligned} m_1 &= x(1 + \delta \overline{y^1}), \\ m_2 &= x\delta \overline{y^2} + m_1^2, \\ m_3 &= \overline{y^3} x\delta + 3\overline{y^2} x\delta m_1 + m_1^2, \end{aligned} \tag{3.52}$$

with the squared coefficient of variation

$$\operatorname{Cv}_C^2 = x\delta \frac{\overline{y^2}}{m_1^2}. \tag{3.53}$$

One has to be cautious in interpreting the results when (3.52) is used in a phase-type approximation since the resulting phase-type distribution is a continuous distribution that may yield particular realizations smaller than x. In fact, the true distribution of C is shifted to the right by x, and a continuous approximation becomes debatable in this case.

To further investigate the structure of (3.49), let us consider even a simpler case in which the processing and the repair times are both exponentially distributed with means μ^{-1} and γ^{-1}, respectively. For $n = 1$ in (3.49), we have

$$F_C^*(s) = \frac{\mu(s + \gamma)}{s^2 + s(\mu + \delta + \gamma) + \mu\gamma}, \tag{3.54}$$

which reminds us of the LST of the density function of the MGE distribution. An important observation at this point is that there should exist an MGE-2 distribution, shown in Figure 3.10, that is equivalent to the distribution of C. Matching the coefficients of s in (3.54) with the ones in (2.45), that is, the LST of the MGE-2 distribution, we obtain

$$\begin{aligned} \mu_1 &= \frac{\mu}{1 - a}, \\ \mu_2 &= \gamma(1 - a), \end{aligned} \tag{3.55}$$

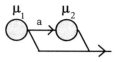

Figure 3.10. MGE-2 completion time.

$$1 - a = \frac{(\mu + \delta + \gamma) - \sqrt{(\mu + \delta + \gamma)^2 - 4\mu\gamma}}{2\gamma}.$$

Notice that $(\mu + \delta + \gamma)^2 - 4\mu\gamma$ is always positive. Thus, in case the processing time, the time to failure, and the repair time are all exponentially distributed, the process completion time becomes an MGE-2 random variable with the parameters given by (3.55). Based on this argument, one can interpret (3.49) as follows: In the case of Erlang-n processing times and exponentially distributed repair times, the process completion time is the sum of n of the above MGE-2 random variables. Consequently, due to the convolution property of the phase-type distributions (Section 2.8), the process completion time itself becomes an MGE random variable. If the resulting distribution has too many phases, one may want to resort to moment approximations to reduce the number of phases (if possible) for implementation purposes.

3.7.2 Process Restarts After Repair

After the repair process is completed, a processing time of the same duration as the one interrupted may start again from scratch. Then, a job leaves the work station only when its processing is completed without interruption, as shown in Figure 3.11. Let us assume that the processing and the repair times are arbitrary and the time to failure is exponentially distributed. Again, using arguments similar to the ones employed in Section 3.7.1, one can obtain an expression for $F_C^*(s)$. First, let $F_{Z|Z \leq X}(z \mid z \leq x) = P(Z \leq z \mid Z \leq X, X = x)$ be the cumulative probability distribution of the time until a failure occurs given that the failure occurs before the process is completed. Then,

$$F_{Z|Z \leq X}(z \mid z \leq x) = \frac{1 - e^{-\delta z}}{1 - e^{-\delta x}}, \qquad 0 \leq z \leq x, \qquad (3.56)$$

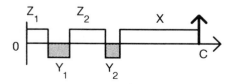

Figure 3.11. Completion time when the process restarts from scratch after repair.

3.7. Failure-Repair Policies

and the derivative of (3.56), that is, the conditional density function of Z, can be used to obtain the following conditional LST of the density function of Z:

$$E[e^{-sZ} \mid Z \leq X, X = x] = \frac{\delta}{\delta + s}\left[\frac{1 - e^{-(s+\delta)x}}{1 - e^{-\delta x}}\right]. \quad (3.57)$$

On the other hand, the probability that there are k failures before the process completion is geometrically distributed since the failures are independent of each other. That is,

$$P(K = k \mid X = x) = (1 - e^{-\delta x})^k e^{-\delta x}, \quad k = 0, 1, \ldots. \quad (3.58)$$

If k failures occur, the process completion time consists of the time until a failure occurs, for k times, k repair times, and the processing time. Then, the conditional LST of the density function of C becomes

$$E[e^{-sC} \mid K = k, X = x] = e^{-sx}\left(\frac{\delta}{\delta + s}\right)^k\left[\frac{1 - e^{-(s+\delta)x}}{1 - e^{-\delta x}}\right]^k [F_Y^*(s)]^k. \quad (3.59)$$

Now, let us first remove the condition on K,

$$E[e^{-sC} \mid X = x] = \frac{e^{-(s+\delta)x}}{1 - F_Y^*(s)(\frac{\delta}{\delta+s})[1 - e^{-(s+\delta)x}]},$$

which becomes the unconditional $F_C^*(s)$ in case the processing time is deterministic. Finally, removing the condition on X leaves us with

$$F_C^*(s) = \int_0^\infty \frac{e^{-(s+\delta)x} f_X(x)\, dx}{1 - F_Y^*(s)\left(\frac{\delta}{\delta+s}\right)[1 - e^{-(s+\delta)x}]}, \quad (3.60)$$

which is the unconditional LST of the density function of the process completion time.

The first two moments of the process completion time are

$$m_1 = \{E[e^{\delta X}] - 1\}\left[\overline{y^1} + \frac{1}{\delta}\right], \quad (3.61)$$

$$m_2 = 2\left[\overline{y^1} + \frac{1}{\delta}\right]^2 \{E[e^{\delta X} - 1]^2\} + \left[\overline{y^2} + 2\overline{y^1}\frac{1}{\delta} + \frac{2}{\delta^2}\right]\{E[e^{\delta X}] - 1\}$$

$$- 2\left[\overline{y^1} + \frac{1}{\delta}\right]E[Xe^{\delta X}].$$

Let us derive the above expression for m_1 using the mean-value arguments, which will make it clear how long the processor spends on each job in addition

3. Analysis of Single Work Stations

to the processing time. For convenience, let us assume that the processing time is deterministic, that is, $X = x$. Clearly, the probability that a failure occurs before a process is completed is $1 - e^{-\delta x}$. Then, the expected value of the time until a failure occurs, given that it occurs before the process is completed, is

$$E[Z \mid Z < x] = \frac{\delta}{1 - e^{-\delta x}} \int_0^x z e^{-\delta z} dz = \frac{1}{\delta} - \frac{x e^{-\delta x}}{1 - e^{-\delta x}}, \quad (3.62)$$

which can also be obtained by taking the derivative of (3.57). Also, from (3.58), the expected number of failures before the process is completed is

$$E[K \mid Z < x] = \frac{1 - e^{-\delta x}}{e^{-\delta x}}. \quad (3.63)$$

Then, the expected process completion time is given by (why?)

$$E[C] = x + \frac{1 - e^{-\delta x}}{e^{-\delta x}} \left(\frac{1}{\delta} - \frac{x e^{-\delta x}}{1 - e^{-\delta x}} + \overline{y^1} \right) \quad (3.64)$$

$$= (e^{\delta x} - 1) \left(\overline{y^1} + \frac{1}{\delta} \right),$$

which is equivalent to m_1 when X is deterministic.

The utilization of the imaginary processor is $\rho = \lambda E[C]$. Clearly, part of its busy period is spent on the reworks due to failures. Therefore, in (3.64), the terms after x (excluding $\overline{y^1}$ of course) represent the average time the processor spends on each job in addition to x. Then, the actual utilization of the processor becomes

$$\rho_a = \frac{\lambda (e^{\delta x} - 1)}{\delta},$$

However, we know that the imaginary processor is effectively producing only during part of its busy period. This leads to the effective actual utilization given by $\rho_e = \lambda x < \rho_a$.

EXAMPLE 3.4. Consider a work station with Poisson job arrivals with rate $\lambda = 1$ per unit time and a fixed processing time that is equal to 0.5 time units. The station is subject to operation-dependent Poisson failures with rate $\delta = 0.02$ per unit time, and the expected repair time is 15 time units. Given this scenario, the expected time to failure given that it occurs before the process is completed is

$$E[Z \mid Z < x] = 0.25.$$

The expected number of failures before a job departure is

$$E[K \mid Z < x] = 0.01005,$$

and the expected process completion time becomes

$$E[C] = 0.5 + 0.153 = 0.653.$$

That is, on the average, the processor spends an extra 0.153 time units on each departing job due to failures and reprocessing. Notice that the long-run average output rate is $\lambda = 1$, which would still be the same even if the station did not experience any failures. This is due to the "average input rate=average output rate" in the long run, as long as the system is stable. The utilization of the imaginary processor is $\rho = \lambda E[C] = 0.653$; the actual utilization is $\rho_a = 0.503$, as opposed to the effective actual utilization, which is $\rho_e = \lambda x = 0.50$, the percentage of the time the processor should have been busy in the case of no failures.

3.7.3 The Single-Failure Assumption

In practical applications, it may sometimes be appropriate to assume that at most one failure may occur during the processing time of a job, with a constant probability q. Let us assume that the process resumes on the same job, after the repair is completed, as shown in Figure 3.12. In this case, the LST of the density function of C is given by

$$F_C^*(s) = F_X^*(s)[1 - q + q F_Y^*(s)]. \tag{3.65}$$

For Erlang processing times with n phases and μ being the phase rate, the moments of the process completion time are given by

$$m_1 = \frac{n}{\mu} + q \overline{y^1},$$

$$m_2 = \frac{2nm_1}{\mu} - \frac{n(n-1)}{\mu^2} + q \overline{y^2}, \tag{3.66}$$

$$m_3 = \frac{n(n-1)(n-2)}{\mu^3} - \frac{3n(n-1)m_1}{\mu^2} + \frac{3nm_2}{\mu} + q \overline{y^3}.$$

The processing and repair times may be thought of as phases 1 and 2 of an MGE-2 distribution with q being the branching probability from the first to the

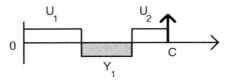

Figure 3.12. The process completion time in the case of a single failure.

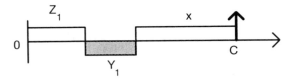

Figure 3.13. The process restarts in the case of single failure.

second phase if the processing and repair times are exponentially distributed. In this case, it is not difficult to find equivalence between (3.54) and (3.65).

If the policy is to restart the process after the repair is completed, as shown in Figure 3.13, then we need the information of when the machine fails during processing. In this case, we may assume that the time to failure is uniformly distributed during the processing time. This is a special case of the multiple-failures scenario described in the previous section, where $F_{Z|Z<x}(z \mid z \leq x) = z/x, 0 < z \leq x$.

3.7.4 Failures Causing Scrapping

In some cases, especially in high-speed manufacturing, it is likely that the job being processed is scrapped upon a failure. In this case, the time a job spends on the machine is the minimum of the processing time and the time until a failure occurs. Upon failure, the work station goes through a repair before the process of the next job starts. Although the process completion time of this case does not include the repair time, it is necessary to assume that the job is scrapped at the end of the repair rather than the beginning, as shown in Figure 3.14. This is to make the machine unavailable for processing until the end of the repair and make the analysis consistent with the completion-time approach for the purpose of obtaining $\overline{W}$ and $\overline{N}$. Thus, our pseudo-process completion time can be expressed as

$$C = \begin{cases} X & \text{w.p. } P(X < Z), \\ Z + Y & \text{w.p. } P(X > Z). \end{cases}$$

Let us again assume Poisson failures, arbitrary repair times, and, for simplicity, fixed processing times, that is, $X = x$. Then, the LST of the density function of

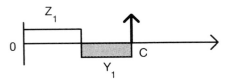

Figure 3.14. The process completion time if $Z < x$, and the job is scrapped after repair.

3.7. Failure-Repair Policies

the process completion time is given by

$$F_C^*(s) = F_X^*(s)P(Z \geq x) + F_Y^*(s)E[e^{-sZ} \mid Z \leq x]P(Z \leq x),$$

where $E[e^{-sZ} \mid Z \leq x]$ is given by (3.57). Hence,

$$F_C^*(s) = e^{-(s+\delta)x} + F_Y^*(s)\left\{\frac{\delta}{\delta+s}[1 - e^{-(s+\delta)x}]\right\}. \quad (3.67)$$

The first two moments of the completion time are

$$m_1 = \left[\overline{y^1} + \frac{1}{\delta}\right](1 - e^{-\delta x}), \quad (3.68)$$

$$m_2 = \left[\frac{2}{\delta^2} + \overline{y^2} + \frac{2\overline{y^1}}{\delta}\right](1 - e^{-\delta x}) - \left[\overline{y^1} + \frac{1}{\delta}\right]2xe^{-\delta x}.$$

The utilization of the imaginary processor whose processing time is C is $\rho = \lambda E[C]$. On the other hand, the expected time that a job is being processed is $(1 - e^{-\delta x})/\delta$. Then the actual machine utilization becomes $\rho_a = \lambda(1 - e^{-\delta x})/\delta$. However, this actual utilization is not effective, since some of the jobs—those that are processed to a point—are scrapped. The actual effective utilization is given by $\rho_e = \lambda x e^{-\delta x}$, that is, the percentage of the time the processor is busy with the unscrapped jobs.

EXAMPLE 3.5. If we apply the scrapping policy to the work station analyzed in Example 3.4, the measures of interest would be as follows: $E[C] = 0.647$, while $x = 0.5$. The expected time a job is processed is 0.498 time units. The actual utilization is $\rho_a = 0.498$, whereas the effective actual utilization is $\rho_e = 0.495$, which is slightly less than ρ_a. This is expected since, to produce one unit, the processor works more in the scenario with failures than in the one with no failures.

EXAMPLE 3.6. Let us consider a work station experiencing operation-dependent failures, with the policy that the process resumes on the interrupted job after the repair is completed. The job arrival process is Poisson with rate $\lambda = 1$ per unit time. The processing time is an Erlang random variable with two phases and a phase rate of 3.0. The failure process is Poisson with a rate of .05 failures per unit time, and the repair-time distribution is assumed to be exponential with rate 0.45 repairs per unit time.

From (3.50), the first three moments of the process completion time are

$$m_1 = 0.740741,$$
$$m_2 = 1.152263,$$
$$m_3 = 4.511507,$$

102   3. Analysis of Single Work Stations

and Cv_C^2 is 1.1.

Let us first compare the CDFs of the process completion time and its approximate MGE equivalent. The exact cumulative distribution function can be found using the convolution argument since the process completion time is the sum of two i.d.d. MGE-2 random variables, as mentioned in Section 3.7.1:

$$P_{EX}(C \leq c) = 1 - \left(\mu_1 cc_1^2 + c_1^2 - \frac{2c_1 c_2 \mu_2}{\mu_1 - \mu_2}\right) e^{-\mu_1 c}$$
$$- \left(\mu_2 cc_2^2 + c_2^2 + \frac{2c_1 c_2 \mu_1}{\mu_1 - \mu_2}\right) e^{-\mu_2 c}, \qquad c \geq 0, \quad (3.69)$$

where $c_1 = (\mu_1(1-a) - \mu_2)/(\mu_1 - \mu_2)$, $c_2 = 1 - c_1$, and μ_i is the phase rate at the ith phase and a is the branching probability in the MGE-2 random variable. In our example,

$$\mu_1 = 3.0586252, \qquad c_1 = 0.9776005,$$
$$\mu_2 = 0.4413747, \qquad c_2 = 0.0223995.$$

The above process completion-time distribution can be approximated by a two-phase MGE distribution using all three moments. The corresponding approximate MGE-2 distribution and the values of its parameters are

$$P_{APX}(C \leq c) = 1 - r_1 e^{-\mu_1' c} - r_2 e^{-\mu_2' c}, \qquad c \geq 0, \qquad (3.70)$$

where

$$\mu_1' = 1.3548373, \qquad r_1' = 0.9997452,$$
$$\mu_2' = 0.0899782, \qquad r_2' = 0.0002548.$$

Comparison of the exact and the approximate process completion-time distributions for a certain range of the completion time is given in Figure 3.15. The comparison supports the use of the simplified MGE approximation.

Next, let us compute the steady-state probability distribution of the number of jobs in the above work station. Without loss of generality, we assume that the work station has a buffer with a finite capacity of six jobs making it a loss system. The buffer capacity includes the place on the machine. We compute the probability distribution of the number of jobs in two ways: (i) considering (3.69) as the CDF of the process completion time, which is the exact approach; and (ii) considering (3.70) as the CDF of the proces completion time, which is the approximate approach. The underlying M/PH/1/N work station can be analyzed using the method presented in Section 3.3, where the process completion time is treated as the processing time. The results are shown in Figure 3.16. The approximation again seems to be reasonable and suggests the use of simplified forms of the process completion time in the analysis of a work station.

3.7. Failure-Repair Policies 103

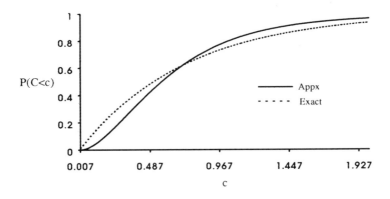

Figure 3.15. CDFs of the exact and approximate process completion times.

EXAMPLE 3.7. Consider a work station where the processor expreriences operation-independent failures. Upon a failure, the job being processed (if there is one) is scrapped. Job arrivals are assumed to come from a Poisson distribution with $\lambda = 0.2$ per unit time. The processing time is constant, and $x = 5$ time units. The first two moments of the repair time are $E[Y] = \gamma^{-1} = 1.0$ and $E[Y^2] = 5.0$. Finally, the failure process is assumed to be Poisson with rate $\delta = 0.15$. Let us find the long-run average number of delayed jobs in the work station.

The moments of the process completion time can be found using (3.68) and are as follows:

$$m_1 = 4.045,$$
$$m_2 = 20.359,$$

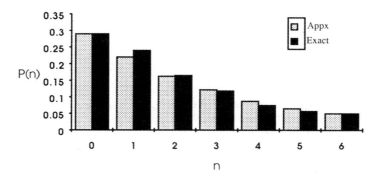

Figure 3.16. The probability distribution of the buffer contents.

and the corresponding squared coefficient of variation is $Cv^2 = 0.224$.

Now we have an imaginary processor with the moments of its processing time equal to m_1 and m_2. This imaginary processor will not fail while processing; however, it may fail during the idle period. Consequently, we can write

$$P(B) = \lambda m_1 = 0.809,$$

$$P_d^o = \frac{\delta}{\gamma + \delta}[1 - \lambda m_1] = 0.0249.$$

The average delay experienced by a job before processing starts is given by (3.44), that is,

$$\overline{W} = P(B)E[C_r] + \overline{N}E[C] + P_d^o E[Y_r],$$

where $E[C_r] = m_2/2m_1 = 2.517$ and $E[Y_r] = 2.5$. Using $\overline{N} = \lambda \overline{W}$, the above expression for $\overline{W}$ gives us $\overline{W} = 10.987$ and $\overline{N} = 2.197$.

The probability that the original work station is down is $P_d = \delta/(\delta + \gamma) = 0.130$. The actual processor utilization is $\rho_a = \lambda(1 - e^{-\delta x})/\delta = 0.704$. The output rate of the unscrapped jobs is $\lambda e^{-\delta x} = 0.0945$. However, the effective actual utilization is $\rho_e = 0.472$. That is, the processor is spending 23 percent of its time on the jobs that are to be scrapped.

So far, we have considered cases where the immediate attention of the repair crew is available when machines fail. However, in a job-shop environment, the repair facility may have a limited capacity and a failed machine may therefore wait to receive service. In other words, machines may interfere with each other in case they experience failures. This problem is known as the **machine interference** or the machine–repairman problem. It can be studied under a variety of assumptions and within different contexts. We have already introduced the problem with simple assumptions in Example 2.2. In the next section, we elaborate further on the machine interference problem.

3.8 The Machine Interference Problem

In the basic machine interference problem, there are N machines subject to failures and there is a repair crew who services them. Machines are assumed to be in constant operation unless they are down. When a machine fails, it receives service from a repairman, if any are available. Otherwise, it waits for its turn. Once a machine is repaired, it immediately starts operating again. Machines experience delays when the repairmen are busy. That is, machines interfere with each other. In this problem, the performance measures of our interest usually consist of the long-run average number of down machines, the average delay in starting a repair, and the repairman utilization. There have been deterministic as well as probabilistic treatments of the problem using both analytical and simulation approaches depending on the assumptions. Probabilistic treatments have usually consisted of queueing theory and queueing network approaches, as well as simulation. Next we introduce the machine interference problem, focusing on queueing and queueing network models, and raise some new research issues.

3.8.1 Exponential Failure and Repair Times

Consider a shop floor with N machines and S repairmen, where $S < N$. The time until a failure occurs on each machine is independent of the others and exponentially distributed with mean δ^{-1}. The repair times are also independent and exponentially distributed with mean γ^{-1}. Each machine operates independently until a failure occurs and then joins an imaginary queue to wait for the service of a repairman. The machines are taken into service according to the FIFO (first-in–first-out) rule. It is equally likely for each available repairman to work on a failed machine. The repair facility resembles a multiserver work station.

Let N_t be the number of down machines at time t. At any moment, the time until a failure occurs is the minimum of the times to failure on all machines that are operational, and it is exponentially distributed. The time to a repair completion is the minimum of the remaining repair times of busy repairmen and is also exponentially distributed. Hence, at any moment, the time until a change occurs in N_t is exponentially distributed. Clearly, $\{N_t, t \geq 0\}$ becomes a continuous-time, discerete-state Markov chain. Let P_n be the steady-state probability that n machines are down. The steady-state flow-balance equations of $\{N_t\}$ can be written as

$$[(N-n)\delta + v(n)\gamma]P_n = (N-n+1)\delta P_{n-1} + u(n+1)\gamma P_{n+1}, \quad (3.71)$$

$$n = 0, 1, \ldots, S, \ldots, N,$$

where

$$v(n) = \begin{cases} n & \text{if } n < S, \\ S & \text{if } n \geq S, \end{cases}$$

$$u(n+1) = \begin{cases} n+1 & \text{if } n < S, \\ S & \text{if } n \geq S, \end{cases} \quad \text{and} \quad P_{-1} = P_{N+1} = 0.$$

These equations can be recursively solved to obtain

$$P_n = \begin{cases} \binom{N}{n}\left(\frac{\delta}{\gamma}\right)^n P_0, & 0 \leq n \leq S, \\ \binom{N}{n}\left(\frac{n!}{S!S^{n-S}}\right)\left(\frac{\delta}{\gamma}\right)^n P_0, & S \leq n \leq N. \end{cases} \quad (3.72)$$

Finally, P_0 can be obtained through normalization. Measures of interest can be easily computed from $\{P_n\}$. A special case of this problem, where $S = 1$, is presented in Example 2.2.

3.8.2 Exponential Repair Times: A Queueing Network Approach

The shop floor together with the maintenance facility in the machine interference problem can be viewed as a queueing network with two stations, where the machines represent the customers in the system, as shown in Figure 3.17. The first queue represents the shop floor, the second queue represents the repair facility, and N machines circulate in the system. The first node is an infinite server (or N server) work station since all the up machines are running simultaneously and the processing time at this station is the time until a failure occurs.

The single-repairman machine interference problem can be treated as a closed network of quasi-reversible queues[1] resulting in the following steady-state probability of n machines being down:

$$P_n = \frac{B\gamma^{-n}\delta^{N-n}}{(N-n)!}, \qquad n = 0, \ldots, N, \qquad (3.73)$$

where B is found through normalization. When the time until a failure occurs is arbitrary, yet identical on all the machines, the first node in Figure 3.17 behaves as an M/G/∞ queue and (3.73) still remains valid since M/G/∞ is a quasi-reversible queue with δ^{-1} being the expected processing time. The basic argument here is that a closed queueing network is quasi-reversible if the individual queues or stations (with Poisson arrival processes) in isolation are quasi-reversible. A quasi-reversible network of queues has a steady-state probability distribution that can

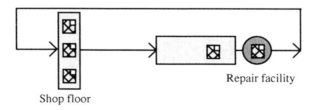

Repair facility

Shop floor

Figure 3.17. Queueing network representation of the machine interference problem.

---

[1] A queue is a quasi-reversible queue if its state forms a stationary Markov process with the property that the system state is independent of both the arrival and the departure times. M/M/1 and M/G/∞ are examples of quasi-reversible queues. We refer the reader to Kelly (1979) for details in reversible queueing networks.

be expressed as the product of the individual steady-state probability distributions (product-form solution).

It is possible to extend the above product-form solution to the case where the machines differ. That is, machine i, $i = 1, 2, \ldots, N$, remains in station i until a failure occurs, and it goes back to station i after its repair is completed. The stochastic process $\{X_0(t), X_1(t), X_2(t), \ldots, X_N(t)\}$ with $X_i(t) = 0$ or 1, $i > 0$, representing the status of the ith machine and $\underline{X}_0(t) = (t(1), t(2), \ldots, t(n))$ being the sequence of failed machines at time t, represents the system behavior. If the time to failure is exponentially distributed with mean δ_i^{-1} on machine i, then the above stochastic process becomes a Markov chain with a steady-state probability distribution having the following product form:

$$P(\underline{X}_0, X_1, X_2, \ldots, X_N) = B'\gamma^{-n} \prod_{i=1}^{N} \delta_i^{-X_i}, \qquad (3.74)$$

where $B' = BN!$. Notice that when the machines are different, the order in which they fail becomes important. For instance, when n machines are down, there are $n!$ possible orders in which they may fail, Consequently, the probability that n machines are down becomes the sum of the joint probabilities of all these possibilities, resulting in

$$P_n = B'n!\gamma^{-n} \prod_{i=n+1}^{N} \delta_i^{-1}.$$

Again, (3.74) remains valid for general failure times due to the quasi-reversibility of queues $i = 1, 2, \ldots, N$. Clearly, the general failure times bring higher complexity to the state definition, since additional information is necessary on how long each machine has been operating. However, the above expression for P_n remains the same for the general failure case. The repair times have been assumed to be exponentially distributed in all the cases since the M/M/1 queue is the only single-server quasi-reversible queue with the FIFO discipline.

3.8.3 General Repair Times

In the case of arbitrary repair times, the repair facility can be viewed as an M/G/1 work station with state-dependent Markovian arrivals. Let us assume that the machines are identical and the time to failure is exponentially distributed with mean δ^{-1}. The repair times are arbitrarily distributed with mean γ^{-1}. Then the arrival process at the repair facility becomes Poisson with rate

$$\lambda_j = \begin{cases} (N - j)\delta, & 0 \le j \le N, \\ 0, & j \ge N, \end{cases}$$

indicating that the machines fail with rate λ_j when j machines are already down. Hence, the repair facility can be modeled as a single-repairman M/G/1 work station with state-dependent arrivals.

Again, let P_n be the steady-state probability that n machines are down. Let T be the length of the time between two successive arrivals of machines finding the repair facility empty and idle. Then each entry into the empty-and-idle state initiates a regeneration cycle. Using this regeneration argument, we can write $P_n = E[T_n]/E[T]$, for all n.

Next, let us define A_{jk} as the amount of time k machines are down during a repair time that started with j down machines, $j \leq k$. Then, we can write

$$E[T_n] = E[A_{1n}] + \sum_{j=1}^{n} E[M_j]E[A_{jn}], \qquad n = 1, 2, \ldots. \qquad (3.75)$$

and

$$E[T_0] = \frac{1}{\lambda_0},$$

where M_j is the number of repair completions in a cycle at which j down machines are left behind, $j = 0, 1, 2, \ldots$. Notice that the very first repair, which initiates the busy period of the repairman, starts with one down machine. Also, there should be $j(j \leq n)$ down machines at the start of a repair to reach the state of n down machines before the repair ends. The number of "down" machines can only increase during a repair time.

As mentioned earlier in Section 3.1, in an M/G/1 work station, the long-run average measures as observed at the departure points (repair completions) are the same as the ones observed at the arrival points (machine failures) and the same as the arbitrary-time averages. Therefore, we can write

$$\frac{E[T_j]}{E[T]} = \frac{E[M_j]}{E[M]},$$

where M is the total number of arrivals in a cycle, resulting in

$$E[M_j] = \lambda_j E[T_j]. \qquad (3.76)$$

Dividing both sides of (3.75) by $E[T]$ and using (3.76) with $1/E[T] = \lambda_0 P_0$, we obtain

$$P_n = \lambda_0 P_0 E[A_{1n}] + \sum_{j=1}^{n} \lambda_j P_j E[A_{jn}], \qquad n = 1, 2, \ldots. \qquad (3.77)$$

which enables us to compute P_n, $n \geq 1$, as a function of P_0 that in turn is obtained through normalization.

The computation of $E[A_{jk}]$ is somewhat involved:

$$E[A_{jk}] = \int_0^\infty \binom{N-j}{k-j}(1 - e^{-\delta t})^{k-j}(e^{-\delta t})^{N-k}[1 - F_Y(t)]dt, \quad (3.78)$$
$$1 \leq j \leq k \leq N,$$

where $F_Y(t)$ is the CDF of the arbitrary repair time. From the definition of A_{jk}, there are j down machines and $N - j$ operating machines at the beginning of the repair. There have to be $k - j$ machines failing during this repair time with probability $(1 - e^{-\delta t})^{k-j}$ to have k down machines before the repair ends. Notice the binomial distribution that takes care of all the possibilities in which the system goes from state j to k down machines during the time interval $(0, t)$. Consequently, $P_n, n = 0, 1, \ldots, N$, can be obtained using (3.77) and (3.78). Measures such as the average down time and the repairman utilization are obtained using P_n. Notice that this is the first time we have employed the above approach where the repair facility is observed at repair completion epochs. As mentioned in Section 3.1, this is one way to analyze the M/G/1-type work stations. In case the repair times are MGE type, one can study the arbitrary-time Markov chain, as we have done in Chapters 2 and 3.

For the case with general repair times and S repairmen, we can extend the argument of "the effective input rate=the effective output rate" at the repair facility, introduced in Example 2.2 in Chapter 2. Again, let us assume that the time to failure is exponentially distributed. Since a Poisson arrival (arrival of a failed machine) sees on the average $\overline{N}_q$ machines waiting to be repaired and $\overline{N}_r$ machines being repaired, the expected arrival rate of failed machines at the repair facility is $\delta(N - \overline{N}_q - \overline{N}_r)$. The effective output rate of the repair facility is $\overline{N}_r \gamma$. Thus, we have

$$\delta(N - \overline{N}_q - \overline{N}_r) = \overline{N}_r \gamma,$$

resulting in

$$\overline{N}_q = N - \overline{N}_r \frac{\delta + \gamma}{\delta}, \quad (3.79)$$

which is an explicit relationship between the number of machines receiving repair and the number of machines waiting for repair. It is also possible to obtain the following important design measure, that is, given that N, δ, and γ are fixed and $r = \overline{N}_r/S$ and $q = \overline{N}_q/N$, the number of repairmen S is

$$S = \frac{(1 - q)\delta N}{r(\delta + \gamma)}, \quad (3.80)$$

for fixed values of q and r. Hence, the necessary value for S can be obtained from the above equation for given values of r and q.

Clearly, extensive research is being done on the machine interference problem, with a variety of assumptions. It would be quite rewarding to consider this problem within the context of job shops or transfer lines. The machine interference problem does not deal with the jobs on the machines or with the fact that machines may be idle for a period of time during which failures may not occur. Moreover, in this type of environment, machines are usually different from one another and have different failure characteristics. Thus, performance issues relevant to the shop floor are actually inseparable from those relevant to the machines. However, for the purpose of tractability, the two issues have been separated in all the related studies in the literature. Future research in the failure-repair models of shop floors will hopefully deal with these issues simultaneously.

Related Literature

Analytical queueing-theoretic models of manufacturing work stations with various assumptions have been extensively studied in the literature. Books such as Kleinrock (1975), Gross and Harris (1974), Çinlar (1975), Neuts (1981), Tijms (1986), and more recent ones as Wolff (1989), Cooper (1990), Buzacott and Shanthikumar (1993), and Perros (1994) have many examples and cases of single work stations and their analysis to obtain the steady-state probabilities and related measures. Among the specific problems mentioned in this chapter, the approximation for the average number of jobs waiting in the buffer of a GI/G/m work station is based on the heavy-traffic limit theorems by Whitt (1983) and reported by Suri et al. (1993). Work stations with different types of products fall into the category of priority systems and have been studied by Cobham (1954), Jaiswal (1968), Neuts (1981), and Shanthikumar (1989), among many others. The analysis of M/G/1 queues where units are served according to the SPT rule is due to Shanthikumar (1982). Matrix-geometric analysis of stochastic service systems was introduced by Neuts (1981).

The problem of machine failures has been studied by various authors within the context of queueing systems; see, for instance, White and Christie (1958), Keilson (1962), Gaver (1962), Avi-Itzhak and Naor (1963), Thiruvengadam (1963), Mitrani and Avi-Itzhak (1968), and more recently by Federgruen and Green (1986), and Altiok (1989). The process completion-time approach was due to Gaver (1962) and Avi-Itzhak and Naor (1963). Mixed types of failures were studied by Nicola (1986).

The earliest study of the machine interference problem is due to Palm (1947). It has also been introduced in many books such as Feller (1968), Kelly (1979), Neuts (1981), Bhat (1984), Heyman and Sobel (1982), Tijms (1986), and Kashyap and Chaudhry (1988), with varying levels of complexity. Kelly (1979) extended the product-form solution to the case where the machines differ. Also, the single-work-station approach with general repair times is due to Tijms (1986). Finally, Heyman and Sobel (1982) proposed the design measure for the number of repairmen given in (3.80). A review of the literature can be found in Stecke and Aronson (1985).

References

Altiok, T., 1989, Queueing Modeling of a Single Processor with Failures. *Performance Evaluation*, Vol. 9, pp. 93–102.

Avi-Itzhak, B., and P. Naor, 1963, Some Queueing Problems with the Service Station Subject to Breakdown. *Operations Research*, Vol. 10, pp. 303–320.

Bhat, U. N., 1984, *Elements of Applied Stochastic Processes*. John Wiley, New York.

Buzacott, J. A., and J. G. Shanthikumar, 1993, *Stochastic Models of Manufacturing Systems*. Prentice Hall, Englewood Cliffs, NJ.

Çinlar, E., 1975, *Introduction to Stochastic Processes*. Prentice Hall, Englewood Cliffs, NJ.

Cobham, A., 1954, Priority Assignment in Waiting Lines. *Operations Research*, Vol. 2, pp. 70–76.

Cooper, R. B., 1990, *Introduction to Queueing Theory.*, 2nd ed. North Holland, Amsterdam.

Federgruen, A., and L. Green, 1986, Queueing Systems with Service Interruptions. *Operations Research*, Vol. 34, pp. 752–768.

Feller, W., 1968, *Introduction to Probability Theory and Its Applications*. John Wiley, New York.

Gaver, D. P., 1962, A Waiting Line with Interrupted Service Including Priorities. *J. Royal Stat. Soc.*, B24, pp. 73–90.

Gross, D., and C. M. Harris, 1974, *Fundamentals of Queueing Theory*. John Wiley, New York.

Heyman, D. P., and M. J. Sobel, 1982, *Stochastic Models in Operations Research.*, Vol. 1. McGraw Hill, New York.

Jaiswal, N. K., 1968, *Priority Queues*. Academic Press, New York.

Kashyap, B. R. K., and M. L. Chaudhry, 1988, *An Introduction to Queueing Theory*. A & A Publications, Kingston, Ontario.

Keilsen J., 1962, Queues Subject to Service Interruption. *Ann. Math. Stat.*, Vol. 33, pp. 1314–1322.

Kelly, F. P., 1979, *Reversibility and Stochastic Networks*. John Wiley, New York.

Kleinrock, L., 1975, *Queueing Systems, Vol. 1: Theory*. John Wiley, New York.

Kraemer, W., and M. Langenbach-Belz, 1976, Approximate Formulae for the Delay in the Queueing System G/G/1. *Proc. 8th Intl. Teletraffic Congress*, Melbourne.

Mikou, N., O. Kacimi, and S. Saadi, 1994, Two Processors Interacting Only During Breakdown: The Case Where the Load Is Not Lost. to appear in *Queueing Systems*.

Mitrani, I. L., and B. Avi-Itzhak, 1968, A Many Server Queue with Service Interruptions. *Operations Research*, Vol. 16, pp. 628–638.

Neuts, M. F., 1981, *Matrix-Geometric Solutions in Stochastic Models—An Algorithmic Approach*. Johns Hopkins University Press, Baltimore, MD.

Nicola, V. F., 1986, A Single Server Queue with Mixed Types of Interruptions. *Acta Informatica*, Vol. 23, pp. 465–486.

Palm, C., 1947, The Distribution of Repairmen in Servicing Automatic Machines. *Industritidningen Norden*, Vol. 75, pp. 75–80, 90–94, 119–123.

Perros, H. G., 1994, *Queueing Networks with Blocking*. Oxford University Press, New York.

Ross, S. M., 1980, *Introduction to Probability Models*. Academic Press, New York.

Shanthikumar, J. G., 1982, On Reducing Time Spent in M/G/1 Systems. *European J. Operations Research*, Vol. 9, pp. 286–294.

Shanthikumar, J. G., 1989, Level Crossing Analysis of Priority Queues and a Conservation Identity for Vacation Models. *Naval Research Logistics*, Vol. 36, pp. 797–806.

Stecke, K. E. and J. E. Aronson, 1985, Review of Operator/Machine Interference Models. *Intl. J. Production Research*, Vol. 23, pp. 129–151.

Suri, R., J. L. Sanders, and M. Kamath, 1993, Performance Evaluation of Production Networks. *Handbooks in OR and MS* (S. Graves, et al., eds.), Elsevier Science Publishers.

Thiruvengadam, K., 1963, Queueing with Breakdowns. *Operations Research*, Vol. 11, pp. 62–71.

Tijms, H. C., 1986, *Stochastic Modeling and Analysis—A Computational Approach*. John Wiley, New York.

White, H.C., and L. S. Christie, 1958, Queueing with Preemptive Priorities or with Breakdown. *Operations Research*, Vol. 6, pp. 79–95.

Whitt, W., 1983, The Queueing Network Analyzer. *Bell Syst. Technical J.* Vol. 62, pp. 2779–2815.

Wolff, R. W., 1989, *Stochastic Modeling and the Theory of Queues*. Prentice Hall, Englewood Cliffs, NJ.

Problems

3.1. Consider a work station where jobs arrive according to a Poisson process with rate $\lambda = 0.1$ per unit time. The following data on the processing time (time units) is collected:

6.5 5.2 5.9 8.2 9.3 5.7 5.9 7.1 6.9 8.5 8.2 5.9 6.3 5.1 6.4 6.9 4.3 8.7 9.1

(a) Based on the above data, compute the average number of jobs waiting to be processed at the work station.

(b) Compute the steady-state probabilities P_0, P_1, P_2, and P_3. You may use MGE-type representations of the processing time.

(c) If we know that the job is arriving at a busy system, how many jobs on the average will there be in the system at the point of arrival?

(d) Once the processor starts a busy period, on the average how long does it take to clear the system (expected length of the production (busy) period)?

3.2. Let us consider a work station with hyperexponentially distributed (H_2=type) processing times and a finite buffer capacity of five jobs, including the position on the machine. The parameters of the processing time are $(\mu_1, \mu_2) = (2, 7)$ and $(p_1, p_2) = (0.3, 0.7)$. Jobs arrive according to a Poisson process with rate $\lambda = 2$ per unit time. Jobs that arrive while the buffer is full are diverted to other stations. Compute the output rate and the average number of jobs waiting to be processed at the work station.

3.3. Consider a work station with three identical parallel machines. Each available machine has the same chance to be assigned to an arriving job. The mean and the squared coefficient of variation of the processing time on any machine are $m_1 = 15$ minutes and $Cv^2 = 0.15$, respectively. The job interarrival times have a mean of 10 minutes and a squared coefficient of variation of 1.7. Obtain the expected number of jobs in the system in steady state (perhaps an approximation to it).

3.4. Consider a work station where jobs arrive according to a Poisson process with rate $\lambda = 1$ per unit time. The processing time is Erlang type with $K = 5$ phases and a mean $m_1 = 0.75$. After clearing all the jobs, if the processor remains idle for 5 or more time units, it has to go through a set-up process before it starts processing another job. The set-up starts at the arrival instance of the first job. The set-up time is uniformly distributed between 10 and 20 time units. Hint: Due to the memoryless property of the exponential distribution (interarrival time), this case is equivalent to the case with a fixed probability of setting up the system after every idle time. Consequently, the regeneration argument may still be used.

(a) What is the stability condition for this work station? Is it satisfied?
(b) What are the steady-state probabilities that the station is empty and idle, producing, and in set-up?
(c) What is the average number of jobs waiting in the buffer to be processed?
(d) If the processing time is a decision variable, say x (and deterministic), what is the optimal value of x that minimizes the average total cost per unit time for a holding cost of $h = 1$/unit per unit time, and a set-up cost of $\Theta = 15$/set-up?
(e) What happens to (d) if a set-up occurs every time a busy period starts?

3.5. Consider a work station equipped to process three types of jobs. Type 1 job is the highest-priority job, and type 3 is the lowest. Each job type arrives according to a Poisson process with rates 2, 5, and 1 per unit time, respectively. The first two moments of the processing times are given below:

Moments	Job Type		
	1	2	3
m_1	0.1	0.05	0.3
m_2	0.0125	0.003	0.12

(a) Obtain the average number of jobs of each type in the buffer.
(b) If each class has the same priority, and the jobs are processed according to the FIFO rule, suggest an approach to find the average total number of jobs in the system.

3.6. Consider a work station with an Erlang-2 processing time with rate 5 units per unit time and a Poisson job arrival process with rate λ per unit time. The input buffer cannot house any of the arriving jobs when the station is busy. These jobs are diverted to an overflow station.

(a) For what value of the arrival rate λ is the output rate of the system equal to 4 units per unit time?
(b) Assume that the overflow station has an exponential processing time with rate $\mu = 20$ per unit time and a buffer of infinite capacity. Suggest a method to study the overflow station to obtain its steady-state probabilities and the average number of jobs in the overflow station for the arrival rate found in part (a).

114   3. Analysis of Single Work Stations

3.7. Develop a matrix-recursive approach similar to the one in Section 3.3 for a work station with an MGE-2 job arrival process with parameters $(\lambda_1, \lambda_2, a)$ and an exponential processing time with rate μ.

3.8. Extend the matrix-geometric approach (3.28) to analyze a work station where the processing time is Erlang type with $Cv^2 = 0.1$. Test your results for numerical instability with $\lambda = 3$ and the mean processing time of 0.25.

3.9. Consider a work station where the processing time is of MGE-2 type with parameters $\mu_1 = 2, \mu_2 = 4$ and $a = 0.8$. Jobs arrive according to a Poisson process with rate 0.75 per unit time. Let us assume that the processing times are assigned from the given distribution at the points of arrival and the jobs are processed according to the SPT rule. If an arriving job is assigned 0.50 time units, what is the average number of jobs that are ahead of this job, and what is its expected delay going to be in the buffer?

3.10. Consider a work station subject to failures. Jobs arrive according to a Poisson process with rate $\lambda = 0.60$ per unit time. The time until a failure occurs is exponentially distributed with rate $\delta = 0.15$. The repair time and the processing times are also exponentially distributed with rates $\gamma = 0.5$, and $\mu = 1.0$ per unit time, respectively. The process resumes after the repair is completed.

(a) Obtain the steady-state probability that the processor is down and the average number of jobs waiting to be processed, assuming operation-dependent failures. What happens if the job arrival process is turned off during the repair times?
(b) Repeat (a) for operation-independent failures.
(c) Repeat (a) for more general processing and the repair times with means 1.0 and 4.0, and the squared coefficient of variations 0.25 and 1.5, respectively.

3.11. Consider stations 1 and 2 shown below. The processing times and the job interarrival times are exponentially distributed with respected rates. Station 1 experiences operation-independent failures, and station 2 never fails. The time to failure is exponentially distributed with rate δ. Upon a failure, all the jobs including the one being processed are transferred to station 2 and the repair process starts immediately.

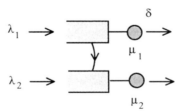

The repair time is also exponentially distributed with rate γ. During the down time, the arriving jobs at station 1 are sent to station 2. The job arrival stream is diverted back to station 1 as soon as it is repaired. The analysis of station 1 in isolation is given in Example 2.3. Here, using that analysis, develop an approach to obtain the average number of jobs in station 2. What is the output rate of station 2? A more involved version of this problem was studied by Mikou, et al. (1994).

3.12. Show that the work station experiencing operation-independent failures with the arrival process turned off during the repair times has exactly the same marginal probabilities of the number of jobs as it would be in the same work station had it not experienced any failures.

3.13. At a particular station, the processing time, the time to failure, and the repair time are all exponentially distributed with rates μ, δ, and γ, respectively. The process resumes after the repair is completed. Represent the process completion time by the following phase structure:

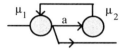

Identify μ_1, μ_2, and a in terms of μ, δ, and γ such that the two process completion times are equivalent.

3.14. Assume that the processing time is hyperexponentially distributed with parameters $(\mu_1, \mu_2) = (2, 4)$ and $(p_1, p_2) = (0.3, 0.7)$. The time until a failure occurs is exponentially distributed with rate $\delta = 0.25$. The repair time is exponentially distributed with rate $\gamma = 1.0$. The job arrival process is Poisson with rate $\lambda = 1.5$. The process resumes after the repair is completed.

(a) Develop an expression for the LST of the process completion time. Compute its mean and the squared coefficient of variation.
(b) Compute $\overline{N}$ and P_d assuming operation-independent failures.

3.15. A work station is subject to operation-dependent failures and the process restarts again after the repair is completed. The job arrival process is Poisson with rate $\lambda = 5.0$. The processing time is $x = 0.10$ time units. The time to failure is exponentially distributed with rate $\delta = 0.05$, and the repair-time distribution is MGE type with $(\mu_1, \mu_2, a) = (8.0, 2.0, 0.05)$.

(a) Compute the mean and the squared coefficient of variation of the process completion time.
(b) Compute $\overline{N}$, P_d, and the average time a job spends in the system.

3.16. A work station experiences operation-independent failures and the interrupted job is scrapped. The processing time, the time to failure, and the repair time are all exponentially distributed with rates μ, δ, and γ, respectively. Jobs arrive according to a Poisson process with rate λ.

(a) Develop an expression for the LST of the density function of the process completion time. Obtain its mean and its squared coefficient of variation.
(b) Develop an expression for $\overline{W}$ and $\overline{N}$.
(c) Obtain the average actual processor utilization and the percentage of the scrapped jobs.

116   3. Analysis of Single Work Stations

3.17. Consider a work station with random processing times with a mean of 0.6 and a variance of 0.25. Jobs arrive following a Poisson distribution with rate $\lambda = 0.6$. Failures are operation-dependent and cause scrapping of the current job. The time to failure is exponentially distributed with rate $\delta = 0.15$. The repair time has a mean of 2.5 and a variance of 32.0. Assume that the processing and the repair time distributions are of MGE type.

(a) Find the mean and the squared coefficient of variation of the process completion time.
(b) Find the steady-state down-time probability and the actual server utilization.
(c) Find the average number of jobs waiting to be processed.
(d) What is the percentage of the scrapped jobs?
(e) What percentage of the processor's busy time is spent on the scrapped jobs?

3.18. Consider a work station that is subject to operation-dependent failures. Assume that the processing time is uniformly distributed in (x_1, x_2). We expect that only one failure may occur during the processing time of a job with a fixed probability q, and this failure may occur uniformly during the processing time. The repair time is exponentially distributed with rate γ. The process resumes after the repair is completed.

(a) Develop an expression for the LST of the process completion time. Obtain its mean and its squared coefficient of variation.
(b) Develop an expression for the steady-state average number of jobs waiting to be processed.
(c) Find the mean process completion time for the case where the process restarts after the repair is completed.

3.19. Consider a work station subject to a possible single failure during the processing time with a fixed probability q. The unit processing time is x time units. Assume that the time to failure is uniformly distributed in $(0, x]$. The process restarts when a failure occurs. The repair time is exponentially distributed with rate γ.

(a) Develop an expression for the LST of the process completion time, and find its first two moments.
(b) What is the actual processor utilization?
(c) What percentage of the processor's busy time is spent on reruns?
(d) Compute $\overline{N}$ and P_d for $q = 0.20$, $x = 2.0$, $\gamma = 5.0$, and $\lambda = 0.3$.

3.20. Consider a work station that may fail once during a processing time with probability q. The processing time is Erlang with K phases with a phase rate of μ. The time to failure is uniform within the processing time. The repair time is exponentially distributed with rate γ. Develop an expression for the mean process completion time if

(a) the process restarts for the same duration,
(b) the process restarts for an independent processing time from the same distribution,
(c) the interrupted job is scrapped.

3.21. Consider a work station experiencing operation-dependent failures. Jobs arrive according to a Poisson process with rate $\lambda = 0.004$ per unit time. The processing time is fixed at $x = 25$ time units. The time to failure is exponentially distributed with rate $\delta = 0.01$, and the repair time is uniform between 40 and 60 time units.

(a) Obtain the mean and the squared coefficient of variation of the process completion time for the case of process resumption after repair.
(b) Repeat (a) for the policy of scrapping upon a failure.
(c) If p is the price of a finished unit, and h is the unit holding cost per unit time, for what values of p and h is scrapping justified? [Hint: Develop an average profit expression and compare the two alternatives.]

3.22. The processor restarts after the repair is completed. The processing time is fixed at x, the time to failure is exponentially distributed with rate δ, and the repair time is exponentially distributed with rate γ. Assume that the failures are operation-dependent. Jobs arrive according to a Poisson process with rate λ. For $\lambda = 0.50$, $\delta = 0.10$ $\gamma = 1.5$, and $x = 1.0$,

(a) compute the mean and the squared coefficient of variation of the process completion time,
(b) compute $\overline{N}$, P_d, and the actual processor utilization,
(c) find the percentage of the processor's busy time spent in reruns.

3.23. There are four identical machines in a shop floor. The time to failure and the repair time are exponentially distributed with rates δ and γ, respectively. There are two repairmen in the repair facility. For $\delta = 0.50$ and $\gamma = 2.0$ for each machine,

(a) compute the utilization of the repair facility and the expected number of machines waiting to be repaired.
(b) What should be the number of repairmen if one plans for a repairman utilization of 0.75 and on the average one machine waiting to be repaired at any point in time?

3.24. There are three identical machines in a shop floor with a single repairman. The time to failure is exponentially distributed with rate $\delta = 0.25$. The repair-time distribution is weighted hyperexponential with a mean of 2.5 and a variance of 10.0. Compute the steady-state repairman utilization and the average number of down machines waiting to be repaired.

4

Analysis of Flow Lines

4.1 Introduction

Various manufacturing facilities in different industrial sectors have some type of series arrangements of production stages. Jobs in a variety of forms are transferred from one stage to another to be processed in some order and leave the system as finished or semifinished products. The flexibility generated by computer-controlled machinery enables production stages to handle a variety of operations. This in turn creates a sequence of intelligent work stations in the shop floor, processing or assembling different types of products. Each work station may consist of one or more machines, one or more operators (robots may also be involved), and a work-in-process buffer, as depicted in a rather abstract form in Figure 4.1. Upon process completion at a work station, jobs join the work-in-process buffer at the next work station, if space is available. Conceivably, some cases may require repetition of a certain process or a set of processes for a number of times. Some of the work stations may process jobs in batches. Transfer of jobs from one station to another is usually done by means of vehicles or conveyors. Eventually, jobs depart from the system, which may happen practically at any of the work stations. In general, flow lines are **push-type** systems where little attention is given to the finished-product inventory. The line is to produce as much as it can with the assumption that all finished products are either used or demanded. Otherwise, it has to stop producing when there is an excessive accumulation of finished products. Those systems are known as **pull-type** systems and will be discussed in Chapter 7.

Due to space limitations and cost considerations, there exist explicit or implicit target levels for storage between stages in a flow line. Consequently, the flow of jobs is likely to experience stoppages due to the limited space between work stations. This type of stoppage propagates backward among the stations, where a longer stoppage at a station may cause the upstream stations to stop due to excessive accumulation of the work-in-process inventories in between. Similarly, starvation may be experienced, causing idleness in stations due to a lack of jobs to process from the upstream stages. Starvation propagates forward in such a way that a station that remains idle for a long time causes idleness and therefore stoppage in the downstream stations due to a lack of jobs in the system.

4.1. Introduction

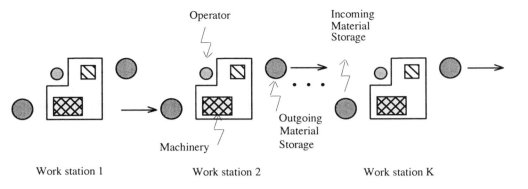

Figure 4.1. Work stations in series with buffers in between.

Another type of idleness in flow lines is caused by failures. Usually, a failed station is taken into repair as soon as possible, and it resumes its operation once it is back up again. Clearly, longer down times cause idleness in the downstream machines. The failure/repair policies and the policies of handling the interrupted job introduced in Chapter 3 can be implemened in any of the work stations. Also, the process completion-time approach is convenient to model failures in flow lines.

The purpose in the analysis of flow lines is to be able to evaluate their performance for a given set of system parameters. The most widely sought-after measures of performance in flow lines are the average throughput, average inventory levels in the buffers, downtime probabilities, bottleneck (blocking) probabilities, and the average system flow time. These measures will lead to better designs of these systems by showing those areas where loss of productivity is experienced. The loss of productivity occurs in those periods where machines are idle due to either failures or bottlenecks which result in excessive accumulation of inventories. The uncertainty in flow lines is mainly due to variable processing times, and more so to failure occurrence and repair times. The randomness makes it difficult to control these systems or to predict their behavior and this is where algorithm development for performance modeling becomes important.

Design problems in flow lines include the work-load allocation and the allocation of buffer capacities for a given set of work stations with processing times. In general, design problems are quite difficult to solve in manufacturing systems. This is due in part to the combinatorial nature of the problems and in part to the lack of explicit differentiable equations for the measures of performance involved in the design problems. We will visit these problems later in the chapter.

Storage limitations in work stations give rise to a concept of bottleneck called **blocking**. It causes stoppages at work stations due to lack of space or excessive accumulation of in-process inventory in the downstream stages. There are various scenarios in which a station gets blocked as well as various events that occur during blocking. Blocking policies differ depending on whether or not a work station starts processing a new job upon filling up the immediate downstream buffer. When a process completion occurs at a work station, the immediate downstream buffer may become full. If it is not full, and if there are jobs to process, the work station will start processing a new job. Even when the next buffer is full, since the work station is operational, it may start processing a new job. In this case, when the process is completed, if the next buffer is still full, the work station gets blocked. This type of blocking policy is called **blocking-after-processing** (**BAP**). The blocking job remains at the work station, keeping the machine idle, until space is available in the next buffer. That is, blocking will be over when the process at the next station is over. Another scenario is that the work station may not start processing a new job until space becomes available in the next buffer. This is referred to as **blocking-before-processing** (**BBP**). Note that during a BBP blocking, a new job may be placed on the machine (not to process but to store). The BAP policy is more likely to be encountered in flow lines.

4.2 Characteristics of Flow Lines

Consider the flow line in Figure 4.2, where M_i denotes the ith machine and B_i denotes the buffer between M_{i-1} and M_i. Buffer B_i has a finite-capacity N_i, which includes the position on M_i. Let X_i be the random variable representing the processing time on M_i. The first machine may have a sufficient amount of raw material so that it does not experience starvation. Another possibility is that M_1 may have a buffer to which jobs arrive in a particular manner (either randomly or according to a schedule).

Let $P_i(n_i)$ be the steady-state probability distribution of the number of jobs (or units) in B_i, where $n_i = 0, \ldots, N_i$. Note that $P_i(0)$ is the probability that M_i is idle. Also, let $P_i(B)$ be the probability that M_i is blocked at any point in time. Then, the utilization of M_i, that is, the percentage of time M_i is producing, is given by

$$P_i(U) = 1 - P_i(0) - P_i(B), \qquad (4.1)$$

indicating that M_i produces when it is not idle and not blocked. One can include the downtime probability into (4.1) in case M_i experiences failures. However, in this chapter we will not deal with failures; we will assume that failures are handled using the concept of process completion time introduced in Chapter 3.

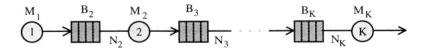

Figure 4.2. A pictorial representation of a production line.

While M_i is actually busy (with probability $P_i(U)$), it is producing with rate $1/E[X_i]$, yielding a throughput $\bar{o}_i$ given by

$$\bar{o}_i = \frac{P_i(U)}{E[X_i]} \tag{4.2}$$

In case jobs are not lost or scrapped, the long-run flow rate of jobs in a line with K machines is the same at every station or machine, that is,

$$\bar{o}_1 = \bar{o}_2 = \cdots = \bar{o}_K. \tag{4.3}$$

That is, the station throughputs are the same and equal to the line throughput, $\bar{o}_\ell$. The throughput can be obtained accordingly if the jobs are scrapped with a certain probability at any station in the line.

Let us next look at some equivalence relations between flow lines with different blocking policies. Consider two flow lines, namely L1 and L2, that have identical work stations. In L1, the BAP policy is in effect with $(N_1, N_2, \ldots, N_K)$ being the vector of buffer capacities. In L2, the BBP policy is in effect with the machines not holding jobs during blocking, with buffer capacities $(N_1+1, N_2+1, \ldots, N_K+1)$. Then, the two lines L1 and L2 have the same throughput. Intuitively, the position on M_i in line L1 and the $(N_i + 1)$st position in buffer B_i in line L2 function in the same manner.

Furthermore, in two-station flow lines, the one with the BAP policy and a buffer of capacity $N - 1$ has the same behavior as the one with the BBP policy with a buffer of capacity N. This is valid only for two-station lines, yet it provides the following bounds for the throughput in lines with multiple machines:

$$\bar{o}_\ell(\text{BBP}, \underline{N}) \leq \bar{o}_\ell(\text{BAP}, \underline{N}) \leq \bar{o}_\ell(\text{BBP}, \underline{N}+1). \tag{4.4}$$

It is also possible to write

$$\bar{o}_\ell(\text{BAP}, \underline{N}-1) \leq \bar{o}_\ell(\text{BAP}, \underline{N}) \leq \bar{o}_\ell(\text{BAP}, \underline{N}+1)$$

due to the monotonic behavior of the line throughput as a function of the buffer capacities. However, (4.4) provides a tighter bound due to the following relation:

$$\bar{o}_\ell(\text{BAP}, \underline{N}-1) \leq \bar{o}_\ell(\text{BBP}, \underline{N}) \leq \bar{o}_\ell(\text{BAP}, \underline{N}).$$

As the buffer capacities increase, the line throughput converges to a fixed value, that is, $\lim_{N \to \infty} \bar{o}_\ell = \bar{o}_\ell(\infty)$, which happens to be the throughput of the line with infinitely large buffer capacities. In this case, the throughput is the production rate of the slowest station, which must be the first station due to the stability argument. That is, if M_1 is not the slowest machine (having the maximum expected processing time), then the upstream buffers of the slowest machine will keep building large accumulations, causing instability in the flow line. Therefore, the slowest machine has to be the first machine in flow lines with large or infinite buffer capacities.

Finally, let us discuss the **reversibility property**. A flow line operating with the BAP policy retains its throughput if it is reversed. That is, M_K becomes M_1, B_K becomes B_2, and so on. In the case of BBP policy, the two lines not only maintain the same throughput, but also have the probability distributions of the buffer contents complementing each other. That is, $P(n)$ in a buffer in the original line is the same as $P(N - n)$ in the same buffer in the reversed line, resulting in the sum of the average contents of the same buffer in the two lines being equal to its capacity, for each buffer. The reversed line can be thought of as the system of holes or the empty cells moving in the opposite direction of the physical flow in the original line. The movement of the holes and the physical units are synchronized under the BBP policy. These relationships play an important role in understanding the behavior of flow lines.

4.3 Analysis of Two-Station Flow Lines

In this section, we focus on two-station lines with a finite-capacity buffer in between and with the first station having an infinite supply of raw material. Two-station lines are essential in introducing solution methodologies as well as behavioral aspects of flow lines. They also form building blocks in the analysis of longer lines.

4.3.1 Exponential Processing Times

Let us first consider a two-station flow line, as shown in Figure 4.3, with exponential processing times and a finite-capacity buffer in between. The first station, M_1, has a sufficient amount of raw material such that it does not experience starvation. It gets blocked when a departing job finds the intermediate buffer full, that is. the blocking policy is of BAP type. Blocking is over as soon as the second station, M_2, completes processing its current job. During blocking, M_1 simply holds the blocking job and remains idle. The second station does not experience blocking. The intermediate buffer, B, has a finite capacity N, including the position on M_2.

Generally, the processing times in manufacturing systems tend to have low variability (low Cv^2), and therefore they may not be properly characterized by an exponential distribution. However, the assumption of exponentiality forms a good basis to introduce the Markovian approach for the analysis of flow lines. We will later extend the processing-time assumptions to include phase-type distributions

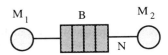

Figure 4.3. A two-station flow line.

4.3. Analysis of Two-Station Flow Lines

that can assume any variability. Let λ and μ be the processing rates at stations 1 and 2, respectively. Due to the exponential processing times, the number of units observed in the buffer is sufficient to describe the system state at any point in time. The stochastic process $\{N_t, t \geq 0\}$, where $N_t = 0, 1, 2, \ldots, N, N+1$, representing the level of in-process inventory at time t, is a continuous-time, discrete-state Markov chain with $N+2$ states. $N_t = N+1$ is the state where the first station is blocked. The possible events causing a state change in the above Markov chain are the process completions. The level of in-process inventory increases by one upon a process completion at station 1 and decreases by one upon a process completion at station 2. Note that the first station is always busy unless it is blocked. At any point in time, the time until the next state change occurs is either the remaining processing time at station 1 (when the second station is idle), or the remaining processing time at station 2 (when the first station is blocked), or the minimum of the remaining processing times at the two stations. The time until any one of these events occurs is exponentially distributed, as is the time until the next event occurs. Due to the memoryless property of the exponential distribution, $\{N_t\}$ gives rise to a Markov chain. Let n be a particular state of $\{N_t\}$ and P_n be its steady-state probability. Then, the steady-state flow-balance equations of $\{N_t\}$ can be written as follows:

$$\lambda P_0 = \mu P_1, \tag{4.5a}$$

$$(\lambda + \mu) P_n = \lambda P_{n-1} + \mu P_{n+1}, \quad 1 \leq n \leq N, \tag{4.5b}$$

$$\mu P_{N+1} = \lambda P_N. \tag{4.5c}$$

The structure of the above equations makes the following simple recursive scheme possible. Letting $\rho = \lambda/\mu$, from (4.5a) we have

$$P_1 = \rho P_0,$$

and from (4.5b),

$$P_2 = (\rho + 1) P_1 - \rho P_0 = \rho^2 P_0.$$

Similarly,

$$P_n = \rho^n P_0, \quad n \leq N,$$

and from (4.5c), we have

$$P_{N+1} = \rho P_N = \rho^{N+1} P_0,$$

resulting in

$$P_n = \rho^n P_0, \quad 1 \leq n \leq N.$$

Then, $\sum_{n=0}^{N+1} P_n = 1$ yields

$$P_o = \frac{1-\rho}{1-\rho^{N+2}},$$

thus providing the explicit expression for P_n:

$$P_n = \frac{(1-\rho)\rho^n}{1-\rho^{N+2}}, \qquad n = 0, 1, \ldots, N, N+1, \qquad (4.6)$$

where P_{N+1} is the blocking probability. Note that the probability that the buffer is full is $P_N + P_{N+1}$. The above two-station system is equivalent to an $M/M/1/N+1$ finite-capacity work station with an arrival rate λ and a processing rate μ. This reduction to a single queue is a consequence of the exponential processing time at M_1. Having M_1 blocked is the same as having the $(N+1)$st position filled in the $M/M/1/N+1$ queue. After the process is completed at M_2, the time until the first station completes its process is the same as the time until an arrival occurs in the single queue system. The performance measures of interest to us, such as the line throughput $\overline{o}_\ell$, utilizations $P_i(U)$, and average number of jobs in the buffer, $\overline{N}$, are

$$\overline{o}_\ell = (1-P_0)\mu, \qquad P_1(U) = 1 - P_{N+1}, \qquad P_2(U) = 1 - P_0, \qquad (4.7)$$

and

$$\overline{N} = \sum_{n=1}^{N} nP_n + NP_{N+1}. \qquad (4.8)$$

There are N units in the buffer during blocking, and $\overline{N}$ has to be computed accordingly. The system time of a job includes the processing time at M_1, the time it spends while waiting in B, and the processing time at M_2. Since $\overline{N}$ is the average number of jobs in B including M_2, we can use Little's formula to find the time in B and M_2. Thus, the expected system time of a job is given by

$$E[T_s] = \frac{1}{\mu_1} + \frac{\overline{N}}{\overline{o}_\ell}. \qquad (4.9)$$

4.3.2 MGE-2 Processing Times

Let us continue our analysis of the two-station system with another special case that is simple enough to easily comprehend the problem formulation and the method of analysis. Assume that stations 1 and 2 have two-phase MGE distributions with parameters $(\lambda_1, \lambda_2, a_1)$ and (μ_1, μ_2, a_2), respectively, as shown in Figure 4.4. Jobs may depart from either phase at either of the stations.

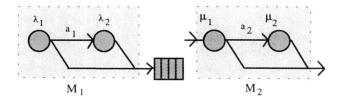

Figure 4.4. Stations with MGE-2 processing time distributions.

The first station is again assumed always to be busy except when blocked. Consider the stochastic process $\{I_t, J_t, N_t, t \geq 0\}$, where $I_t = 0, 1, 2, B$ represents the phase of the first station, $J_t = 0, 1, 2$ represents the phase of the second station, and $N_t = 0, 1, \ldots, N$ represents the work-in-process inventory level including the job being processed at station 2 at time t. $I_t = B$ implies that station 1 is blocked. Then, $\{I_t, J_t, N_t\}$ is a continuous-time, discrete-state Markov chain with a finite number of states, for which the steady-state probabilities uniquely exist. Let (i, j, n) be a particular state of the above stochastic process and $P(i, j, n)$ be its steady-state probability. Possible state transitions in the above Markov chain can be listed as follows. As soon as M_1 finishes processing a job and places it into the buffer, it starts processing a new job in its first phase. M_2 behaves the same way, provided that there are jobs in the buffer waiting to be processed. Otherwise, M_2 remains idle until M_1 completes its process. In any one of the stations, from the first phase a job either goes to the second phase of the same station or departs from the station. From the second phase, a job definitely departs from the station. When M_1 is blocked, it does not matter from which phase of M_1 the blocking job departs. Using these possible transitions with their associated transition rates, we can write the steady-state flow-balance equations of $\{I_t, J_t, N_t\}$ as shown below. We write them in such a way that there is a set of equations for each value of n.

$n = 0$

$$\lambda_1 P(1, 0, 0) = \mu_1(1 - a_2)P(1, 1, 1) + \mu_2 P(1, 2, 1) \quad (4.10a)$$

$$\lambda_2 P(2, 0, 0) = \mu_1(1 - a_2)P(2, 1, 1) + \mu_2 P(2, 2, 1)$$
$$+ \lambda_1 a_1 P(1, 0, 0) \quad (4.10b)$$

$n = 1$

$$(\lambda_1 + \mu_1)P(1, 1, 1) = \lambda_1(1 - a_1)P(1, 0, 0) + \lambda_2 P(2, 0, 0)$$
$$+ \mu_1(1 - a_2)P(1, 1, 2) + \mu_2 P(1, 2, 2) \quad (4.11a)$$

$$(\lambda_1 + \mu_2)P(1, 2, 1) = \mu_1 a_2 P(1, 1, 1) \quad (4.11b)$$

$$(\lambda_2 + \mu_1)P(2, 1, 1) = \lambda_1 a_1 P(1, 1, 1) + \mu_1(1 - a_2)P(2, 1, 2)$$
$$+ \mu_2 P(2, 2, 2) \quad (4.11c)$$

$$(\lambda_2 + \mu_2)P(2, 2, 1) = \lambda_1 a_1 P(1, 2, 1) + \mu_1 a_2 P(2, 1, 1) \qquad (4.11d)$$

$\underline{2 \leq n \leq N - 1}$

$$(\lambda_1 + \mu_1)P(1, 1, n) = \lambda_1(1 - a_1)P(1, 1, n - 1) + \lambda_2 P(2, 1, n - 1)$$
$$+ \mu_1(1 - a_2)P(1, 1, n + 1)$$
$$+ \mu_2 P(1, 2, n + 1) \qquad (4.12a)$$

$$(\lambda_1 + \mu_2)P(1, 2, n) = \lambda_2 P(2, 2, n - 1) + \lambda_1(1 - a_1)P(1, 2, n - 1)$$
$$+ \mu_1 a_2 P(1, 1, n) \qquad (4.12b)$$

$$(\lambda_2 + \mu_1)P(2, 1, n) = \lambda_1 a_1 P(1, 1, n) + \mu_1(1 - a_2)P(2, 1, n + 1)$$
$$+ \mu_2 P(2, 2, n + 1) \qquad (4.12c)$$

$$(\lambda_2 + \mu_2)P(2, 2, n) = \lambda_1 a_1 P(1, 2, n) + \mu_1 a_2 P(2, 1, n) \qquad (4.12d)$$

$\underline{n = N}$

$$(\lambda_1 + \mu_1)P(1, 1, N) = \lambda_1(1 - a_1)P(1, 1, N - 1)$$
$$+ \lambda_2 P(2, 1, N - 1) + \mu_1(1 - a_2)P(B, 1, N)$$
$$+ \mu_2 P(B, 2, N) \qquad (4.13a)$$

$$(\lambda_1 + \mu_2)P(1, 2, N) = \lambda_2 P(2, 2, N - 1) + \lambda_1(1 - a_1)P(1, 2, N - 1)$$
$$+ \mu_1 a_2 P(1, 1, B) \qquad (4.13b)$$

$$(\lambda_2 + \mu_1)P(2, 1, N) = \lambda_1 a_1 P(1, 1, N) \qquad (4.13c)$$

$$(\lambda_2 + \mu_2)P(2, 2, N) = \lambda_1 a_1 P(1, 2, N)$$
$$+ \mu_1 a_2 P(2, 1, N) \qquad (4.13d)$$

$\underline{i = B}$

$$\mu_1 P(B, 1, N) = \lambda_1(1 - a_1)P(1, 1, N) + \lambda_2 P(2, 1, N) \qquad (4.14a)$$

$$\mu_2 P(B, 2, N) = \mu_1 a_2 P(B, 1, N) + \lambda_1(1 - a_1)P(1, 2, N)$$
$$+ \lambda_2 P(2, 2, N) \qquad (4.14b)$$

The transition rate matrix **Q** of $\{I_t, J_t, N_t\}$, or the coefficient matrix of (4.10)–(4.14), is given below:

(100)	$-\lambda_1$	$\lambda_1 a_1$	$\lambda_1(1-a_1)$									
(200)		$-\lambda_2$	λ_2									
(111)	$\mu_1(1-a_2)$		$-(\lambda_1+\mu_1)$	$\mu_1 a_2$	$\lambda_1 a_1$	$\lambda_1(1-a_1)$						
(121)	μ_2			$-(\lambda_1+\mu_2)$		$\lambda_1 a_1$	$\lambda_1(1-a_1)$					
(211)		$\mu_1(1-a_2)$			$-(\lambda_2+\mu_1)$	$\mu_1 a_2$	λ_2					
(221)		μ_2				$-(\lambda_2+\mu_2)$	λ_2					
(112)			$\mu_1(1-a_2)$				$-(\lambda_1+\mu_1)$	$\mu_1 a_2$	$\lambda_1 a_1$	$\lambda_1(1-a_1)$		
(122)			μ_2					$-(\lambda_1+\mu_2)$		$\lambda_1 a_1$	$\lambda_1(1-a_1)$	
(212)					$\mu_1(1-a_2)$				$-(\lambda_2+\mu_1)$	$\mu_1 a_2$	λ_2	
(222)						μ_2				$-(\lambda_2+\mu_2)$		λ_2
(113)							$\mu_1(1-a_2)$				$-(\lambda_1+\mu_1)$	$\mu_1 a_2$	$\lambda_1 a_1$	$\lambda_1(1-a_1)$
(123)							μ_2					$-(\lambda_1+\mu_2)$		$\lambda_1 a_1$	$\lambda_1(1-a_1)$
(213)									$\mu_1(1-a_2)$				$-(\lambda_2+\mu_1)$	$\mu_1 a_2$	λ_2
(223)										μ_2				$-(\lambda_2+\mu_2)$		λ_2
(B13)											$\mu_1(1-a_2)$				$-\mu_2$	$\mu_1 a_2$
(B23)											μ_2					$-\mu_2$

That is, we have

$$\mathbf{Q}^T \underline{\mathbf{P}} = \underline{\mathbf{0}}, \tag{4.15}$$

where $\underline{\mathbf{P}}$ is the vector of all the steady-state probabilities. There are various ways of solving the above set of homogenous, linear equations. These include direct methods such as the Gaussian elimination, iterative methods such as the Gauss-Seidel, the Jacobi, the power method, and the matrix-recursive methods. Next, we introduce a matrix-recursive approach to obtain the steady-state probabilities. In the next section, we will generalize the processing times to accommodate large numbers of phases and introduce an iterative method to solve for the probabilities.

A Matrix-Recursive Approach

The matrix-recursive approach takes advantage of the special structure of $\mathbf{Q}$. Let us first define the following probability vectors:

$$\tilde{\underline{\mathbf{P}}}(0) = \begin{bmatrix} P(1,0,0) \\ P(2,0,0) \end{bmatrix}, \quad \tilde{\underline{\mathbf{P}}}(n) = \begin{bmatrix} P(1,1,n) \\ P(1,2,n) \\ P(2,1,n) \\ P(2,2,n) \end{bmatrix},$$

and

$$\tilde{\underline{\mathbf{P}}}(B) = \begin{bmatrix} P(B,1,N) \\ P(B,2,N) \end{bmatrix}.$$

This approach is based on grouping the equations in such a way that the probabilities of having n jobs in the buffer, that is, $\tilde{\underline{\mathbf{P}}}(n)$, are obtained in terms of the probabilities of having n-1 and n-2 jobs, that is, $\tilde{\underline{\mathbf{P}}}(n-1)$ and $\tilde{\underline{\mathbf{P}}}(n-2)$. Thus, we construct a set of equations to solve for the joint probabilities for each n. This is a matter of selecting appropriate equations such that the desired recursion is obtained and the resulting coefficient matrices are invertible. For instance, consider

Eqs. (4.10a), (4.10b), (4.11b), and (4.11d) written in matrix form as follows:

$$\begin{bmatrix} \mu_1(1-a) & \mu_2 & 0 & 0 \\ -\mu_1 a_2 & (\lambda_1+\mu_2) & 0 & 0 \\ 0 & 0 & \mu_1(1-a_2) & \mu_2 \\ 0 & -\lambda_1 a_2 & -\mu_1 a_2 & (\lambda_1+\mu_2) \end{bmatrix} \begin{bmatrix} P(1,1,1) \\ P(1,2,1) \\ P(2,1,1) \\ P(2,2,1) \end{bmatrix}$$

$$= \begin{bmatrix} \lambda_1 & 0 \\ 0 & 0 \\ -\lambda_1 a_1 & \lambda_2 \\ 0 & 0 \end{bmatrix} \begin{bmatrix} P(1,0,0) \\ P(2,0,0) \end{bmatrix},$$

which may be rewritten as

$$\mathbf{R}\tilde{\mathbf{P}}(1) = \mathbf{A}\tilde{\mathbf{P}}(0),$$

or

$$\tilde{\mathbf{P}}(1) = \mathbf{R}^{-1}\mathbf{A}\tilde{\mathbf{P}}(0) = \mathbf{Z}(1)\tilde{\mathbf{P}}(0), \qquad (4.16)$$

where $\mathbf{Z}(1) = \mathbf{R}^{-1}\mathbf{A}$ and $\mathbf{R}$ and $\mathbf{A}$ are coefficient matrices.

Now, consider eqs. (4.11a), (4.11c), (4.12b), and (4.12d), for $n = 2$, put into the same matrix-equation form:

$$\begin{bmatrix} \mu_1(1-a_2) & \mu_2 & 0 & 0 \\ -\mu_1 a_2 & (\lambda_1+\mu_2) & 0 & 0 \\ 0 & 0 & \mu_1(1-a_2) & \mu_2 \\ 0 & -\lambda_1 \alpha_2 & -\mu_1 a_2 & (\lambda_2+\mu_2) \end{bmatrix} \begin{bmatrix} P(1,1,2) \\ P(1,2,2) \\ P(2,1,2) \\ P(2,2,2) \end{bmatrix}$$

$$= \begin{bmatrix} \lambda_1+\mu_1 & 0 & 0 & 0 \\ 0 & \lambda_1(1-a_1) & 0 & \lambda_2 \\ -\lambda_1 a_1 & 0 & \lambda_2+\mu_1 & 0 \\ 0 & 0 & 0 & 0 \end{bmatrix} \begin{bmatrix} P(1,1,1) \\ P(1,2,1) \\ P(2,1,1) \\ P(2,2,1) \end{bmatrix}$$

$$+ \begin{bmatrix} -\lambda_1(1-a_1) & -\lambda_2 \\ 0 & 0 \\ 0 & 0 \\ 0 & 0 \end{bmatrix} \begin{bmatrix} P(1,0,0) \\ P(2,0,0) \end{bmatrix}$$

that is,

$$\mathbf{R}\tilde{\mathbf{P}}(2) = \mathbf{B}\tilde{\mathbf{P}}(1) + \mathbf{C}\tilde{\mathbf{P}}(0),$$

resulting in

$$\tilde{\mathbf{P}}(2) = \mathbf{R}^{-1}[\mathbf{B}\tilde{\mathbf{P}}(1) + \mathbf{C}\tilde{\mathbf{P}}(0)] = \mathbf{R}^{-1}[\mathbf{B}\mathbf{Z}(1)\tilde{\mathbf{P}}(0) + \mathbf{C}\tilde{\mathbf{P}}(0)]$$
$$= \mathbf{R}^{-1}[\mathbf{B}\mathbf{Z}(1) + \mathbf{C}]\tilde{\mathbf{P}}(0).$$

4.3. Analysis of Two-Station Flow Lines

We can rewrite $\tilde{\underline{P}}(2)$ as

$$\tilde{\underline{P}}(2) = \mathbf{Z}(2)\tilde{\underline{P}}(0), \qquad (4.17)$$

where $\mathbf{Z}(2) = \mathbf{R}^{-1}[\mathbf{B}\mathbf{Z}(1) + \mathbf{C}]$.

Similarly, using (4.12a), (4.12c), (4.13b), and (4.13d), we can write

$$\mathbf{R}\tilde{\underline{P}}(3) = \mathbf{B}\tilde{\underline{P}}(2) + \mathbf{D}\tilde{\underline{P}}(1) = [\mathbf{B}\mathbf{Z}(2) + \mathbf{D}\mathbf{Z}(1)]\tilde{\underline{P}}(0),$$

or

$$\tilde{\underline{P}}(3) = \mathbf{R}^{-1}[\mathbf{B}\mathbf{Z}(2) + \mathbf{D}\mathbf{Z}(1)]\tilde{\underline{P}}(0) = \mathbf{Z}(3)\tilde{\underline{P}}(0), \qquad (4.18)$$

where $\mathbf{Z}(3) = \mathbf{R}^{-1}[\mathbf{B}\mathbf{Z}(2) + \mathbf{D}\mathbf{Z}(1)]$. The structure of (4.18) remains the same for $n = 3, \ldots, N$, that is,

$$\tilde{\underline{P}}(n) = \mathbf{R}^{-1}[\mathbf{B}\mathbf{Z}(n-1) + \mathbf{D}\mathbf{Z}(n-2)]\tilde{\underline{P}}(0) = \mathbf{Z}(n)\tilde{\underline{P}}(0). \qquad (4.19)$$

Finally, using Eqs. (4.14a) and (4.14b), we obtain

$$\begin{bmatrix} \mu_1 & 0 \\ -\mu_1 a_2 & \mu_2 \end{bmatrix} \begin{bmatrix} P(\mathrm{B}, 1, N) \\ P(\mathrm{B}, 2, N) \end{bmatrix}$$

$$= \begin{bmatrix} \lambda_1(1 - a_1) & 0 & \lambda_2 & 0 \\ 0 & \lambda_1(1 - a_1) & 0 & \lambda_2 \end{bmatrix} \begin{bmatrix} P(1, 1, N) \\ P(1, 2, N) \\ P(2, 1, N) \\ P(2, 2, N) \end{bmatrix},$$

which can be rewritten as

$$\mathbf{E}\tilde{\underline{P}}(\mathrm{B}) = \mathbf{F}\tilde{\underline{P}}(N),$$

or

$$\tilde{\underline{P}}(\mathrm{B}) = \mathbf{E}^{-1}\mathbf{F}\tilde{\underline{P}}(N) = \mathbf{E}^{-1}\mathbf{F}\mathbf{Z}(N)\tilde{\underline{P}}(0),$$

resulting in

$$\tilde{\underline{P}}(\mathrm{B}) = \mathbf{Z}(\mathrm{B})\tilde{\underline{P}}(0), \qquad (4.20)$$

where $\mathbf{Z}(B) = \mathbf{E}^{-1}\mathbf{F}\mathbf{Z}(N)$. Thus, $\mathbf{Z}(n)$s are obtained recursively. Once $\tilde{\underline{P}}(0)$ is found, $\tilde{\underline{P}}(1), \ldots, \tilde{\underline{P}}(\mathrm{B})$ can easily be obtained using the above scheme. Notice that (4.13a) and (4.13b) are not used yet. These equations can be used to obtain $\tilde{\underline{P}}(0)$,

4. Analysis of Flow Lines

as shown below:

$$\begin{bmatrix} \lambda_1 + \mu_1 & 0 & 0 & 0 \\ -\lambda_1 a_1 & 0 & \lambda_2 + \mu_1 & 0 \end{bmatrix} \begin{bmatrix} P(1,1,N) \\ P(1,2,N) \\ P(2,1,N) \\ P(2,2,N) \end{bmatrix}$$

$$= \begin{bmatrix} \lambda_1(1-a_1) & 0 & \lambda_2 & 0 \\ 0 & 0 & 0 & 0 \end{bmatrix} \begin{bmatrix} P(1,1,N-1) \\ P(1,2,N-1) \\ P(2,1,N-1) \\ P(2,2,N-1) \end{bmatrix}$$

$$+ \begin{bmatrix} \mu_1(1-a_2) & \mu_2 \\ 0 & 0 \end{bmatrix} \begin{bmatrix} P(B,1,N) \\ P(B,2,N) \end{bmatrix},$$

or

$$\mathbf{G}\tilde{\mathbf{P}}(N) = \mathbf{H}\tilde{\mathbf{P}}(N-1) + \mathbf{M}\tilde{\mathbf{P}}(B),$$

which can be rewritten as

$$\mathbf{GZ}(N)\tilde{\mathbf{P}}(0) = \mathbf{HZ}(N-1)\tilde{\mathbf{P}}(0) + \mathbf{MZ}(B)\tilde{\mathbf{P}}(0). \tag{4.21}$$

Equation (4.21) can be solved by setting $P(1,0,0) = 1$ and finding $P(2,0,0)$ accordingly. The joint probabilities are normalized after $\tilde{\mathbf{P}}(i)$, $i = 1, \ldots, B$, are obtained. The above recursive procedure can be summarized as follows:

$$\tilde{\mathbf{P}}(n) = \mathbf{Z}(n)\tilde{\mathbf{P}}(0), \quad n = 1, \ldots, B \tag{4.22}$$

where

$$\mathbf{Z}(1) = \mathbf{R}^{-1}\mathbf{A}, \quad \mathbf{Z}(2) = \mathbf{R}^{-1}[\mathbf{BZ}(1) + \mathbf{C}],$$
$$\mathbf{Z}(n) = \mathbf{R}^{-1}[\mathbf{BZ}(n-1) + \mathbf{DZ}(n-2)], \quad n = 3, \ldots, N, \tag{4.23}$$

and

$$\mathbf{Z}(B) = \mathbf{E}^{-1}\mathbf{FZ}(N).$$

$\tilde{\mathbf{P}}(0)$ is obtained using

$$\mathbf{GZ}(N)\tilde{\mathbf{P}}(0) = \mathbf{HZ}(N-1)\tilde{\mathbf{P}}(0) + \mathbf{MZ}(B)\tilde{\mathbf{P}}(0). \tag{4.24}$$

The steady-state joint probabilities are normalized using

$$\sum_{i=1}^{B} \sum_{j=0}^{2} \sum_{n=0}^{N} P(i,j,n) = 1.$$

4.3. Analysis of Two-Station Flow Lines

The marginal probabilities of the buffer content are given by

$$P(0) = \sum_{i=1}^{2} P(i, 0, 0), \qquad P(n) = \sum_{i}\sum_{j} P(i, j, n), \qquad n = 1, \ldots, N,$$

and the blocking probability is $P_1(B) = \sum_{j=1}^{2} P(B, j, N)$.

The measures of interest to us, such as the line throughput ($\bar{o}_\ell$), the utilization $P_i(U)$, and the average buffer contents are as follows:

$$\bar{o}_\ell = \frac{1 - P(0)}{E[X_2]}, \qquad \bar{N} = \sum_{n=1}^{N} n P(n),$$

$$P_1(U) = 1 - P_1(B), \quad \text{and} \quad P_2(U) = 1 - P(0),$$

where X_i is the random variable representing the processing time at station i. As mentioned earlier, an important characteristic of the above production line is the equality of the long-run throughputs at both stations, that is,

$$\frac{1 - P_1(B)}{E[X_1]} = \frac{1 - P(0)}{E[X_2]}. \tag{4.25}$$

EXAMPLE 4.1. Consider two work stations with the following MGE-2 processing times: $(\lambda_1, \lambda_2, a_1) = (2, 1, 0.25)$ and $(\mu_1, \mu_2, a_2) = (2.5, 2, 0.5)$, with a buffer capacity of $N = 3$. For illustration purposes, let us implement the above matrix-recursive algorithm to obtain $\tilde{\mathbf{P}}(0), \tilde{\mathbf{P}}(1), \tilde{\mathbf{P}}(2), \tilde{\mathbf{P}}(3)$, and $\tilde{\mathbf{P}}(B)$. Since the matrices **A**, **B**, **C**, and so on are all defined earlier, we will not write them explicitly, but rather show only the **Z** matrices:

$$\mathbf{Z}(1) = \begin{bmatrix} 1.0667 & 0 \\ 0.3333 & 0 \\ -0.2933 & 0.4800 \\ -0.0667 & 0.2000 \end{bmatrix}, \quad \mathbf{Z}(2) = \begin{bmatrix} 1.6444 & -0.5867 \\ 0.6222 & -0.1333 \\ -0.8484 & 0.8277 \\ -0.2498 & 0.3227 \end{bmatrix},$$

$$\mathbf{Z}(3) = \begin{bmatrix} 3.0675 & -1.6967 \\ 1.1295 & -0.4996 \\ -2.0006 & 1.6113 \\ -0.6453 & 0.5881 \end{bmatrix}, \quad \mathbf{Z}(B) = \begin{bmatrix} 1.0402 & -0.3735 \\ 1.1746 & -0.3140 \end{bmatrix}.$$

Having $P(1, 0, 0) = 1$, Eq. (4.23) yields $P(2, 0, 0) = 1.3103$. With these values of **Z**s, and $\tilde{\mathbf{P}}(0)$, we use (4.22) to obtain $\tilde{\mathbf{P}}(1), \tilde{\mathbf{P}}(2), \tilde{\mathbf{P}}(3)$, and $\tilde{\mathbf{P}}(B)$ and normalize all the so-called probabilities to find the actual probabilities. The results are given below:

$$\tilde{\underline{P}}(0) = \begin{bmatrix} 0.1130 \\ 0.1486 \end{bmatrix} \quad \tilde{\underline{P}}(1) = \begin{bmatrix} 0.1205 \\ 0.0377 \\ 0.0382 \\ 0.0222 \end{bmatrix} \quad \tilde{\underline{P}}(2) = \begin{bmatrix} 0.0986 \\ 0.0505 \\ 0.0272 \\ 0.0197 \end{bmatrix}$$

$$\tilde{\underline{P}}(3) = \begin{bmatrix} 0.0944 \\ 0.0534 \\ 0.0135 \\ 0.0145 \end{bmatrix} \quad \tilde{\underline{P}}(B) = \begin{bmatrix} 0.0620 \\ 0.0860 \end{bmatrix},$$

$$P_0 = 0.2616, \quad P_1 = 0.2159, \quad P_2 = 0.1960,$$
$$P_3 = 0.1757 + 0.1480 = 0.3237.$$

The same example with $N = 30$ would yield the probabilities given in Table 4.1. Unfortunately, for large values of N ($N > 40$), the entries of $\mathbf{Z}(n)$ reach large values, causing numerical instabilities. Thus, it is advisable to use the recursive approach for smaller values of N.

EXAMPLE 4.2. Let us implement the algorithm to the case where the processing times are exponentially distributed with rates 2 and 4 at stations 1 and 2, respectively. Here, we use the two-phase minimal representation of the exponential distribution with rates $(\lambda_1, \lambda_2, a_1) = (4, 2, 0.5)$ for station 1, for instance. The resulting probability distribution is given in Table 4.2. Clearly, this example can be used to verify the matrix-recursive algorithm. In this case, the steady-state probabilities of the process $\{I_t, J_t, N_t\}$ coincide with those generated by (4.6), given in Table 4.2.

TABLE 4.1. $\{P_n\}$ for Example 4.1 with $N = 30$

n	P_n
0	0.13481
1	0.11180
2	0.09929
5	0.06592
10	0.03264
15	0.01649
20	0.00799
25	0.00396
30	0.00374
Blocking	0.00171
($\bar{o}_\ell$)	1.33106

4.3. Analysis of Two-Station Flow Lines   133

TABLE 4.2. Probabilities of the number of jobs in the buffer in Example 4.2

n	P_n
0	0.50012
1	0.25006
2	0.12503
3	0.06252
4	0.03126
5	0.01563
6	0.00781
7	0.00391
8	0.00195
9	0.00098
10	0.00048
Blocking	0.00024

4.3.3 MGE-k Processing Times: A Numerical Approach

In principle, the recursive approach summarized in (4.22)–(4.24) can be applied to flow lines with more than two stations and/or with processing times having more than two phases (accommodating lower variability). However, when the state space grows, it is highly likely that one encounters numerical instabilities due to very large entries of the Z matrices. Therefore, in this section, we emphasize a numerical approach in which the states are generated in a systematic manner and the transition rate matrix Q is stored in compact form, where only the nonzero elements are stored, with their row and column numbers. This approach significantly benefits from the savings in computer memory requirements and execution time. The steady-state probabilities are obtained using an iterative scheme, carried out in compact form.

Thus, consider a two-station flow line where the processing times are MGE random variables of the form shown in Figure 2.7. M_1 and M_2 have ℓ_1 and ℓ_2 phases, respectively, with N being the buffer capacity including the position on M_2. Thus, the processing times may assume practically any variability. The stochastic process $\{I_t, J_t, N_t\}$ is again a continuous-time Markov chain with the states listed in lexicographic order in Table 4.3. The total number of states is given by

$$s = N\ell_1\ell_2 + \ell_1 + \ell_2.$$

The transition rate matrix Q can be generated by considering the following possible transitions:

- A job may go from one phase of the processing time to the next phase.

TABLE 4.3. Lexicographic ordering of the states of $\{I_t, J_t, N_t\}$

i	j	n	$V(i, j, n)$	
1	0	0	1	Set 1: M_2 is idle.
$\vdots$	$\vdots$	$\vdots$	$\vdots$	
ℓ_1	0	0	ℓ_1	
1	1	1	$\ell_1 + 1$	
$\vdots$	$\vdots$	$\vdots$	$\vdots$	
1	ℓ_2	1	$\ell_1 + \ell_2$	
2	1	1		
$\vdots$	$\vdots$	$\vdots$	$\vdots$	
2	ℓ_2	1	$\ell_1 + 2\ell_2$	
$\vdots$	$\vdots$	$\vdots$	$\vdots$	
ℓ_1	1	1		
$\vdots$	$\vdots$	$\vdots$	$\vdots$	Set 2: M_1 and M_2 are both busy.
ℓ_1	ℓ_2	1		
1	1	2		
$\vdots$	$\vdots$	$\vdots$	$\vdots$	
1	ℓ_2	2		
$\vdots$	$\vdots$	$\vdots$	$\vdots$	
ℓ_1	1	N		
$\vdots$	$\vdots$	$\vdots$	$\vdots$	
ℓ_1	ℓ_2	N	$\ell_1 + N\ell_1\ell_2$	
0	1	N		Set 3: M_1 is blocked.
$\vdots$	$\vdots$	$\vdots$	$\vdots$	
0	ℓ_2	N	$\ell_1 + N\ell_1\ell_2 + \ell_2$	

- A job may depart from any one of the phases at M_1 to the first phase at M_2, provided that there is space in buffer B.
- A job may depart from any one of the phases of M_2. The job from M_1 may join B synchronously in case M_2 blocks M_1.

To be able to generate Q in compact form, the position of each state in the list in Table 4.3 has to be known. It is the variable $V(i, j, n)$ that keeps the position of each state $v = (i, j, n)$.

$$V(i, j, n) = \begin{cases} i & \text{if } v \text{ is in Set 1,} \\ (n-1)\ell_1\ell_2 + \ell_1 + (i-1)\ell_2 + j & \text{if } v \text{ is in Set 2,} \\ N\ell_1\ell_2 + \ell_1 + j & \text{if } v \text{ is in Set 3.} \end{cases}$$

4.3. Analysis of Two-Station Flow Lines

During the generation process of $\mathbf{Q}$, let the current state be $(1,0,0)$ with $V(1, 0, 0) = 1$. A transition is possible from $(1,0,0)$ to $(1,1,1)$, for which $V(1, 1, 1) = \ell_1 + 1$ with transition rate λ_1. The corresponding positive element of $\mathbf{Q}$ is λ_1 with a row number of 1 and a column number of $\ell_1 + 1$. Through this approach, all the positive elements of $\mathbf{Q}$ can be generated by a procedure that goes through the list of states asking to which states a transition may occur from the current state, and stores the transition rates, their row numbers, and the column numbers in three respective vectors. Values are assigned to the diagonal elements of $\mathbf{Q}$ so that the row sums become zero, making it a transition rate matrix.

So far, what we have is (4.15), that is, $\mathbf{Q}^T \underline{\mathbf{P}} = \underline{\mathbf{0}}$. To implement an iterative scheme to obtain $\underline{\mathbf{P}}$, one needs to formulate the problem as an eigenvalue problem in which $\underline{\mathbf{P}}$ becomes the eigenvector. Using (4.15), it is possible to proceed as follows:

$$\mathbf{Q}^T \Delta t \underline{\mathbf{P}} + \underline{\mathbf{P}} = \underline{\mathbf{P}},$$

where Δt is arbitrary;

$$(\mathbf{Q}^T \Delta t + \mathbf{I})\underline{\mathbf{P}} = \underline{\mathbf{P}},$$

$$\tilde{\mathbf{Q}} \underline{\mathbf{P}} = \underline{\mathbf{P}}, \qquad (4.26)$$

where $\tilde{\mathbf{Q}} = \mathbf{Q}^T \Delta t + \mathbf{I}$.

If Δt is chosen such that $\Delta t \leq [\max |q_{ii}|]^{-1}$, then the matrix $\tilde{\mathbf{Q}}$ becomes a stochastic matrix (row sums equal to 1) and in fact it becomes the transition probability matrix of $\{I_t, J_t, N_t\}$. From the fundamental properties of the stochastic matrices, $\tilde{\mathbf{Q}}$ has a unit eigenvalue and no other eigenvalues can exceed it. If Δt is further restricted so that $\Delta t < [\max |q_{ii}|]^{-1}$, then the application of Gershgorin's theorem (see Varga (1963)) shows that the eigenvalues that are not of unit value must be strictly less than 1. The vector $\underline{\mathbf{P}}$ then becomes the left-hand eigenvector of $\tilde{\mathbf{Q}}$. This approach is equivalent to the discretization of the Markov chain in which transitions occur at intervals of Δt, where Δt is sufficiently small to ensure that the possibility of two or more state changes within this interval is negligible.

The iterative scheme called the power method is based on (4.26) and has the following iteration equation:

$$\tilde{\mathbf{Q}} \underline{\mathbf{P}}_{k-1} = \underline{\mathbf{P}}_k, \qquad (4.27)$$

where $\underline{\mathbf{P}}_0$ is chosen arbitrarily. Notice that $\tilde{\mathbf{Q}}$ has the same nonzero entries as $\mathbf{Q}$, except with different values. Then, $\underline{\mathbf{P}}$ is obtained by carrying out the multiplication of $\tilde{\mathbf{Q}} \underline{\mathbf{P}}_{k-1}$ in compact form, utilizing the columnwise storage of $\tilde{\mathbf{Q}}$.

The convergence of the power method can be explained as follows: Let α_i and $\underline{\mathbf{u}}_i$ be the ith eigenvalue and the ith eigenvector of $\tilde{\mathbf{Q}}$, respectively. Let us assume that $|\alpha_1| > |\alpha_2| \geq \cdots \geq |\alpha_S|$, making sure that α_1 is the dominant eigenvalue.

Then, we can write

$$\tilde{Q}\underline{u}_i = \alpha_i \underline{u}_i, \qquad i = 1, \ldots, S.$$

From (4.27), we have $\underline{P}_1 = \tilde{Q}\underline{P}_0$, and $\underline{P}_0$ can be expressed as $\underline{P}_0 = \sum_{i=0}^{S} c_i \underline{u}_i$. Then

$$\underline{P}_1 = \sum_{i=1}^{S} c_i \tilde{Q}\underline{u}_i = \sum_{i=1}^{S} c_i \alpha_i \underline{u}_i.$$

Also,

$$\underline{P}_2 = \tilde{Q}\,\underline{P}_1 = \sum_{i=1}^{S} c_i \alpha_i^2 \underline{u}_i.$$

Generalizing to $\underline{P}_k$, we have

$$\underline{P}_k = \sum_{i=1}^{S} c_i \alpha_i^k \underline{u}_i = \alpha_1^k \left[c_i \underline{u}_1 + \sum_{i=2}^{S} \left(\frac{\alpha_i}{\alpha_1}\right)^k \underline{u}_i \right]. \qquad (4.28)$$

Since, α_1 is the dominant eigenvalue, as k tends to infinity, the second group of terms in (4.28) tends to zero, yielding

$$\underline{P}_k = \alpha_1 \underline{P}_{k-1},$$

and making $\underline{P}_{k-1}$ and $\underline{P}_k$ converge, for $\alpha_1 = 1$. Clearly the convergence rate depends on the magnitude of the eigenvalues. That is, convergence will be slow if α_2 is close to 1.

EXAMPLE 4.3. Consider a two-station system with a buffer capacity of $N = 25$. The first station has an E-10 processing time (Erlang with 10 phases) with phase rate $\lambda = 10$. The mean processing time is $E[X_1] = 1$, and $Cv_{X_1}^2 = 0.1$. The second station has an E-8 processing time with phase rate $\mu = 10$. Its mean processing time is $E[X_2] = 0.8$, and $Cv_{X_2}^2 = 0.125$. The performance measures and the probability distribution of the buffer contents are given in Table 4.4. The total number of states is 2,018. We have applied the power method presented earlier. The results after 2,000 iterations are given in Table 4.4. The goodness of these results can be tested by computing the throughput of M_1 and M_2, which are supposed to be the same. We have

$$\bar{o}_{M_1} = 1.00000,$$

$$\bar{o}_{M_2} = 1.01449,$$

TABLE 4.4. Probabilities of the buffer content in Example 4.3

n	P_n
0	0.18841
1	0.55937
2	0.17908
5	0.0079
10	1.1518×10^{-3}
15	6.7889×10^{-5}
20	1.3789×10^{-6}
24	1.9004×10^{-8}
25	4.2252×10^{-9}
Blocking	2.6166×10^{-10}
$\bar{N}$	1.24542

with an absolute difference of 0.0145. The closeness of these throughputs depends on the number of iterations. Clearly, the state space has a significant impact on the execution time.

4.3.4 Line Behavior

Before we proceed any further, let us briefly discuss the impact of the line parameters, such as the buffer capacities and the moments of the processing times, on the measures of the production line. Buffers with larger capacities cause less blocking and therefore result in higher throughput. On the other hand, buffers with larger capacities will have a larger accumulation of work-in-process inventory, since an upstream machine continues processing jobs until either it is starved or the downstream buffer is full. Figure 4.5 shows the impact of the buffer capacity on $\bar{o}_\ell$ and $\bar{N}$ for increasing values of N. The parameters of M_1 and M_2 are $(\lambda_1, \lambda_2, a_1) = (2, 2, 1)$ and $(\mu_1, \mu_2, a_2) = (2.5, 2.5, 1)$, respectively. $\bar{N}$ and $\bar{o}_\ell$ are monotone nondecreasing functions of N, and as N gets larger, both $\bar{o}_\ell$ and $\bar{N}$ converge to what their values would be when $N = \infty$. That is, $\bar{o}_\ell$ converges to the minimum of the processing rates of the two stations. For instance, at $N = 15$, $P(B) \cong 0$, yielding $\bar{o}_\ell \cong 1.0$, which is the throughput of the first station (see (4.25)). As N tends to infinity, $\bar{N}$ will also do so if $E[X_1] \leq E[X_2]$, and the Markovian model becomes unstable. Thus, for higher values of N, the first station must be slower than the second to prevent a large accumulation in the buffer.

The flow line is said to be balanced if $E[X_1] \cong E[X_2]$. The imbalance in the line has a direct impact on the measures of performance. For instance, higher work-in-process inventory tends to accumulate in the buffer if M_1 is faster than M_2. On the other hand, there will be a smaller accumulation if M_1 is slower than M_2. The line balance is a consequence of the workload allocation problem that is to be discussed in Section 4.7.1.

138   4. Analysis of Flow Lines

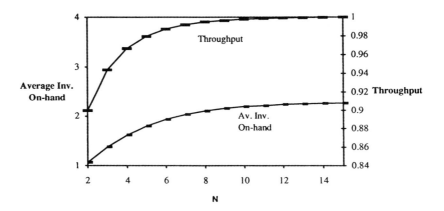

Figure 4.5. $\bar{N}$ and $\bar{o}_\ell$ as a function of N.

Since the processing-time distributions have a direct impact on the behavior of $\{I_t, J_t, N_t\}$, their moments have a direct impact on $\bar{o}_\ell$ and $\bar{N}$. For instance, it is clear that an increase in any of the mean processing times will decrease $\bar{o}_\ell$ because of higher starvation or higher blocking. Also, as a rule of thumb, when M_1 does not experience starvation,[1] an increase in the variability of the processing time at any of the stations reduces the line throughput. Figure 4.6 shows the behavior of $\bar{o}_\ell$ with respect to the Cv^2 of the processing time at M_2 when N=10. In the balanced line, we have $E[X_1] = E[X_2] = 1.0$, $Cv^2_{X_1} = 0.5$, and $Cv^2_{X_2}$ ranging from 0.5 to 3.5. Note that the squared coefficient of variation of the processing times may take large values due to failures, in which case they are modeled as the process completion times (see Chapter 3). In the unbalanced line, we have $E[X_1] = 1.0$, $E[X_2] = 0.5$ and $Cv^2_{X_1} = 0.5$. The processing-time parameters can be obtained using (2.67) and (2.72). In general, the variability of the processing times has a bigger impact on $\bar{o}_\ell$ when the line is balanced. The impact gets larger toward the downstream end of the line. As shown in Figure 4.6, the slope of the throughput is larger in the balanced line. However, the impact of the processing-time variability on $\bar{N}$ is more involved and depends on the mean processing times also. That is, the Cv^2 of the processing times at M_1 and M_2 may affect $\bar{N}$ in different ways depending on the system parameters.

As mentioned in Section 4.2, **reversibility** is another property of the flow lines. Consider two lines, namely L1 and L2, that are designed in such a way that the processing time at station 1 of line L1 is identical to that at station 2 of line L2. Also, M_2 of line L1 is identical to M_1 of line L2. Having the same buffer capacity,

---

[1] In case $N_1 = \infty$, and there exists a random job arrival process at M_1 with rate λ, such that M_1 remains idle at times, then $\bar{o}_\ell = \lambda$ no matter what the moments of the processing times are. If $N_1 < \infty$, then $\bar{o}_\ell$ is affected by the processing-time distributions.

4.3. Analysis of Two-Station Flow Lines

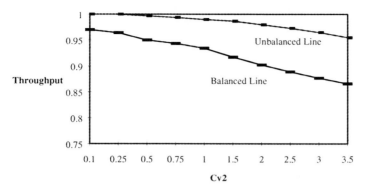

Figure 4.6. $\bar{o}_\ell$ as a function of the Cv^2 of the processing time at station 2.

same. Furthermore, it is possible to obtain the probability distribution of the work-in-process inventory in one system using the distribution of the one in the other. In two-station lines, this is true regardless of the type of blocking mechanism.

To supplement the above argument of throughput reversibility, let us focus on some special cases. For instance, let us first look at a two-station system with exponential processing times and an intermediate buffer of capacity 1 (the position on the second processor). The processing rates of the stations are λ and μ, respectively. Due to the memoryless property of the exponential distribution, this two-station line is equivalent to a single work station of type M/M/1/2, as explained in Section 4.3.1. The output rate of this system is given by

$$\bar{o}_\ell = \frac{\lambda \mu}{\lambda^2 + \lambda \mu + \mu^2},$$

which clearly shows the symmetry of $\bar{o}_\ell$ with respect to λ and μ.

For a two-station buffered line, Table 4.5 shows the probability distribution of the number of jobs in the system and those in the buffer in the original and the reversed lines. The BAP policy is assumed, and the system parameters are $(\lambda_1, \lambda_2, a_1) = (1.333, 2, 0.5)$ and $(\mu_1, \mu_2, a_2) = (4, 1.333, 0.333)$, with the buffer capacity of $N = 5$. First of all, we have

$$P_1(0) = P_{L2}(\text{blocking}), \qquad P_{L1}(\text{blocking}) = P_{L2}(0).$$

Also, having n jobs waiting in the buffer in line L1 is equivalent to having $N-n-1$ jobs waiting in the buffer in line L2; that is,

$$\Pr(n \text{ waiting in L1}) = \Pr(N - n - 1 \text{ waiting in L2}), \qquad n = 0, 1, \ldots, N - 1. \tag{4.29}$$

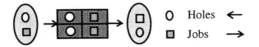

Figure 4.7. Jobs and holes move in opposite directions.

resulting in $\overline{N}_{L1,q} = N - 1 - \overline{N}_{L2,q}$ as shown in Table 4.5. Another way to interpret (4.29) is that the number of jobs waiting in line L2 is equal to the number of holes (or empty cells) waiting in line L1. The jobs and the holes in the buffer move in opposite directions, and their movement is synchronized. A hole leaves the buffer (moves to M_1) as soon as a job joins the buffer, as depicted in Figure 4.7 for $N = 5$. The concept of holes helps in the analysis of tandem systems. We will refer to them whenever possible in this and the following chapters.

Again in the two-station system of Table 4.4, if M_1 is considered a part of the buffer space, it automatically increases the buffer capacity to $N + 1$ and converts the blocking policy to the BBP type. In this case, we have

$$P_{L1}(n) = P_{L2}(N - n), \qquad (4.30)$$

where N is already augmented (i.e., $N = 6$ for the BBP policy in Table 4.5), providing $\overline{N}_{L1} = N - \overline{N}_{L2}$. The concept of holes applies in this case also.

In the case of symmetric lines with $K = 2$, we have

$$\bar{N}_q = \frac{N-1}{2} \quad \text{for the BAP policy,}$$

TABLE 4.5. Probability distributions of the number of jobs in the system and the jobs waiting in the buffer in the original and the reversed systems ($N = 5$)

n	Original System (L1)		Reversed System (L2)	
	Pr(n in sys.)	Pr(n waiting)	Pr(n in sys.)	Pr(n waiting)
0	0.5052	0.7573	0.0104	0.0278
1	0.2521	0.1203	0.0174	0.0327
2	0.1203	0.0619	0.0327	0.0619
3	0.0619	0.0327	0.0619	0.1203
4	0.0327	0.0278	0.1203	0.7573
5 w/o B	0.0174		0.2521	
$P(B)$	0.0104		0.5052	
$\bar{N}_q$(BAP)	0.9482	0.4534	4.5362	3.5466
$\bar{N}$(BBP)	0.9586		5.0414	

and

$$\overline{N} = N/2 \text{ for the BBP policy.}$$

The throughput equivalence also holds for both BAP and BBP policies in lines with $K > 2$ stations. However, the correspondence between the distributions of the number of jobs in the two systems with $K > 2$ holds only for the BBP policy. We will refer to the reversibility property in this and later chapters.

4.4 Analysis of Flow Lines With More Than Two Stations

It is of genuine interest to us to study longer flow lines, since in practice these systems have more than two stations. In principle, one can apply the direct, the iterative, or the recursive methods presented in Section 4.3 to longer systems. For instance, a K-station flow line with phase-type processing times can be analyzed by studying the continuous-time Markov chain represented by

$$\{N_1(t), J_1(t); N_2(t), J_2(t); \ldots; N_K(t), J_K(t), t \geq 0\}, \quad (4.31)$$

where $N_i(t) = 0, 1, \ldots, N_i$ is the number of jobs in B_i at time t, and $J_i(t)$ is the index of the phase the processor is in at station i at time t. $J_i(t) = B$ implies that station i is blocked. The capacity of B_1 may be infinite or finite, or M_1 may be exhausted (in which case $N_1(t)$ drops out of the above stochastic process).

4.4.1 A Special Case: $N_i = 1$

A special case is a three-station flow line with exponential processing times and with intermediate buffers of capacity 1. This problem has been attributed to Hunt (1956), who derived an explicit expression for the line throughput. Let μ_i be the processing rate at station i. Then the stochastic process $\{J_1(t), J_2(t), J_3(t), t \geq 0\}$ is a continuous-time Markov chain where $J_1(t) = $ U, B, $J_2(t) = $ I, U, B, and $J_3(t) = $ I, U, with I being idle, U being processing, and B being blocked. The steady-state flow-balance equations of the Markov chain are given below:

$$\mu_1 P(\text{U, I, I}) = \mu_3 P(\text{U, I, U})$$
$$(\mu_1 + \mu_2) P(\text{U, U, I}) = \mu_1 P(\text{U, I, I}) + \mu_3 P(\text{U, U, U})$$
$$\mu_2 P(\text{B, U, I}) = \mu_1 P(\text{U, U, I}) + \mu_3 P(\text{B, U, U})$$
$$(\mu_1 + \mu_3) P(\text{U, I, U}) = \mu_2 P(\text{U, U, I}) + \mu_3 P(\text{U, B, U})$$
$$(\mu_1 + \mu_2 + \mu_3) P(\text{U, U, U}) = \mu_1 P(\text{U, I, U}) + \mu_2 P(\text{B, U, I})$$
$$+ \mu_3 P(\text{B, B, U}) \quad (4.32)$$

$$(\mu_2 + \mu_3)P(B, U, U) = \mu_1 P(U, U, U)$$
$$(\mu_1 + \mu_3)P(U, B, U) = \mu_2 P(U, U, U)$$
$$\mu_3 P(B, B, U) = \mu_1 P(U, B, U) + \mu_2 P(B, U, U)$$

The solution to the preceding set of equations (utilizing the normalization equation) yields the following throughput:

$$\overline{o}_\ell = \frac{\mu_1(A + 1) + \mu_2 + \mu_3}{A\left(1 + \frac{\mu_1}{\mu_2} + \frac{\mu_1}{\mu_3}\right) + B\left(1 + \frac{\mu_2^2 + \mu_3^2}{\mu_2 \mu_3}\right)}, \qquad (4.33)$$

where

$$A = \frac{\mu_2 \mu_3^2 \left(1 + \frac{\mu_1 + \mu_2}{\mu_1 + \mu_3}\right)}{\mu_1^2(\mu_1 + \mu_2 + \mu_3)} \quad \text{and} \quad B = \frac{\mu_1 + \mu_2 + \mu_3}{\mu_2 + \mu_3}.$$

As the system size increases (i.e., having more machines and larger buffer capacities), the use of the exact methods in general becomes infeasible due to the magnitude of the computational effort and the computer memory requirements. Moreover, the numerical instabilities resulting from the mathematical structure of the problem and the round-off errors in computers contribute to the infeasibility of the exact approaches.[1] Hence, the only remaining viable approach for the analysis of longer production lines appears to be the use of approximation techniques and computer simulation. In the search for fast and reliable methods to compute the performance measures in flow lines, various approximation methods have been developed during the past few decades. A widely used approach has been to decompose the system into several smaller subsystems that can be analyzed in isolation. Usually, an iterative procedure relates these subsystems to each other. The main theme of the decomposition is to create smaller subsystems that can be associated to the original flow line and can be analyzed in isolation with lesser difficulty. For instance, the behavior of the work-in-process inventory in a subsystem should resemble the behavior of the contents of the same buffer in the original system. Similarly, the throughput of a processor in a subsystem should be sufficiently close to that of the same processor in the original system. A subsystem may consist of a single station, that is, a machine and a buffer with an appropriate job arrival process, or it may consist of two or more stations with intermediate buffers. The single-station approach may not be preferred, because of the lack of knowledge about the job arrival process. On the other hand, subsystems with more than two stations are also difficult to analyze. We showed earlier in Section 4.3 that

---

[1] On the other hand, we do not want to totally disregard the numerical approaches (Gaussian elimination and its derivatives and others), since the advances in the computer technology such as parallel processing will gradually make these techniques feasible also.

it is possible to solve two-node systems with phase-type service times efficiently. Therefore, in the next section, we present a two-node decomposition approach for the approximate analysis of flow lines. Clearly, as the subsystems have more stations, they capture the behavior of the flow line better. This brings the argument of accuracy versus computational efficiency in approximations, where a balance is needed between the two.

4.4.2 An Approximate Decomposition Method

Let us consider the flow line with K machines and $K - 1$ intermediate buffers shown in Figure 4.8. Let M_i represent the machine at station i and N_i denote the capacity of the buffer B_i that is between stations $i - 1$ and i. The BAP-type blocking is in effect. M_1 is assumed to be busy except when blocked. The processing time, X_i at station i, is assumed to be of phase type with an $(\underline{\alpha}_i, S_i)$ representation with ℓ_i phases. $\underline{S}_i^o$ is the vector of process completion rates from the phases of X_i. Define $P_i(n)$ to be the steady-state probability of having n jobs at station i at any point in time. Our objective here is to approximate $P_i(n)$ for all i and n, from which most of the performance measures of interest can be obtained.

In a two-node decomposition approach, a two-node subsystem exists for every buffer in the flow line, as shown in Figure 4.9. For clarity, let Ω stand for the flow line shown in Figure 4.8, and let $\Omega(i)$ represent the subsystem for buffer i, $i = 2, \ldots, K$. Clearly, the behavior of a station in a flow line, that is, the length of its idle time, blocked time, and so forth, depends on the behavior of the upstream as well as the downstream stations in the system. Therefore, for the buffer in $\Omega(i)$ to replicate the behavior of B_i in Ω, the upstream node in $\Omega(i)$, represented by M_i', should model the behavior of station M_{i-1} as it is affected by the behavior of the upstream part of the flow line. Similarly, the downstream node in $\Omega(i)$, that is, M_i'', should model the behavior of station M_i as it is affected by the downstream part of the flow line. Clearly, nodes M_i' and M_i'' are fictitious machines modeling the impacts of starvation (M_i') and blocking (M_i'') phenomena on the behavior of the work-in-process in B_i.

The two main steps in the decomposition of flow lines are the characterization of the subsystems, that is, the identification of the processing times of M_i' and M_i'' for all i, and linking the subsystems to each other. Next we discuss these steps in detail.

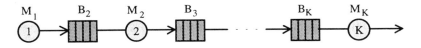

Figure 4.8. A flow line with K stations.

144   4. Analysis of Flow Lines

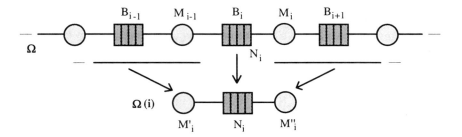

Figure 4.9. Two-node decomposition.

Characterization of $\Omega(i)$

Let us first discuss how the upstream and the downstream parts of the line affect the behavior of M_{i-1} and M_i. As soon as a job departure occurs at M_{i-1}, M_{i-1} starts processing a new job, if there are any in B_{i-1}. Otherwise, M_{i-1} starts an idle period and awaits a new job. This idle period ends upon the arrival of a new job, whose process at M_{i-1} starts immediately. On the other hand, M_i may be blocked at a departure instant due to lack of space in B_{i+1}. Otherwise, or after blocking is over, M_i starts processing a new job if there are any in B_i. Thus, the processing times in $\Omega(i)$ should be modeled to reflect these scenarios. The blocking of M_{i-1} and the starvation of M_i are issues to be handled within the modeling of $\Omega(i)$. The following parameters are necessary to formalize the processing times at M' and M''. Let

Δ_{i-1} = Pr(a departing job leaves M_{i-1} empty and idle),
V_{i-1} = the length of the idle period at M_{i-1},
Π_i = Pr(a departing job blocks station i),
Z_i = the length of the period during which M_i remains blocked.

Using these parameters and the possible scenarios mentioned above, one can develop a representation of $\Omega(i)$ as shown in Figure 4.10. M'_i is always busy

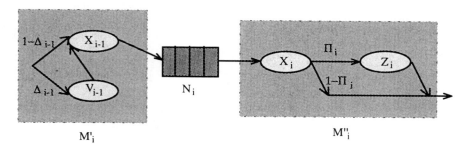

Figure 4.10. A pictorial representation of subsystem $\Omega(i)$.

except when blocked. The structure of the processing time at M'_i emerges out of the scenario mentioned earlier. First of all, M'_i is always busy except when blocked. Upon a process completion at M'_i, it is immediate that a new job arrives. This new job goes through the processing time X_{i-1} with probability $1 - \Delta_{i-1}$. With probability Δ_{i-1}, the job goes through the idle period V_{i-1}, and then the process starts for a period of X_{i-1}. Upon process completion, M'_i starts processing a new job or gets blocked.

On the other hand, M''_i operates as long as there are jobs in the buffer. Each job goes through the processing time X_i. After the process is completed, with probability Π_i each job goes through Z_i, which is the period M_i is blocked. With probability $1 - \Pi_i$, a job departs without causing blocking. Clearly, $\Omega(i)$ will be completely characterized when the probabilities Δ_{i-1} and Π_i, and the distributions of the random variables V_{i-1} and Z_i, are known.

EXAMPLE 4.4. Let us implement the two-node decomposition principle in a three-station flow line with exponentially distributed processing times with rates μ_1, μ_2 and μ_3. The intermediate buffers in Ω are B_2 and B_3, and the corresponding subsystems in the approximation are $\Omega(2)$ and $\Omega(3)$, as shown in Figure 4.11. Note that M'_2 and M''_3 are identical to M_1 and M_3, respectively. Thus, we have only M''_2 and M'_3 to characterize. M''_2 is the downstream machine to buffer B_2 in $\Omega(2)$, and M'_3 is the upstream machine to buffer B_3 in $\Omega(3)$. That is, both of these fictitious machines replicate the behavior of M_2 of Ω in their respective subsystems. Therefore, the behavior of M_2 becomes critical in characterizing their processing times.

Let us start with M''_2. Upon a process completion at M_2 in Ω, the job joins B_3 with probability $1 - \Pi_2$, or it remains on M_2 with probability Π_2, keeping it blocked until the process at M_3 is completed. In case M_2 is blocked, the time until blocking is over (that is Z_2) is the remaining processing time on M_3, which is also exponentially distributed with rate μ_3. The phase-type representation of the processing time on M''_2 is familiar to us: MGE-2 distribution! The time a job spends on M''_2 approximates the interdeparture times from M_2 with the blocking effect incorporated into it. The idle state of M_2 is modeled by the idleness of M''_2 in $\Omega(2)$.

The processing time on M'_3 is obtained by going through similar arguments. Upon a process completion M_2 may start processing a new job with probability $1 - \Delta_2$, or it may remain idle, with probability Δ_2, until the next job arrives from M_1. The idle period of M_2 (that is, V_2) is the remaining processing time on M_1 and is also exponentially distributed with rate μ_1. Thus, M'_3 is always busy, and the time a job spends on M'_3 approximates the interdeparture times from M_2 with the starvation effect incorporated into it. The blocked state of M_2 is modeled by the blocking of M'_3 in $\Omega(3)$.

The two-phase processing time at M'_3 is also MGE-2 distributed due to the equivalence shown in Figure 4.12. The two phase-structures have the same distribution.

146 4. Analysis of Flow Lines

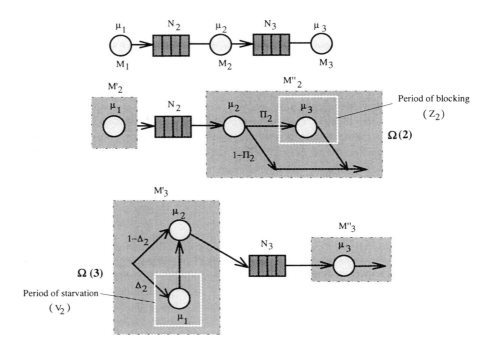

Figure 4.11. Decomposition of a three-station flow line.

Let us now continue with the details of the decomposition method. In Example 4.4, since the processing times are exponentially distributed and there are only three stations, the random variables V_2 and Z_2 are also exponentially distributed. However, if there were more than three stages in Ω, or if the processing times had more than one phase then V_2 and Z_2 would be a bit more involved. For instance, since the processing times in Ω are of phase type, the random variables V_i and Z_i are also phase-type distributed. The remaining processing time at station $i+1$ at the moment it blocks station i, that is, Z_i, is the sum of the remaining time in

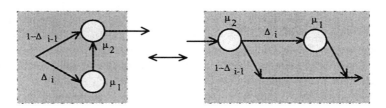

Figure 4.12. The processing time at M'_3 in Example 4.4.

the current exponential phase and the time that the residing job on M_{i+1} spends in what is left of the processing time. This is the convolution of an exponential and a phase-type random variable resulting in a phase-type random variable (see Chapter 3). The same argument is valid for V_i also. Then, let V_i have a $(\underline{\gamma}_i, D_i)$ representation with ℓ'_i phases, and let Z_i have a $(\underline{\beta}_i, B_i)$ representation with ℓ''_i phases.

The Pseudo Processing Times in $\Omega(i)$

The processing times in $\Omega(i)$ are mixtures of the random variables X, V and Z as shown in Figure 4.10. Let U'_i and U''_i represent the processing times on M'_i and M''_i, respectively. Then, we can write

$$U'_i = \begin{cases} X_{i-1} & \text{wp } 1 - \Delta_{i-1}, \\ V_{i-1} + X_{i-1} & \text{wp } \Delta_{i-1}, \end{cases} \quad (4.34)$$

where V_{i-1} and X_{i-1} are assumed to be independent of each other (an approximation). U'_i is of phase-type since it is a mixture of two phase type random variables. Then, U'_i can be represented by the pair $(\underline{\alpha}'_i, S'_i)$. The processing time on M'_i may start from X_{i-1} as well as V_{i-1}, having a vector of the start-up probabilities,

$$\underline{\alpha}'_i = \begin{bmatrix} (1 - \Delta_{i-1})\underline{\alpha}_{i-1} \\ \Delta_{i-1}\underline{\gamma}_{i-1} \end{bmatrix}.$$

That is, the start-up probabilities of U'_i are the ones of X_{i-1} with probability $(1 - \Delta_{i-1})$ and the ones of V_{i-1} with probability Δ_{i-1}. When the process starts in one of the phases of X_{i-1}, it experiences transitions associated with the rate matrix S_{i-1}. If it starts in one of the phases of V_{i-1}, its transitions will be according to the matrix D_{i-1}. Consequently, the matrix of transition rates of U'_i, that is, S'_i, is given by

$$S'_i = \begin{bmatrix} S_{i-1} & 0 \\ T'_{i-1} & D_{i-1} \end{bmatrix},$$

with a dimension of $m_i = \ell_{i-1} + \ell'_{i-1}$, which is the sum of the number of phases in the processing and the idle times. T'_{i-1} is the matrix of transition rates from the phases of V_{i-1} to the phases of X_{i-1}. It designates all the possible routes through which a job may arrive from the idle period to start the process at M_{i-1}, with the associated arrival (transition) rates. If we let $\underline{D}^o_{i-1}$ be the vector of departure rates for V_{i-1} and $\underline{\alpha}_{i-1}$ the start-up probabilities for X_{i-1}, we can write $T'_{i-1} = \underline{D}^o_{i-1}\underline{\alpha}^T_{i-1}$.

The job departures from M'_i can only occur at the end of the processing time X_{i-1}; therefore, the vector of departure rates for U'_i is given by

$$\underline{S}^{o'}_i = \begin{bmatrix} \underline{S}^o_{i-1} \\ \underline{0} \end{bmatrix}.$$

For instance, in Example 4.4, U_3' has the following $(\underline{\alpha}_3', \mathbf{S}_3')$ representation:

$$\underline{\alpha}_3' = \begin{bmatrix} 1 - \Delta_2 \\ \Delta_2 \end{bmatrix}, \qquad \mathbf{S}_3' = \begin{bmatrix} -\mu_2 & 0 \\ \mu_1 & -\mu_1 \end{bmatrix}, \qquad \underline{\mathbf{S}}_3^{o'} = \begin{bmatrix} \mu_2 \\ 0 \end{bmatrix}.$$

Each job moves from the idle period to the busy period with rate μ_1, and from the busy period to the buffer with rate μ_2.

Similarly, the processing time on M_i'', denoted by U_i'', can be expressed as

$$U_i'' = \begin{cases} X_i & \text{wp } 1 - \Pi_i, \\ X_i + Z_i & \text{wp } \Pi_i, \end{cases} \qquad (4.35)$$

where X_i and Z_i are assumed to be independent (an approximation). U_i'' is also a phase-type random variable for reasons stated for U_i'. Let the pair $(\underline{\alpha}_i'', \mathbf{S}_i'')$ represent U_i'' with m_i'' phases. Since every job starts with X_i on M_i'', the vector of the start-up probabilities $\underline{\alpha}_i''$ is given by

$$\underline{\alpha}_i'' = \begin{bmatrix} \underline{\alpha}_i \\ \mathbf{0} \end{bmatrix}.$$

$\mathbf{S}_i''$, the matrix of transition rates of U_i'', is given by

$$\mathbf{S}_i'' = \begin{bmatrix} \mathbf{S}_i & \mathbf{T}_i'' \\ 0 & \mathbf{B}_i \end{bmatrix},$$

having a dimension of $m_i'' = \ell_i + \ell_i''$. $\mathbf{T}_i''$ is the matrix of transition rates from the phases of the processing time to the phases of Z_i, the time spent in blocking. Since the vector of departure rates for X_i is $\underline{\mathbf{S}}_i^o$, and the vector of start-up probabilities for Z_i is $\underline{\beta}_i$, $\mathbf{T}_i''$ is given by $\mathbf{T}_i'' = \Pi_i \underline{\mathbf{S}}_i^o \underline{\beta}_i^T$.

The job departures from M_i'' may occur from both X_i and Z_i, therefore, the vector of departure rates of U_i'' is given by

$$\underline{\mathbf{S}}_i^{o''} = \begin{bmatrix} (1 - \Pi_i)\underline{\mathbf{S}}_i^o \\ \mathbf{B}_i^o \end{bmatrix}.$$

Note that the pseudo processing times in $\Omega(i)$ are modeled assuming fixed starvation and blocking probabilities, namely Δ_{i-1} and Π_i. For instance, in the construction of the processing time on M_i'', the argument is that every job has probability Π_i to go through the blocking period. However, only those jobs that depart while B_{i+1} is full get blocked. That is, these probabilities depend on the number of jobs in buffers B_{i-1} and B_{i+1}. This argument, coupled with the independence assumptions, forms the main thrust of the approximation.

In Example 4.4, the $(\underline{\alpha}_2'', \mathbf{S}_2'')$ representation of U_2'' is given by

$$\underline{\alpha}_2'' = \begin{bmatrix} 1 \\ 0 \end{bmatrix}, \qquad \mathbf{S}_2'' = \begin{bmatrix} -\mu_2 & \Pi_2\mu_2 \\ 0 & -\mu_3 \end{bmatrix}, \qquad \underline{\mathbf{S}}_2^{o''} = \begin{bmatrix} (1 - \Pi_2)\mu_2 \\ \mu_3 \end{bmatrix}.$$

4.4. Analysis of Flow Lines With More Than Two Stations   149

Thus, we have completed the approximate phase type representations of the processing times in $\Omega(i)$. This is convenient because we can use any procedure, including the matrix-recursive procedure or any of the numerical procedures, for the analysis of $\Omega(i)$ for all i. Clearly, we want to compute $P_i(k, j, n)$, the steady-state probability that M_i' is in phase k, M_i'' in phase j, and the number of jobs in the buffer, including the one on M_i'', is n. $P_i(\bullet)$ is referred to as the steady-state joint probabilities of $\Omega(i)$ in the remainder of this chapter.

Linking $\Omega(i)$s

Having the phase-type representations of the processing times in $\Omega(i)$ laid out, the next step is to relate these subsystems to each other. Notice that the behavior of M_i is modeled by M_i'' in $\Omega(i)$ and by M_{i+1}' in $\Omega(i+1)$. Consequently, the period of starvation experienced by M_i should be the same as the one experienced by M_i'' in $\Omega(i)$. Also, the period of blocking experienced by M_i should be the same as the one experienced by M_{i+1}' in $\Omega(i+1)$. This suggests that γ_i and $\mathbf{D}_i$ may be identified using the parameters of $\Omega(i)$, and $\underline{\beta}_i$ and $\mathbf{B}_i$ using the parameters of $\Omega(i+1)$. Let us pursue this idea starting with γ_i and $\mathbf{D}_i$. The period of starvation in $\Omega(i)$ starts with a job departure, leaving M_i'' idle and the buffer empty. At this moment, M_i' may be in any phase of its processing time. Remember that M_i' is always busy unless it is blocked. Then, the remaining processing time on M_i', at the time M_i'' becomes idle, is in fact the period of starvation for M_i''. Let

$$\varpi_i'(j) = \Pr(M_i' \text{ is in phase } j \text{ at the time a departing job leaves } M_i'' \text{ idle}).$$

$\varpi_i'(j)$ can be computed as the ratio of the number of job departures leaving M_i'' idle and M_i' in phase j of its processing time to the total number of departures leaving M_i'' idle per unit time. That is,

$$\varpi_i'(j) = \frac{\sum_{k=1}^{\ell_i''} \mathbf{S}_i^{\circ''}(k) P_i(j, k, 1)}{\sum_{m=1}^{\ell_i'} \sum_{k=1}^{\ell_i''} \mathbf{S}_i^{\circ''}(k) P_i(m, k, 1)}, \quad j = 1, \ldots, \ell_i', \quad (4.36)$$

which provides the percentage of time a departing job leaves M_i'' idle, the buffer empty, and M_i' in phase j. Notice that $P_i(j, k, 1)$ is the steady-state probability that the first machine is in phase j and the second node is in phase k of their processing times (that is, both are busy), and there is one job in the buffer. $\mathbf{S}_i^{\circ''}(k)$ is the rate at which jobs are departing from phase k on M_i'' of $\Omega(i)$. Then, V_i has a $(\underline{\gamma}_i, \mathbf{D}_i)$ representation, where

$$\underline{\gamma}_i^T = [(\varpi_i'(1), \varpi_i'(2), \ldots, \varpi_i'(\ell_i')], \quad \text{and} \quad \mathbf{D}_i = \mathbf{S}_i'.$$

The period of time M_i remains blocked can be obtained from $\Omega(i+1)$. The period of time M_{i+1}' remains blocked depends on what phase M_{i+1}'' is in at the moment

150   4. Analysis of Flow Lines

M'_{i+1} gets blocked. Let

$\varpi''_{i+1}(j) = \Pr(M''_{i+1}$ is in phase j at the time when a departing job blocks $M'_{i+1})$.

Again, $\varpi''_{i+1}(j)$ can be computed as the ratio of the number of job departures seeing M''_{i+1} in phase j and blocking M'_{i+1} to the total number of departures blocking M'_{i+1}. That is,

$$\varpi''_{i+1}(j) = \frac{\sum_{k=1}^{\ell'_{i+1}} \mathbf{S}^{o'}_{i+1}(k) P_{i+1}(k, j, N)}{\sum_{m=1}^{\ell''_{i+1}} \sum_{k=1}^{\ell'_{i+1}} \mathbf{S}^{o'}_{i+1}(k) P(k, m, N)}, \quad j = 1, \ldots, \ell_{i+1''}. \quad (4.37)$$

Then, Z_i has a $(\underline{\beta}_i, \mathbf{B}_i)$ representation, where

$$\underline{\beta}_i^T = [\varpi''_{i+1}(1), \varpi''_{i+1}(2), \ldots, \varpi''_{i+1}(\ell''_{i+1})],$$

and $\mathbf{B}_i = \mathbf{S}''_{i+1}$. Notice that $\varpi'(\cdot)$ and $\varpi''(\cdot)$ are the probabilities of where the machines are (in which phase of their processing time) at the instants of blocking and starving. They are essential elements of the phase-type representations of the periods of starvation and blocking. It may also be possible (and simpler) to approximate $\varpi'_{i+1}(j)$ by

$$\varpi'_i(j) = \frac{P_i(j, 0, 0)}{\sum_{k=1}^{\ell'_i} P_i(k, 0, 0)}, \quad (4.38)$$

that is, the conditional probability that M'_i is in phase j given that M''_i is idle. $\varpi''_i(j)$ may also be approximated similarly.

We have yet to provide expressions for Δ_i and Π_i. Let us start with Δ_i. Starvation starts as soon as a job departs from the ith station, leaving it empty, and lasts until another job enters the station. Since a departing job causes starvation with probability Δ_i, the rate of departures causing starvation is $\bar{o}_\ell \Delta_i$, where $\bar{o}_\ell$ is the line throughput. The expected length of the starvation period is $E[V_i]$, and the steady-state probability that M_i is starving is $P_i(0) \cong \sum_{j=1}^{\ell'_i} P_i(j, 0, 0)$. Then, we can write

$$\bar{o}_\ell \Delta_i E[V_i] = P_i(0), \quad (4.39)$$

which is merely Little's relationship applied at the empty-and-idle state.

A similar argument can be applied to compute Π_i. The rate at which blocking occurs is $\bar{o}_\ell \Pi_i$ since a departing job blocks station i with probability Π_i. The expected period of time station i remains blocked is $E[Z_i]$. The steady-state probability that M_i is blocked is $P_i(B) \cong \sum_{j=1}^{\ell''_{i+1}} P_{i+1}(B, j, N)$. Then, we can write

$$\bar{o}_\ell \Pi_i E[Z_i] = P_i(B). \quad (4.40)$$

4.4. Analysis of Flow Lines With More Than Two Stations   151

Thus, Δ_i and Π_i can be computed from (4.39) and (4.40), respectively. Before we present a formal algorithm to compute $P_j(n)$, let us demonstrate pictorially how the periods of starving and blocking are characterized.

EXAMPLE 4.5. Consider a flow line consisting of four stations with intermediate buffers of capacities N_2, N_3, and N_4. For simplicity, the processing times are exponentially distributed with rate μ_i at station i. The above approach decomposes the line into three subsystems, $\Omega(2)$, $\Omega(3)$, and $\Omega(4)$, corresponding to the three buffers in the line. The decomposition is shown in Figure 4.13. In $\Omega(2)$, there is no V_1 since M_1 never starves, and consequently the processing time of M_1' is the same as that of M_1.

In $\Omega(2)$, Z_2 models the period of time M_2 remains blocked, which depends on the phase M_3'' is in at the instant of blocking. In $\Omega(3)$, V_2, which is the period of starvation experienced by M_2, is the remaining processing time at M_2'. The period during which M_3 is blocked, Z_3 is the remaining processing time at M_4''. In $\Omega(4)$, V_3 models the idle period of M_3, which is the remaining processing time at M_3''. Hence, $\Omega(2)$, $\Omega(3)$, and $\Omega(4)$ are characterized as shown in Figure 4.13.

EXAMPLE 4.6. Let us next consider a flow line with three work stations and with MGE-2 processing times as shown in Figure 4.14. In $\Omega(2)$, Z_2 depends on which phase M_3'' (or M_3) is in at the instant of blocking. In $\Omega(3)$, V_2 depends on which phase M_2' (or M_1) is in at the moment starvation starts.

Having the flow lines decomposed as shown in Figures 4.13 and 4.14, we need a procedure to relate the subsystems to each other to find the unknown parameters and eventually obtain the steady-state distributions of the buffer contents. For this purpose, we need the following iterative algorithm designed to link $\Omega(\cdot)$s to each other.

An Iterative Algorithm

In principle, what we have is a set of linear equations consisting of the steady-state flow-balance equations of each subsystem and the equations for Δ_j and Π_j for all j. Our final task is to solve these equations to find the unknowns, that is, $P_i(k, j, n)$, namely the steady-state joint probabilities of $\Omega(i)$ for every i. The fact that we have subsystems that supply information to each other and that can be analyzed individually naturally gives rise to an iterative algorithm. For instance, in Figure 4.14, we may start with $\Omega(2)$ assuming some values for Π_2, $\varpi_3''(1)$, and $\varpi_3''(2)$, or ignoring Z_2 completely, and solve for $P_2(k, j, n)$ using any procedure of our choice. We obtain Δ_2, $\varpi_2'(1)$, and $\varpi_2'(2)$ from the analysis of $\Omega(2)$. Now we are ready to analyze $\Omega(3)$ since all the parameters necessary for its analysis are known. We obtain $P_3(k, j, n)$. This in turn enables us to compute Π_2, $\varpi_3''(1)$, and $\varpi_3''(2)$. Then we go back to analyze $\Omega(2)$. This procedure continues in the same manner until the throughputs of $\Omega(2)$ and $\Omega(3)$ are sufficiently close to each other. Although the degree of closeness is a subjective matter, it may be recommended to stop the procedure when $|\bar{o}_2 - \bar{o}_3| < 10^{-5}$ is satisfied.

152   4. Analysis of Flow Lines

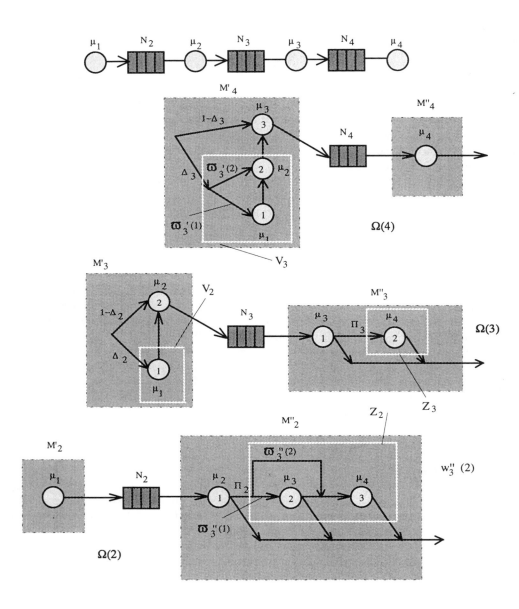

Figure 4.13. Decomposition of the flow line in Example 4.5.

4.4. Analysis of Flow Lines With More Than Two Stations

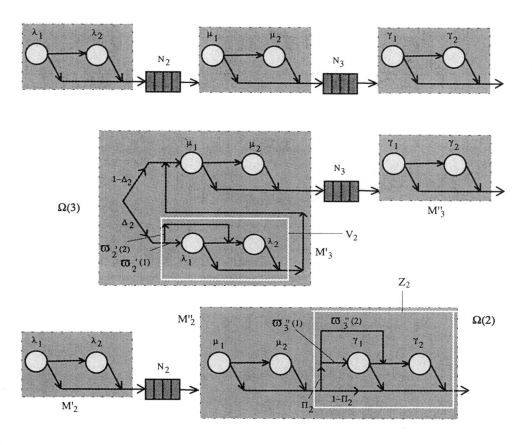

Figure 4.14. Decomposition of the flow line in Example 4.6.

Thus, the algorithm executes a series of forward and backward passes. In a flow line with K stations, in each forward pass, the subsystems $\Omega(2), \ldots, \Omega(K-1)$ are analyzed in that order. The analysis of $\Omega(i-1)$ provides information (that is $\Delta_{i-1}, \varpi'_{i-1}(\cdot)$) necessary for the analysis of $\Omega(i)$. In each backward pass, the subsystems $\Omega(K), \ldots, \Omega(3)$ are analyzed in that order. The analysis of $\Omega(i+1)$ provides information (that is, $\Delta_{i+1}, \varpi'_{i+1}(\cdot)$) necessary for the analysis of $\Omega(i)$. The iterative procedure continues in this manner until the throughputs of all the subsystems, $\bar{o}_2, \ldots, \bar{o}_K$ are sufficiently close to each other. The algorithm is summarized in Table 4.6.

TABLE 4.6. The two-node decomposition algorithm

INITIALIZATION: $k = 1, i = 2, \Delta_1 = 0, \Pi_2 = \Pi_3 = \cdots = \Pi_{K-1} = 0$

STEP 1: FORWARD PASS
- Characterize $(\underline{\alpha}'_i, \mathbf{S}'_i)$ and $(\underline{\alpha}''_i, \mathbf{S}''_i)$ of U'_i and U''_i using (4.34) and (4.35).
- Obtain $P_i(\cdot, \cdot, \cdot)$ and $\overline{o}_i^{(k)}$.
- Obtain $\varpi'_i(j), j = 1, \ldots, \ell'_i$ using (4.36).
- Characterize $(\underline{\gamma}_i, \mathbf{D}_i)$ of V_i.
- Obtain Δ_i using (4.39).
- $i \leftarrow i + 1$, if $i < K$, then repeat Step 1. Else, $k \leftarrow k + 1$ and go to Step 2.

STEP 2: BACKWARD PASS
- Characterize $(\underline{\alpha}'_i, \mathbf{S}'_i)$ and $(\underline{\alpha}''_i, \mathbf{S}''_i)$ of U'_i and U''_i
- Obtain $P_i(\cdot, \cdot, \cdot)$ and $\overline{o}_i^{(k)}$
- Obtain $\varpi''_i(j), j = 1, \ldots, \ell''_i$ using (4.37)
- Characterize $(\underline{\beta}_i, \mathbf{B}_i)$ of $\mathbf{Z}_i$
- Obtain Π_i using (4.40)
- $i \leftarrow i - 1$
- If $i < 2$, then repeat Step 2.
- Else, if $\overline{o}_M^{(k)} \cong \overline{o}_{M-1}^{(k)} \cdots \cong \overline{o}_3^{(k)} \cong \overline{o}_2^{(k-1)}$, then stop; else, $k \leftarrow k + 1$ and go to Step 1.

Although the parameters do not have a superscript k (to indicate their values at iteration k), except $\overline{o}_i^{(k)}$, their most recent values are used in evaluating any expression at any iteration.

In case $N_1 = \infty$, we also have $\Omega(1)$ in the approximate decomposition, which is a •/MGE/1-type work station with a particular job arrival process (either M or MGE) with rate λ. Principles of the decomposition method apply in this case also. The simplification is that there is no need to solve for the steady-state probabilities of $\Omega(1)$ until the algorithm converges. This is because Δ_2 is readily available through $\Delta_2 = P_1(0) = 1 - \lambda E[U''_1]$ to be used in the analysis of $\Omega(2)$ in the forward pass.

Before we present some numerical examples, let us discuss a computational issue. As the number of stations increases, or as the number of phases in the processing times increases, the number of phases to be dealt with in $\mathbf{V}_i$ or in $\mathbf{Z}_i$ also increases. To avoid dealing with a large number of phases, it is advisable to approximate $\mathbf{V}_i$ (or $\mathbf{Z}_i$) by a simpler phase-type distribution using a few moments, and then analyze $\Omega(i)$. This approximation has practically no impact on the accuracy of the results. A minor issue in the analysis of $\Omega(i)$ is that the procedure presented in Section 4.3 handles cases where the vector of start-up probabilities at the first station is of the form of $\underline{\alpha}'^T = (1, 0, 0, \ldots, 0)$. To be able to use this procedure here in $\Omega(i)$, we need to make the transformation shown in Figure 4.15. This transformation does not change the distribution of U' and has no effect on the results of the approximation.

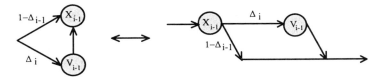

Figure 4.15. Transformation of U' to a convenient form.

EXAMPLE 4.7. Let us first implement the approximation to the simplest case and observe the convergence of the algorithm. Consider a three-station flow line with buffer capacities of $N_2 = N_3 = 1$. That is, we have only the machines and no space in between. Let us assume exponentially distributed processing times with rates $\mu_1 = \mu_2 = \mu_3 = 2$. The subsystems $\Omega(2)$ and $\Omega(3)$ involve Δ_2 and Π_2. As the output rates of the subsystems converge, Δ_2 and Π_2 also converge to their final values. Table 4.7 shows the values of Δ_2 and Π_2, and $\bar{o}_\ell$ as the algorithm executes forward and backward passes. A new value for Δ_2 is obtained at the end of each forward pass (analysis of $\Omega(2)$), and a new value for Π_2 is obtained at the end of each backward pass (analysis of $\Omega(3)$). In each pass, we have a new value for $\bar{o}_\ell$. Convergence is obtained with $|\bar{o}_\ell^{(k)} - \bar{o}_\ell^{(k-1)}| \leq 10^{-6}$. The approximate line throughput is $\bar{o}_\ell(\text{Appx}) = 1.11111$, whereas the exact throughput (Hunt's formula) is $\bar{o}_\ell(\text{Ex}) = 1.12821$.

EXAMPLE 4.8. Consider a production line consisting of three work stations with buffer capacities of $N_2 = 3$ and $N_3 = 3$. The mean processing times are 2, 4, and 3, and the squared coefficient of variations are 2, 1, and 0.25 in stations 1 to 3, respectively. Hence, the line is not balanced in terms of the mean processing times.

TABLE 4.7. Convergence of the approximation in Example 4.7

Iteration No.	Δ_2	Π_2	$\bar{o}_\lambda$
0	0.1	0.1	—
1	0.47499	—	1.26984
2	—	0.38125	1.07744
3	0.40469	—	1.11986
4	—	0.39883	1.10895
5	0.40029	—	1.11165
6	—	0.39993	1.11098
7	0.40002	—	1.11115
8	—	0.39999	1.11110
9	0.40000	—	1.11111
10	—	0.39999	1.11111

156   4. Analysis of Flow Lines

The approximation procedure described above was implemented on a Pyramid 9810 RISC machine with 10 mips using a FORTRAN 77 compiler. The results were tested against the results of a discrete-event simulation model. Each simulation run consisted of 800,000 transactions and generated the probability distribution of the number of units in each queue and a 95 percent confidence interval for the mean queue lengths. The approximation and the simulation results are given in Table 4.8. The arbitrary-time blocking probabilities (simulation) experienced by the first and second stations are 0.549 and 0.051, respectively. The first station experiences a significant level of blocking.

EXAMPLE 4.9. The production line consists of five stations with buffer capacities $N_2 = 4$, $N_3 = 4$, $N_4 = 10$, $N_5 = 4$. The mean processing times are 2.5, 4, 3, 2, and 5, respectively. The squared coefficient of variations are 2.5, 0.25, 3.0, 0.30, and 0.75, respectively. The system is unbalanced and experiences high levels of processing time-variability. The results are shown in Table 4.9. The arbitrary-time blocking probabilities (simulation) for stations 1 through 4 are 0.518, 0.186, 0.226, and 0.551, respectively.

The decomposition algorithm presented above is fast, and it generally produces acceptable results. The factors expected to impact the accuracy of an approximation algorithm for the analysis of flow lines are the processing-time variability, the imbalance in the system, the sizes of the buffer capacities, and the number of stations. For this particular algorithm, the processing-time variability has almost no impact on the accuracy of the approximation. However, the smaller the variability is, the longer the CPU times are, due to the larger number of phases. The imbalance affects the results such that the balanced systems (in terms of the mean processing times) tend to have better approximate results. The negative effect of the imbalance is especially observed in the cases with a larger number of stations. The approximation deteriorates in the middle stations. This may be attributed to the fact that the subsystems have only two nodes. A decomposition with three or more nodes in the subsystems would naturally give better results but at the expense of immense computational effort. Buffer capacities also affect the approximation. The approximation in the steady-state probabilities of the buffer content deterio-

TABLE 4.8.   Results of Example 4.8 (CPU = 0.08 sec.)

		$P_i(0)$	$P_i(N)$	$\bar{N}$
Station 2	App.	0.0476	0.7781	2.6209
	Sim.	0.0433	0.7815	2.6334
				(± 0.0127)
Station 3	App.	0.3239	0.1734	1.2239
	Sim.	0.3214	0.1744	1.2293
				(± 0.0199)

$\bar{o}_\ell$: App. = 0.2254, Sim. = 0.2262

TABLE 4.9. Results of Example 4.9 (CPU = 0.17 sec.)

		$P_i(0)$	$P_i(N)$	$\bar{N}$
Station 2	App.	0.0349	0.7560	3.5095
	Sim.	0.0362	0.7693	3.5293
				(± 0.0179)
Station 3	App.	0.1836	0.3109	2.2104
	Sim.	0.1918	0.3589	2.3098
				(± 0.1085)
Station 4	App.	0.0358	0.3903	7.4080
	Sim.	0.0603	0.3640	6.9328
				(± 0.3072)
Station 5	App.	0.0082	0.8643	3.7905
	Sim.	0.0181	0.8386	3.7151
$\bar{o}_\ell$: App. = 0.1984, Sim. = 0.1964				

rates in large buffers. The average buffer content usually has an acceptable error level (less than 10 percent relative error). The measure that is best approximated is the throughput (less than 4 percent in systems with a large number of stations, such as 10, and 1 percent in systems with three to five stations).

Thus, we have achieved our goal in the approximate analysis of flow lines by developing a fast and fairly accurate method that can be programmed for applications. For those cases where stations are subject to failures, the phase-type representations of the process completion times can be obtained for each station, and the above method can be used to compute the performance measures.

4.5 Deterministic Processing Times

So far we have considered flow lines with processing times that have some degree of variability. The randomness in flow lines with fixed processing times is usually due to station failures. If the system does not experience any randomness, its analysis becomes trivial provided that the first work station does have raw material to process at all times. In the long run, the station with the longest processing time will cause all the upstream buffers to be full and block the upstream stations. Jobs will not experience delays after the slowest station, whose production rate becomes the output rate of the flow line. That is, the line throughput will be

$$\bar{o}_\ell = \frac{1}{\max\{X_1, \ldots, X_K\}},$$

158   4. Analysis of Flow Lines

and will be independent of the order of the stations and the buffer capacities. A by-product of this argument is that the reversed line has the same throughput, which is the reversibility property in this particular flow line.

In the above scenario, the buffers downstream to the station with the longest processing time will always be empty. That is, the flowtime of a job—the time it takes to complete all the processes—depends on where the slowest station is. If the jth station is the slowest one, then the expected total flow time would be $jX_j + \sum_{i=j+1}^{K} X_i$, which clearly depends on the sequence. Thus, the minimum flow time is obtained by a design that allows $j = 1$. This observation suggests for many reasons (less inventory holding, shorter flow times) that, in the case of an exhausted M_1, the slowest station should be placed as far upstream as possible.

4.5.1 The Case with $N_1 = \infty$

The measures of interest in the case where the first station does not have raw material at all times, and instead has jobs arriving on a random basis, require an analysis that we attempt to introduce in this section. Let us consider a flow line consisting of K work stations with finite-capacity intermediate buffers as shown in Figure 4.16. The first station has an infinite-capacity buffer and an arbitrary job arrival process. The processing time, denoted by X_j for station j, is fixed and may differ from one station to another. The blocking policy is assumed to be the BAP type. Notice that $\bar{o}_\ell = \lambda$, and the system is stable if the slowest station is not exhausted. In this particular flow line, there are no known techniques to obtain measures specific to particular buffers, such as the probability distribution of the buffer contents. It is possible for approximations to be developed after approximately representing the fixed processing times by their Erlang counterparts with fairly large numbers of phases. Here, we limit ourselves to the exact analysis and focus on the time each job spends in the entire system. Consider the following two theorems:

Theorem 1 Avi-Itzhak (1965). *The time an arriving job spends in the system is independent of the order of the stations and is also independent of the buffer capacities.*

Theorem 2 Avi-Itzhak (1965). *The time an arriving job spends in the system is equal to $\sum_{i=1}^{K} X_i$ and the waiting time of the same job in the buffer of a work*

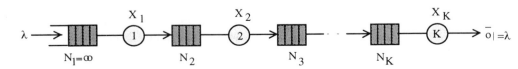

Figure 4.16. Work stations in series with $N_1 = \infty$.

4.5. Deterministic Processing Times

station with the same job arrival process and with a fixed processing time of $X^* = \max\{X_i : i = 1, 2, \ldots, K\}$.

We prove these two theorems for the special case of $N_2 = \cdots N_K = 1$. However, the theorems are immediately applicable to the case with arbitrary capacities in each buffer. This is because of the observation that each buffer location may also be viewed as a station with zero processing time. Let us define k^* as the index for the station with the longest processing time, that is $X^* = X_{k^*}$. Then, the jobs departing from station M_{k^*} will find M_{k^*+1} idle. Similarly, the arriving jobs will always find stations $k^* + 2, \ldots, K$ idle because the time between two successive departures from M_{k^*} is greater than or equal to the processing time at any of the downstream stations. Thus, jobs do not experience blocking at stations $k^*, \ldots, K$, and no matter what the job arrival process is, each job will spend $\sum_{j=k^*+1}^{K} X_j$ time units in the system after station k^*. Let us define the following variables to be able to describe the relationships that will help us prove the above theorems:

a_n = the arrival time of the nth job,
d_n = the departure time of the nth job,
s_n = the time spent in the system by the nth job,
w_n = the total delay (waiting time) of the nth job.

Then, the following relationships exist:

$$s_n = d_n - a_n, \tag{4.41}$$

$$w_n = s_n - \sum_{j=1}^{K} X_j = d_n - a_n - \sum_{j=1}^{K} X_j. \tag{4.42}$$

Let us first show that the following relationship is valid:

$$d_n = \begin{cases} a_n + \sum_{j=1}^{K} X_j, & \text{if } w_n = 0, n \geq 1 \\ d_{n-1} + X_{k^*} & \text{if } w_n > 0. \end{cases} \tag{4.43}$$

The first part of (4.43) is equivalent to (4.42) written for $w_n = 0$, indicating that if a job is not delayed anywhere in the system, it departs from the system after a period of time equal to the sum of the processing times. The second part of (4.43) implies that in case the nth job is delayed, it will depart X_{k^*} time units after the $(n-1)$st job departs. Define D_j to be the difference between the total amounts of processing already received by the $(n-1)$st and the nth jobs at the point of departure of the $(n-1)$st job at station j. The total amount of processing received by the $(n-1)$st job at this point is $\sum_{i=1}^{j} X_i$.

We prove the second part of (4.43) by focusing on where the nth job is delayed for the first time. The nth job, when it is delayed for the first time, may be waiting either in B_1 or on one of processors, that is, blocking that station. Let us first study the case where the nth job is delayed in B_1, as shown in Figure 4.17. This may

Figure 4.17. Delaying of the nth job in B_1.

happen when the nth job arrives at B_1 and finds the $(n-1)$st job either waiting or being processed.

The nth job enters M_1 as soon as the $(n-1)$st job departs from it, making $D_1 = X_1$. D_2 is a little more involved. At the moment the $(n-1)$st job completes its process on M_2, the nth job may have already finished its own process on M_1 or may have yet to finish it. The former case indicates that $X_2 \geq D_1$, and consequently $D_2 = X_2$. That is, the nth job completes its process on M_1 before the $(n-1)$st job completes its own on M_2, and gets blocked. Consequently, the nth job enters M_2 as soon as the $(n-1)$st job departs from it. In the latter case, that is, $X_2 \leq D_1$, the amount of processing the nth job receives by the time the $(n-1)$st job completes its process is simply X_2, and therefore $D_2 = X_1 + X_2 - X_2 = X_1 = D_1$. It is also possible that D_2 may be smaller than D_1 in case there is an intermediate buffer between stations 1 and 2, allowing for the nth job to catch up with the $(n-1)$st job (both being in the buffer), thus reducing D_2 to X_2 and making $D_2 \leq D_1$. On the other hand, $D_2 \geq X_2$ due to a possible time lag between the arrival times of the $(n-1)$st and the nth jobs at B_1. Then we can write $X_2 \leq D_2 \leq D_1$, if $X_2 \leq D_1$. Consequently, $D_j = X_j$ if $X_j \geq D_{j-1}$, and $X_j \leq D_j \leq D_{j-1}$ if $X_j \leq D_{j-1}$ for $j = 1, \ldots, K$. Consider the two examples given in Figure 4.18.

Notice that $X_2 < D_1$ in Figure 4.18(b), where $D_1 = D_2$. If there were a buffer between stations 1 and 2, than D_2 would be smaller due to a possible delay of the $(n-1)$st job in the buffer. We can write $D_{k^*-1} = \max[X_1, \ldots, X_{k^*-1}] \leq X_{k^*}$. It follows that $D_{k^*-1} \leq X_{k^*}$ indicates $D_{k^*} = X_{k^*}$. Finally, since blocking will not be observed in stations $k^*, \ldots, K$, we can write $D_{k^*} = D_{k^*+1} = \cdots = D_K = X_{k^*}$.

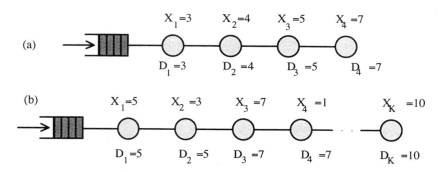

Figure 4.18. Examples of flow lines with fixed processing times.

This indicates that the nth job will depart from the system X_{k^*} time units after the $(n-1)$st job departs.

Next, let us consider the possibility that the nth job is first delayed at the ith station, where $i \le k^* - 1$. This means that the $(n-1)$st job is occupying the $(i+1)$st processor at the time the nth job is delayed at station i and is thus blocking it. This implies that $D_{i+1} = X_{i+1}$, and the earlier relation between D_j, D_{j-1}, and X_j becomes valid for $j = i+2, \ldots, K$. That is, the difference between two consecutive departures is still X_{k^*} even if the nth job is first delayed on a processor. Hence, (4.43) is proven to be valid. Considering the possibility that the nth job arrives sometime after the $(n-1)$st job, we can write $d_n \ge d_{n-1} + X_{k^*}$ if $w_n = 0$. Using (4.43), we have $a_n + \sum_{j=1}^{K} X_j \ge d_{n-1} + X_{k^*}$ for $w_n = 0$, which is equivalent to $a_n + \sum_{j \ne k^*}^{K} X_j \ge d_{n-1}$. Then, (4.43) can be rewritten as follows:

$$d_n = \begin{cases} a_n + \sum_{j=1}^{K} X_j & \text{if } a_n + \sum_{j \ne k^*}^{K} X_j \ge d_{n-1}, \quad n \ge 1 \\ d_{n-1} + X_{k^*}, & \text{otherwise.} \end{cases} \quad (4.44)$$

This verifies that d_n is determined by d_{n-1}, a_n, X_{k^*}, and $\sum_{j=1}^{K} X_j$. Clearly, d_1 is independent of the order of servers in the system since no jobs are ahead of job 1. Since d_n and a_n are independent of the order of servers, s_n is also independent of the order because of (4.41). This completes the proof of Theorem 1. Notice that addition of more stations with zero processing times (buffers!) has no impact on d_n, as is clear in (4.41). Hence, we conclude that the system time is independent of the order of stations as well as the buffer capacities.

Theorem 1 can easily be utilized to prove Theorem 2, which shows how to obtain the system time of a job. Since Theorem 1 allows us to rearrange the stations, we let $k^* = 1$ such that blocking is not experienced at any station in the system. In this case, station 1 behaves as if it is a single station with a fixed processing time, that is, $X_1 = X_{k^*}$. Then, the system time of the nth job is given by

$$s_n = w_n(1) + \sum_{j=1}^{K} X_j, \quad (4.45)$$

where $w_n(1)$ is the delay experienced by the nth job in B_1. This in turn proves Theorem 2. Equation (4.45) can be utilized to actually compute the steady-state average system time of a job, that is, $\bar{s} = \lim_{n \to \infty} 1/n \sum_{j=1}^{n} s_j$. Let $\bar{w}_1 = \lim_{n \to \infty} 1/n \sum_{j=1}^{n} w_j(1)$ be the steady-state average waiting time of a job in the first queue. Then, (4.45) can be rewritten in terms of the averages,

$$\bar{s} = \bar{w}_1 + \sum_{j=1}^{K} X_j, \quad (4.46)$$

which requires the computation of $\bar{w}_1$ to obtain $\bar{s}$.

As we have discussed in Chapter 3, the steady-state average waiting time of a job in a work station depends upon the assumptions about the job interarrival and the processing times. Here, we already made the assumptions that the processing times are deterministic and the arrival process is arbitrary. In case the arrival process is Poisson with rate λ, M_1 becomes an M/D/1 work station, and the Pollaczek-Khinchin formula from Chapter 3 is used to compute $\overline{w}_1$:

$$\overline{w}_1 = \frac{\rho X_1}{2(1-\rho)}, \qquad (4.47)$$

where $\rho = \lambda X_1$ is the traffic intensity. If the arrival process is not Poisson and yet the interarrival times are i.i.d. random variables with an arbitrary distribution, then one can resort to the approximate formula for the average waiting time in GI/GI/1 type systems, proposed by Kramer and Lagenbach-Belz (see (3.13) in Chapter 3).

Finally, the steady-state average number of jobs in the flow line can be obtained using the equality of the throughput to λ at every station:

$$\overline{N}_\ell = \lambda \overline{w}_1 + \lambda \sum_{j=1}^{K} X_j. \qquad (4.48)$$

Clearly, we are considering stable flow lines where the stability condition $\rho = \lambda X^* < 1$ is satisfied.

4.5.2 The Case with $N_1 < \infty$

The theorems of Section 4.5.1 are valid for $N_1 = \infty$. When B_1 has a finite capacity, the system becomes a loss system, indicating that jobs arriving at a full system are lost. In loss systems, the buffer capacities and the order of work stations do affect the proportion of jobs entering B_1 and consequently the time that each job spends in the system. Therefore, Theorems 1 and 2 will no longer be valid if $N_1 < \infty$. Ahead we will introduce lower and upper bounds on the system throughput $\overline{o}_\ell$. We maintain that Ω represents the flow line and k^* represents the station with the longest processing time, that is, X_{k^*}.

Let us consider the special case of Poisson job arrivals with rate λ at B_1, which has a finite capacity N_1. Naturally, we have 0 and λ (or $1/X_{k^*}$) as the lower and upper bounds for $\overline{o}_\ell$, respectively. However, these bounds are usually not tight. Let us study the following tighter bounds:

Lower Bound

The lower bound, $\overline{o}_L$, is the output rate of a single work station that has an incoming-material buffer with finite capacity N_1, a Poisson job arrival process with rate λ, and a fixed processing time of X_{k^*}, the longest processing time in the flow line. Or, equivalently, the lower bound is the throughput of Ω with X_1 replaced by X_{k^*}.

It can be observed that the maximum possible time M_1 remains blocked is $X_{k^*} - X_1$, which happens when M_{k^*} blocks all of the upstream stations, including M_1. Therefore, in case X_1 is replaced by X_{k^*} in Ω, the throughput of M_1 will be the lowest rate at which the jobs depart from the first station. This rate will also be the throughput of the flow line since none of the stations experiences blocking when M_1 has the longest processing time, thus producing the lower bound $\bar{o}_L$ for Ω. To compute the value of $\bar{o}_L$, we can isolate M_1 with its buffer, the job arrival process, and the processing time X_{k^*} and analyze it as a single work station (M/D/1/N queueing system[1]) using the techniques available in queueing theory textbooks. The lower bound is given by $\bar{o}_L = \lambda(1 - P_N)$, where P_N is the probability that an arriving job finds N jobs already in the system and is lost.

Upper Bound

The upper bound, $\bar{o}_U$, is the minimum of $\bar{o}_1$ of M_1 in isolation and $1/X_{k^*}$. This is quite obvious since the throughput of M_1 in isolation with its buffer, the job arrival process, and the processing time X_1 is going to be larger than its throughput when it is part of Ω, where it experiences blocking. However, we also know that the throughput of the network cannot exceed $1/X_{k^*}$, which is the rate of the slowest station in Ω. Consequently, we have $\bar{o}_U = \min(\bar{o}_1, 1/X_{k^*})$. Again, $\bar{o}_U$ can be obtained using the approach suggested for $\bar{o}_L$.

EXAMPLE 4.10. A special case to test the above bounds would be a two-station flow line with $N_1 = N_2 = 1$, as shown in Figure 4.19. The job arrival process is Poisson with rate λ. The jobs that arrive while M_1 is holding a job are lost. M_1 is blocked at every job departure while M_2 is busy. X_1 and X_2 are the processing times with $X_1 < X_2$ (trivial problem otherwise!).

Let Y be the job interarrival time at M_1 and Z be the time M_1 is blocked. U_j and I_j are the busy and the idle times at station j, respectively. Finally, let $D = X_2 - X_1$. Then we can write the following:

$$Z = \max\{0, X_2 - X_1 - Y\} = \max\{0, D - Y\},$$

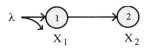

Figure 4.19. A two-station special flow line.

---

[1] An M/D/1/N work station can be studied by focusing on the number of jobs left in the system at job departure instances, which gives rise to a finite-state, continuous-time Markov chain. The steady-state probabilities $d_0, \ldots, d_{N-1}$ are obtained by solving the flow-balance equations. The arrival-point probabilities are given by

$$P_n = \frac{d_n}{d_0 + \rho}, \quad n = 0, \ldots, N - 1 \quad \text{and} \quad P_N = \frac{\rho - 1 + P_0}{\rho}, \quad \rho = \lambda X,$$

due to Cohen (1969).

TABLE 4.10. Numerical results for the bounds in Example 4.10

λ	$\bar{o}_\ell$	$\bar{o}_U$	$\bar{o}_L$
0.1	0.06490	0.06667	0.05556
0.2	0.09307	0.1	0.07692
0.4	0.11425	0.12500	0.09524
0.8	0.12325	0.12500	0.10811
1.6	0.12492	0.12500	0.11594
3.2	0.12499	0.12500	0.12030

implying that $Z > 0$ (blocking occurs) when $Y < D$. For M_1, we have

$$U_1 = \begin{cases} X_1 & \text{if } Y > D, \\ X_1 + Z & \text{if } Y < D, \end{cases}$$

with

$$E[U_1] = X_1 \Pr(Y \geq D) + \int_0^D (X_2 - Y) f_Y(y)\, dy = X_2 - \frac{1 - e^{-\rho}}{\lambda},$$

where $\rho = \lambda D$. Clearly, $E[I_1] = 1/\lambda$, resulting in

$$P_1(0) = \frac{E[I_1]}{E[I_1] + E[U_1]} = \frac{1}{\rho_2 + e^{-\rho}},$$

with $\rho_2 = \lambda X_2$. Then, the line throughput is given by

$$\bar{o}_\ell = \lambda P_1(0) = \frac{\lambda}{\rho_2 + e^{-\rho}}.$$

Let us next focus on the bounds: The lower bound is the throughput of M_1 with X_2 being the processing time. This particular work station goes through cycles of busy and idle times, each having an expected length of X_2 and $1/\lambda$, and produces the following proposed lower bound:

$$\bar{o}_L = \frac{1 - P_L(0)}{X_2} = \frac{\lambda}{\rho_2 + 1}.$$

$P_L(0)$ is the probability that the system producing the lower bound is empty. Similarly, the upper bound is given by

$$\bar{o}_U = \min\{\lambda/(\rho_1 + 1), X_2^{-1}\}.$$

Table 4.10 shows numerical results for $X_1 = 5$ and $X_2 = 8$ for varying values of λ.

TABLE 4.11. Bounds and the simulated throughput for Example 4.11

λ	$\bar{o}_\ell$(Sim.)	$\bar{o}_U$	$\bar{o}_L$
0.5	0.48627	0.48638	0.45639
0.6	0.57053	0.57194	0.51693
0.75	0.65746	0.66667	0.58092
0.9	0.66607	0.66667	0.61924
1.00	0.66655	0.66667	0.63499

EXAMPLE 4.11. Let us next calculate the bounds on the throughput in a four-station line with processing times $X_1 = 1$, $X_2 = 0.75$, $X_3 = 0.40$, and $X_4 = 1.5$. The buffer capacities are $N_1 = 3$, $N_2 = 3$, $N_3 = 5$, and $N_4 = 4$. The bounds and the simulated throughput for varying values of λ are given in Table 4.11.

4.6 Further Bounds on the Output Rate

The approximation algorithm presented earlier yields an approximate value for the output rate, among other performance measures. In the lack of exact results, bounds are useful to test approximations, or they are simply the only information that we can obtain for certain measures. In fact, that is what we have done in Section 4.5.2 due to the lack of other means in lines with deterministic processing times. Furthermore, a bound may exhibit similar behavior to that of the measure itself. Then, at times it may be sufficient to observe the behavior of the bounds for design purposes.

Let us consider bounds on the output rate for those lines whose throughput we do not know (N_1 is finite, or M_1 never starves). It is possible to obtain different bounds on the throughput in a flow line. For instance, (4.4) introduces bounds by shifting the blocking policy from BAP to BBP and augmenting the buffer capacities. However, we still need to compute $\bar{o}_\ell$ (BBP), which is almost as involved as the computation of $\bar{o}_\ell$ (BAP). Also, the notation that we have used in Section 4.5.2 to obtain the upper bound (isolation of the first station with its own processing time) can be generalized to the first two stations of a flow line. That is, since we know how to analyze two-station lines, we can isolate the first two stations as a two-node system with the original intermediate buffer (that is, B_2) and obtain its throughput. It is clear that the line throughput cannot be greater than the throughput of this two-station system, since M_2 gets blocked in the flow line. Thus, we obtain an upper bound on the line throughput at the expense of the computational effort to analyze a two-node system. This upper bound may be loose in those lines where the production rates of the first two stations are much higher than any of the downstream stations. However, it may be possible to improve it if we apply the same argument to the reversed line (that has the same throughput) and choose the minimum of the two upper bounds.

4. Analysis of Flow Lines

For flow lines with low variability in processing times ($Cv^2 < 1$), a lower bound may be conjectured to be the throughput of the isolated two-station system consisting of M_1 and B_2 and M_{k^*}, where M_{k^*} is the station that has the longest mean processing time. If M_1 has the longest processing time, then $k^* = \{k : \max_{k>1}\{X_k\}\}$. This is similar to the argument used in Section 4.5.2 for the deterministic processing times. Unfortunately, this lower bound coincides with the above upper bound in those cases where M_1 and M_2 have the two longest mean processing times, regardless of which one has the longest. Also, this lower bound may exceed the throughput in systems with highly variable processing times ($Cv^2 > 1$). A set of numerical examples for three-station flow lines is provided in Table 4.12, where m and c stand for the mean and the squared coefficient of variation of the processing times, respectively.

Case I is where the two-node systems suggested for the upper and the lower bounds simply coincide. In this case, the upper bound comes from the reverse system. Cases II and IV are the reverse of each other, and consequently the throughput, the upper bounds, and the lower bounds are exactly the same. In these two cases, the bounds come from the line in case II. Cases III, IV, and V have the same mean processing times, but the squared coefficient of variations increase from one case to the next. First, notice that the throughput and the bounds decrease as the processing-time variability increases. Also note that the lower bound behaves in the desirable way in cases III and IV, and it exceeds the throughput in case V, where the processing times have higher variability. As we have mentioned above, these bounds obtained through the analysis of two-station systems may turn out to be loose in cases where there is too much discrepancy among the mean processing times. They may also coincide due to the particular values of the system parameters.

Another set of lower and upper bounds on the throughput in flow lines where the first station never starves, the processing times are arbitrary, and the BAP-type blocking is assumed is given below. These bounds are valid for systems with or without intermediate buffers since again a unit of buffer position can be considered

TABLE 4.12. Numerical results for bounds on the throughput

	Station 1		Station 2		Station 3				
Case	m	c	m	c	m	c	$\bar{o}_\ell$	$\bar{o}_L$	$\bar{o}_U$
I	4	0.5	6	0.5	5	0.5	0.1541	—	0.1555
II	6	1.0	4	1.0	5	1.0	0.1448	0.1443	0.1540
III	5	0.5	4	0.5	6	0.5	0.1584	0.1555	0.1626
IV	5	1.0	4	1.0	6	1.0	0.1448	0.1443	0.1540
V	5	1.5	4	1.5	6	1.5	0.1353	0.1374	0.1479

as a station with zero processing time:

$$\bar{o}_L = \frac{1}{E\{\max(X_1, \ldots, X_M)\}} \qquad (4.49)$$

$$\bar{o}_U = \frac{1}{\max\{E[X_1], \ldots, E[X_M]\}} \qquad (4.50)$$

This upper bound is the mean processing rate of the slowest station. The lower bound is difficult to obtain except in some special cases. Both (4.49) and (4.50) coincide with the throughput when the processing times are fixed.

For the special case of **exponential processing times**, there exist bounds that are easier to compute. Consider a K-station flow line with μ_i being the processing rate and N_i being the buffer capacity at station i. M_1 is always busy except when blocked. The following recursion provides a lower bound $\bar{o}_L(K)$ on the line throughput:

$$\bar{o}_L(j) = \bar{o}_L(j-1) \left(\frac{1 - \rho_j^{N_j+1}}{1 - \rho_j^{N_j+2}} \right) \qquad j = 2, \ldots, K, \qquad (4.51)$$

where $\rho_j = \bar{o}_L(j-1)/\mu_j$ and $\bar{o}_L(1) = \mu_1$. Stage j of the recursion in (4.51) models M_j as an M/M/1 work station with Poisson arrivals with rate $\bar{o}_L(j-1)$, a buffer capacity of $N_j + 1$, and the processing rate of μ_j. The slower arrival rate is compensated for by a fast processing rate (no blocking). The justification of this lower bound is empirical.

A suggested upper bound on the throughput of flow lines with exponential processing times is given by

$$\bar{o}_U = \min_{j \geq 1} \left\{ R_j = \frac{\mu_1 \left[1 - \left(\frac{\mu_1}{\mu_j}\right)^{1+\Sigma N_j} \right]}{1 - \left(\frac{\mu_1}{\mu_j}\right)^{2+\Sigma N_j}} \right\}. \qquad (4.52)$$

R_j is the throughput of an M/M/1 work station with an arrival rate μ_1, a processing rate μ_j, and a buffer capacity of $N_1 + \cdots + N_j + 1$. N_2 is augmented by 1, and μ_2 is assumed to be the arrival rate at B_2. The logic behind (4.52) is that every space in the network prior to M_j is assumed to be a buffer space for M_j. Then the minimum of the output rates of these M/M/1/N work stations becomes the maximum possible rate at which jobs can depart from the network. Table 4.13 provides numerical results for the above bounds and the output rate in three-station flow lines. Notice that (4.52) coincides with the earlier two-node upper bound every time R_2 or R_K is the minimum.

In general, the usefullness of a bound depends on how tight it is and how simple it is to compute. However, bounds are like approximations: The better ones require

168   4. Analysis of Flow Lines

TABLE 4.13. Numerical results for (4.51) and (4.52) ($\bar{o}_\ell$ is obtained numerically)

μ_1	μ_2	μ_3	N_2	N_3	$\bar{o}_\ell$	$\bar{o}_L$	$\bar{o}_U$
1	1	1	1	1	0.56410	0.52632	0.66670
1	1	1	4	4	0.77671	0.74941	0.83333
1	3	2	5	2	0.99475	0.93261	0.99804
4	2	1	3	8	0.99900	0.99873	0.99999
2	1	3	10	3	0.99159	0.99150	0.99975

more computational effort. One can usually find better bounds on the throughput by spending more computational effort in obtaining them. Hopefully, this effort will be less than the one required to obtain the true throughput.

4.7 Design Problems in Flow Lines

Flowline design relates to two main issues, efficiency and effectiveness. The particular performance measure indicative of efficiency is the line throughput. Higher throughput means higher efficiency in flow lines. However, the effectiveness of a flow line's operation can be measured by the inventory accumulation in its WIP buffers. Higher effectiveness implies lower WIP levels. Therefore, efficiency and effectiveness contradict each other in flow lines. A larger throughput is usually the consequence of a larger buffer configuration, which in turn yields a larger inventory accumulation. Thus, the operation of a flow line having a higher throughput obtained by placing buffers between the stations may not be quite effective due to its inventory buildup. Thus, the design problem has to deal with the throughput and the WIP levels in flow lines. The two major design problems are the work-load allocation and the buffer allocation problems.

4.7.1 Work-Load Allocation

The work-load allocation problem is quite similar to the assembly-line balancing problem. It deals with how much of the total work should be done at each station. The lack of information on the variance of the resulting processing times, when some work is allocated to a station, forces us to work with either deterministic or exponential processing times. For the deterministic case, the best work-load assignment strategy is to allocate equal (or almost equal) processing times, which yields the maximum possible throughput with the minimum WIP accumulation. Inventory does not accumulate if M_1 does not starve. Otherwise, the WIP inventory will accumulate in front of M_1. The average WIP level will depend on the assumptions of the job arrival process.

In the case of the exponential processing times, maximum throughput is obtained by allocating smaller work-loads to the middle stations. This strategy was

4.7. Design Problems in Flow Lines

proposed by Hillier and Boling (1967) and is referred to as the *bowl phen* Unfortunately, the bowl phenomenon does not indicate the mean processing time for the maximum throughput or the maximum profit. Therefore, one has to implement a procedure to find the throughput-maximizing values of the processing times. For this purpose, we present the gradient method summarized in Table 4.14.

EXAMPLE 4.12. Consider a three-station flow line with a total work load of 15 time units. The buffer capacities are $N_2 = 10$, $N_3 = 5$. Let us find the throughput-maximizing values of the processing times. Table 4.15 shows the path of the procedure in Table 4.14 while arriving at the optimal solution. It starts with an equal allocation. Notice the bowl-type allocation of the work loads at all intermediate points. The reason for the asymmetry is the uneven buffer configuration. If the stations were reversed, the work load allocation would be completely reversed (also due to the reversibility argument).

4.7.2 Buffer Allocation

Buffers have a significant impact on how a production line operates, and consequently they affect its performance. For instance, the concept of blocking does not exist in production lines with infinite buffer capacities, whereas it becomes an

TABLE 4.14. The gradient search procedure

STEP 1: Begin with an admissible set of work-loads and then compute the associated probabilities and the corresponding value of the objective function.

STEP 2: Compute the partial derivatives of the objective function with respect to the work-loads, and use these in conjunction with the constraints to find the optimal directions, denoted by $v_i, i = 1, \ldots, K$. Because the partial derivatives cannot be computed in closed form, they need to be approximated. Denoting the objective function by $J = (x_1, \ldots, x_K)$, we can write

$$\frac{\partial J}{\partial x_i} = [J(x_1, x_2, \ldots, x_i + \delta, \ldots, x_K) - J(\underline{x})]/\delta, \quad i = 1, 2, \ldots, K,$$

where Δ is the step size and is determined in general by trial and error. We observe that the objective function needs to be evaluated $K + 1$ times in each step of the optimization, which requires the steady-state probabilities to be computed $K + 1$ times as well.

STEP 3: After the optimal direction is determined, the new work-loads $\underline{x}$ are computed using the relation $\underline{x} = \underline{x} + \underline{v}\delta$. Then, the new steady-state probabilities and the new objective function are computed, and the cycle continues until no improvement is obtained in the objective function.

TABLE 4.15. The path of the gradient search procedure

Allocation No.	x_1	x_2	x_3	$\bar{o}_\ell$
1	4.8567	4.8013	5.3419	0.16962
2	4.8422	4.7456	5.4122	0.16950
3	4.9298	4.7893	5.2809	0.16965
4(Opt.)	4.9147	4.7673	5.3180	0.16966

operating policy in those production lines with finite-capacity buffers. In general, the output rate of a production line increases as the buffer capacities (of any of the buffers) increase due to less blocking. When the capacity of a buffer is increased, the output rate increases with a decreasing rate and becomes asymptotic to the output rate of the case with the infinite-capacity buffer. That is, after a certain point, further increases in the buffer capacity will not be effective on the throughput. In those cases where the output rate is already close to that of the infinite-capacity case, placing buffers will not have a major impact on the output rate. On the other hand, if the output rate is already low (e.g., due to failures) the impact of additional buffers on the output rate may be significant. The long-run average work-in-process inventory level in any buffer also has a similar behavior to that of the output rate.

The buffer allocation problem is essentially a combinatorial optimization problem. The decision variables are the buffer capacities, denoted by N_i for the ith buffer. Since buffers include the positions on the machines, N_i will be at least 1. Hence, $N_i > 1$ indicates that there should be a buffer of capacity N_i in station i. On the other hand, $N_i = 1$ indicates that a buffer should not be placed between stations $i - 1$ and i. The objective in the buffer allocation problem may be to optimize (or to improve) either a particular measure such as the output rate or a composite measure such as the profit per unit time.

The output rate is a nondecreasing function of the buffer capacities, and its maximum is attained at $N_i = \infty$, for all i. For this reason, in the output rate maximization, the problem is to determine how to distribute a finite sum of buffer space Γ into $K - 1$ stations in a flow line with K stations. That is,

$$\max_{N_2,\ldots,N_K} \bar{o}_\ell = f(N_2, \ldots, N_K) \quad \text{such that} \quad (4.53)$$

$$\sum_{i=2}^{K} N_i = \Gamma, \quad N_i \geq 1.$$

The drawback of the output rate maximization is that it does not consider the undesirable side of increasing buffer capacities, that is, holding high WIP levels. Holding inventory is costly, and larger buffers incur higher inventory-holding costs. One way to minimize the WIP levels is to minimize the total buffer space for a given desired output rate, $\bar{o}_d$. That is,

$$\min_{N_2,\ldots,N_K} \sum_{i=2}^{K} N_i \quad \text{such that} \tag{4.54}$$

$$\bar{o}_\ell \geq \bar{o}_d, \quad N_i \geq 1.$$

Clearly, $\bar{o}_d$ must be a feasible throughput. Equations (4.53) and (4.54) are related to each other in that the objective function of one is the right-hand side of the constraint in the other. It is conceivable that (4.54) can be solved by solving (4.53) for various values of Γ and choosing the one that provides the smallest $\bar{o}_\ell$ that is greater than $\bar{o}_d$.

As mentioned earlier, the flow-line design can be enhanced by considering an objective function that is more comprehensive than the throughput maximization. For instance, the profit maximization may be preferable since it also deals with the expected WIP levels. In this case, we may assume that every finished product brings in a revenue R ($/unit) and incurs a unit variable production cost V ($/unit). In addition to the above cost parameters, there exists a cost of carrying inventory in the ith buffer, referred to as the unit holding cost per unit time, h_i ($/unit-unit time). Also, one may want to include the cost of placing a unit of buffer space, b_i, at station i. Then, the optimization problem becomes

$$\max_{N_2,\ldots,N_K} (\text{Profit}) F = (R - V)\bar{o}_\ell - \sum_{i=2}^{K} h_i \bar{N}_i - \sum_{i=2}^{K} b_i N_i, \tag{4.55}$$

where $\bar{N}_i$ is the long-run average number of jobs in buffer i. The limitation on the total buffer space can still be included in the problem. Clearly, the profit maximization is an appropriate objective function since it combines not only the benefit of placing buffers but also the cost of doing so.

The difficulty in the above optimization problems is that there are no explicit expressions for $\bar{o}_\ell$ and $\bar{N}_i$ in terms of N_i (discrete decision variables). $\bar{N}_i$ and $\bar{o}_\ell$ are usually obtained from the steady-state distribution of the jobs in the buffers. Thus, due to a lack of differentiable functions that are usually assumed in classical optimization problems, the approach to solve the above optimization problems may be an ad-hoc one. A search procedure may be attached to the algorithm that computes the performance measures for a given buffer configuration, as shown in Figure 1.4. The search procedures by Hooke and Jeeves (1966), Rosenbrock (1960), and Powell (1964) may serve for this purpose. These procedures start with an initial buffer allocation. At any iteration, the search procedure provides values for the buffer capacities, and the performance algorithm provides the values of the measures and the objective function. This procedure continues back and forth until no further improvements can be achieved in the objective function. In Table 4.16 we summarize Hooke and Jeeves search procedure, which is quite useful in discrete optimization problems where the derivatives of the objective function are not available. It has

172   4. Analysis of Flow Lines

TABLE 4.16.   Hooke-Jeeves search procedure

INITIALIZE	$k = 1, \underline{\mathbf{N}}^{(o)}, \Delta$ (increment) $\geq 1, \varepsilon$ (termination criteria, i.e., 1)
STEP 1	*Exploratory move* For each $i = 2, \ldots, K$: Let $N_i^{(k)} = N_i^{(k-1)} + \Delta$. If $f(\underline{N}^{(k)}) > f(\underline{N}^{(k-1)})$, then fix $N_i^{(k)}$ at $N_i^{(k-1)} + \Delta$. Else, fix $N_i^{(k)}$ at $N_i^{(k-1)} - \Delta$.
STEP 2	*Pattern move* Let $\underline{\mathbf{N}}_p^{(k+1)} = \underline{\mathbf{N}}^{(k)} + (\underline{\mathbf{N}}^{(k)} - \underline{\mathbf{N}}^{(k-1)})$.
STEP 3	*Exploratory move* around $\underline{\mathbf{N}}_p^{(k+1)}$ to find $\underline{\mathbf{N}}^{(k+1)}$ If $f(\underline{\mathbf{N}}^{(k+1)}) > f(\underline{\mathbf{N}}^{(k)})$, then $\underline{\mathbf{N}}^{(k-1)} \leftarrow \underline{\mathbf{N}}^{(k)}, \underline{\mathbf{N}}^{(k)} \leftarrow \underline{N}^{(k+1)}$, and go to Step 2. Else, if $\Delta \leq \varepsilon$, then stop (optimality), else reduce Δ and go to Step 1.

two steps, namely the exploratory move and the pattern move. In the exploratory move, each coordinate is tested to establish a base point and a direction to proceed is found. The pattern move accelerates the search in the chosen direction.

EXAMPLE 4.13. Let us consider the buffer allocation problem in a production line with three stations that are subject to operation-dependent failures. Let us assume that the processing time, the time to failure, and the repair time are all exponentially distributed with the data given in Table 4.17.

For a given buffer configuration, this flow line can be studied in various ways. The underlying Markov chain can be analyzed by generating its transition rate matrix and solving for the probabilities using any of the techniques presented earlier. Here we chose to generate the transition rate matrix for the whole production

TABLE 4.17.   Input data for Example 4.13

	Station 1	Station 2	Station 3
Processing rate	0.25	0.20	0.30
Failure rate	0.10	0.15	0.20
Repair rate	0.10	0.30	0.50
Mean process comp. time	8.00	7.50	4.67
MGE-2 parameter (μ_1)	0.38508	0.53860	0.81623
MGE-2 parameter (μ_2)	0.06492	0.11140	0.18377
MGE-2 parameter (a)	0.35078	0.62867	0.63246

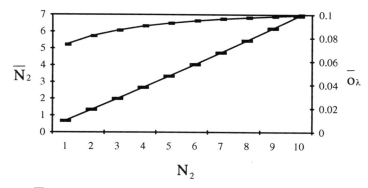

Figure 4.20. $\overline{N}_2$ and $\overline{o}_\ell$ as functions of N_2.

line with a given buffer configuration and employed the power method to compute the steady-state probabilities.

The power method involves only numerical approximations, and therefore the results are considered exact within the present context. Let us first observe the behaviors of the output rate and the average buffer contents at stations 2 and 3 as functions of the buffer capacities. The output rate as a function of N_2 is shown in Figure 4.20. The output ranges from 0.075 to 0.099 and becomes asymptotic to the output rate when $N_2 = \infty$. The output rate increases monotonically with a decreasing marginal rate. That is, additions to the buffer capacity have a diminishing impact on the output rate.

The average buffer contents at stations 2 and 3 also behave in a way that they become asymptotic to their counterparts in the infinite-capacity cases. This is apparent for $\overline{N}_3$, shown in Figure 4.21, and not so for $\overline{N}_2$, shown in Figure 4.20. This is due to the closeness in the mean process completion times at stations 1 and 2. $\overline{N}_2$ will be asymptotic at a very high value. An analogous case is a work station with a traffic intensity close to 1.0.

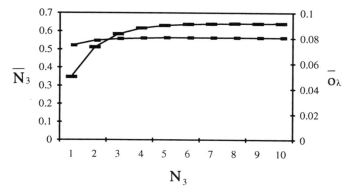

Figure 4.21. $\overline{N}_3$ and $\overline{o}_\ell$ as functions of N_3.

For the optimal buffer allocation problem, let us assume that $h = 0.5$ \$/unit-unit time, $R - V = 30$ \$/unit, and $b_i = 0$, and search for the profit-maximizing buffer allocation. Applying Hooke and Jeeves' pattern search and running the power method every time we need the measures, we obtain the optimal buffer allocation, which is $N_2^* = 1$ and $N_3^* = 10$, with the maximum profit being 1.79909 \$/unit time. The optimal allocation does not assign any additional space to the position on the machine at station 2. The reason for that is the inventory accumulation and the resulting holding cost at station 2 in case the buffer is allocated. If we ignore the holding cost and let the output rate maximization allocate the additional 9 buffer units that the profit maximization came up with, the result would be $N_2^* = 8$ and $N_3^* = 3$. This allocation reduces the blocking probability at station 1 and improves the output rate the most. It makes sense that the two objective functions allocate buffers differently. The output maximization allocates buffers to stations that are likely to accumulate high inventories. However, the profit maximization tends to stay away from these buffers simply because of the average holding cost to be incurred.

Related Literature

The literature on flow lines (production lines or tandem queues with finite interstage buffer capacities) contains various models of performance evaluation emphasizing exact or approximate computation of the steady-state performance measures of the system. The exact approaches are limited to two- or three-station lines. Exact models include Buzacott (1972), Wijngaard (1979), Gershwin and Berman (1981), Berman (1982), Shanthikumar and Tien (1983), Gershwin and Schick (1983), and Jafari and Shanthikumar (1987). These studies mainly focus on the solution of the steady-state flow-balance equations of the underlying Markov chains. Solution methods include recursive, matrix-recursive, as well as matrix-geometric methods. Phase-type processing times are explicitly considered by Buzacott and Kostelski (1987), and Gun and Makowski (1990). The matrix-recursive procedure presented in Section 4.3.2 is due to Buzacott and Kostelski (1987) and Altiok and Ranjan (1989). A matrix-geometric procedure for the same problem with arbitrary phase-type processing times is given by Bocharov (1989), and Gun and Makowski (1990). A good reference on queueing networks with blocking with a variety of assumptions, blocking definitions, and related properties is Perros (1994).

Longer flow lines have been analyzed by using either numerical methods (e.g., the power method) or approximations. Numerical methods rely upon the generation of the rate matrix for the entire flow line and the use of direct or iterative methods to solve for the unknowns. Studies employing these approaches include, Onvural, et al. (1987) and Stewart (1978). The drawback is the size of the statespace of the Markov chains.

Various models of longer lines approximate the measures of performance by decomposing the system into a set of smaller systems that relate to each other within an iterative scheme. These studies include Choong and Gershwin (1987), Gershwin (1987, 1989), Jun and Perros (1987), Kerbache and Smith (1987), De Koster (1988), Murphy (1978), Pollock et al. (1985), Yamashita and Suzuki (1988), Altiok (1989), Altiok and Ranjan (1989), Liu and Buzacott (1989), Gun and Makowski (1989), Takahashi (1989), Mitra and Mitrani (1990), Buzacott and Shanthikumar (1992), and Hong et al. (1992), among many others. These studies include systems with and without failures, and systems with exponential, phase-type, and deterministic processing times. The majority of studies focuses on the impact of failures as well as buffer capacities on the measures of performance. The phase-type modeling of the effective processing times (Section 4.4.2) is due to Perros and Altiok (1986) for exponential processing times, and to Altiok and Ranjan (1989) and Gun and Makowski (1989) for phase-type processing times. The approximation algorithm of Section 4.4.1 is due to Altiok and Ranjan (1989). In queueing theory literature, tandem queues with blocking have attracted a great deal of attention. Queues in series with exponential processing times are decomposed by Hillier and Boling (1967), Caseau and Pujolle (1979), Takahashi et al., Boxma and Konheim (1981), Altiok (1982), Perros and Altiok (1986), and Brandwajn and Jow (1988).

Bounds on the throughput have started to attract increasing attention. Equations (4.49) and (4.50) are proposed by Muth (1973). Further bounds are proposed by Bell (1982) and Kerbache and Smith (1988) and are given by (4.52) and (4.51), respectively. Van Dijk and Lamond (1988) and Van Dijk and van der Wal (1989) also proposed bounds for tandem systems with exponential processing times. Equation (4.4) is a bound involving different definitions of blocking and is due to Dallery et al. (1991).

The reversibility property was first introduced by Yamazaki and Sakasegawa (1975) for tandem queueing systems. Later studies include Dattatreya (1978), Muth (1979), Yamazaki et al. (1985), Melamed (1986), and Dallery et al. (1991), among others.

Models of flow lines with deterministic processing times are either very easy or very difficult to study. In the existence of a random component in the system, such as random job arrivals or failures, the system turns out to be difficult to analyze. The two important theorems in Section 4.5 that enable us to calculate the expected total WIP level in the system are due to Avi-Itzhak (1965). The bounds for the case with $N_1 < \infty$ are given by Altiok and Kao (1989). Other studies assuming deterministic processing times focus on lines with failures, emphasizing fluid models ([see, for instance, Wijangaard (1979), De Koster (1987), and Dallery et al. (1989)). Gun (1987) has given an annotated bibliography of production lines in general.

Design issues in production lines are detailed by Buxey et al. (1973), Buzacott and Hanifin (1978), Sarker (1984), and Gershwin et al. (1986). Hillier and Boling (1966) introduced the bowl phenomenon that allocates less work to the stations in the middle and improves the output rate. Baruh and Altiok (1991) used the first- and second-order numerical perturbations to obtain the actual work load allocation in the bowl phenomenon. Freeman (1964) gave guide lines for buffer capacity allocation. Barten (1962), Young (1967), and Anderson and Moodie (1969) also used simulation and statistical analysis to complement the minimum-cost buffer allocation problem. Ho et al. (1979) developed a gradient technique that is incorporated into a simulation model to maximize the output rate for a given total buffer capacity. Altiok and Stidham (1983), Smith and Daskalaki, (1988) and Smith and Chikhale (1994) utilized a search procedure coupled with the numerical or the approximate analysis of the underlying queueing network to allocate profit-maximizing buffer capacities in three-station and longer production lines. Yamashita and Suzuki (1988) and Jafari and Shanthikumar (1989) have used a dynamic programming approach that incorporates an approximate decomposition scheme to study the throughput maximization problem for a given total buffer capacity in synchronous production lines. Jensen et al. (1988) also used a dynamic programming approach for a cost-minimizing allocation of buffer capacities in production lines where buffers are replenished to the designed levels during down times. Finally, Yamashita and Altiok (1993) proposed a dynamic programming approach that allocates the minimum total buffer space to achieve a desired level of throughput in asynchronous production lines.

References

Altiok, T., 1982, Approximate Analysis of Exponential Tandem Queues with Blocking. *European J. Operations Research*, Vol. 11, pp. 390–398.

Altiok, T., 1989, Approximate Analysis of Queues in Series with Phase-Type Service Times and Blocking. *Operations Research*, Vol. 37, pp. 601–610.

Altiok, T., and Z. Kao, 1989, Bounds for Throughput in Production/Inventory Systems in Series with Deterministic Processing Times. *IIE Transactions*, Vol 21, pp. 82–85.

Altiok, T., and R. Ranjan, 1989, Analysis of Production Lines with General Service Times and Finite Buffers: A Two-Node Decomposition Approach. *Engineering Costs and Production Economics*, Vol. 17, pp. 155–165.

Altiok, T., and S. Stidham, Jr., 1983, The Allocation of Interstage Buffer Capacities in Production Lines. *IIE Transactions*, Vol. 15, pp. 292–299.

Anderson, D.R., and C. L. Moody, 1969, Optimal Buffer Storage Capacity in Production Line Systems. *Int. J. Production Research*, Vol. 7, pp. 233–240.

Avi-Itzhak, B., 1965, A Sequence of Service Stations with Arbitrary Input and Regular Service Times. *Management Science*, Vol. 11, no. 5, pp. 565–571.

Barten, K. A., 1962, A Queueing Simulator for Determining Optimum Inventory Levels in a Sequential Process. *J. Industrial Engineering*, Vol. 13, pp. 245–252.

Baruh, H., and T. Altiok, 1991, Analytical Perturbations in Markov Chains. *European J. Operations Research*, Vol. 51, pp. 210–222.

Bell, P.C., 1982, The Use of Decomposition Techniques for the Analysis of Open Restricted Queueing Networks. *Operations Research Letters*, Vol. 1, pp. 230–235.

Berman, O., 1982, Efficiency and Production Rate of a Transfer Line with Two Machines and a Finite Storage Buffer. *European J. Operations Research*, Vol. 9, pp. 295–308.

Bocharov, P.P., 1989, On the Two-Node Queueing Networks with Finite Capacity. *Proc. 1st Int. Workshop on Queueing Networks with Blocking*, NCSU, Raleigh, NC.

Boxma, O.J., and A.G. Konheim, 1981, Approximate Analysis of Exponential Queueing Systems with Blocking. *Acta Informatica*, Vol. 15, pp. 19–66.

Brandwajn, A., and Y.L. Jow, 1988, An Approximation Method for Tandem Queues with Blocking. *Operations Research*, Vol. 36, pp. 73–83.

Buxey, G.M., N.D. Slack, and R. Wild, 1973, Production Flow Line System Design—A View. *AIIE Transactions*, Vol. 5, pp. 37–48.

Buzacott, J.A., 1972, The Effect of Station Breakdowns and Random Processing Times on the Capacity of Flow Lines with In-Process Storage. *AIIE Transactions*, Vol. 4, pp. 308–312.

Buzacott, J.A., and L.E. Hanifin, 1978, Models of Automatic Transfer Lines with Inventory Banks — A Review and Comparison. *AIIE Transactions*, Vol. 10, pp. 197–207.

Buzacott, J.A,. and D. Kostelski, 1987, Matrix-Geometric and Recursive Algorithm Solution of a Two-Stage Unreliable Flow Line. *IIE Transactions*, Vol. 19, pp. 429–438.

Buzacott, J.A., and J. G. Shanthikumar, 1992, *Stochastic Models of Manufacturing Systems*, Prentice Hall, Englewood Cliffs, N.J.

Caseau, P., and G. Pujolle, 1979, Throughput Capacity of a Sequence of Transfer Lines with Blocking Due to Finite Waiting Room. *IEEE Transactions Software Eng.*, SE-5, pp. 631–642.

Choong, Y.F., and S.G. Gershwin, 1987, A Decomposition Method for the Approximate Evaluation of Capacitated Transfer Lines with Unreliable Machines and Random Processing Times. *IIE Transactions*, Vol. 19, no. 2, pp. 150–159.

Cohen, J. W., 1969, *The Single-Server Queue*. North Holland, Amsterdam.

Dallery, Y., R. David, and X. Xie, 1989, Approximate Analysis of Transfer Lines with Unreliable Machines and Finite Buffers, IEEE Transactions on Automatic Control, Vol. 34, pp. 943-953.

Dallery, Y., Z. Liu, and D. Towsley, 1991, Reversibility in Fork/Join Queueing Networks with Blocking After Service. Laboratoire MASI, Universite P. cf M. Curie Paris, France.

Dattatreya, E.S., 1978, Tandem Queueing Systems with Blocking. Ph.D. dissertation, University of California, Berkeley.

De Koster, M. B. M., 1988, Approximate Analysis of Production Systems. *European J. Operations Research*, Vol. 37, pp. 214–226.

De Koster, M B. M., 1987, Estimation of Line Efficiency by Aggregation. *Int. J. Production Research*, Vol. 25, pp. 615–626.

Freeman, M.C., 1964, The Effect of Breakdowns and Interstage Storage on Production Line Capacity. *J. Industrial Engineering*, Vol. 15, pp. 194–200.

Gershwin, S.B., 1987, An Efficient Decomposition Method for the Approximate Evaluation of Tandem Queues with Finite Storage Space and Blocking. *Operations Research*, Vol. 35, pp. 291–305.

Gershwin, S.B., 1989, An Efficient Decomposition Algorithm for Unreliable Tandem Queueing Systems with Finite Buffers. *Proc. 1st Intl. Workshop on Queueing Networks with Blocking*, Raleigh, NC.

Gershwin, S.B., and O. Berman, 1981, Analysis of Transfer Lines Consisting of Two Unreliable Machines with Random Processing Times and Finite Storage Buffers. *AIIE Transactions*, Vol. 13, pp. 2–11.

Gershwin, S.B., and I.C. Schick, 1983, Modeling and Analysis of Three-Stage Transfer Lines with Unreliable Machines and Finite Buffers. *Operations Research*, Vol. 31, pp. 354–380.

Gun, L., 1987, Annotated Bibliography of Blocking Systems. Systems Research Center, SRC-TR-87-187, University of Maryland, College Park, MD.

Gun, L., and A. Makowski, 1989, An Approximation Method for General Tandem Queueing Systems Subject to Blocking. *Proc. 1st Intl. Workshop on Queueing Networks with Blocking*, Raleigh, NC.

Gun, L., and A. Makowski, 1990, Matrix-Geometric Solution for Two-Node Tandem Queueing Systems with Phase-Type Servers Subject to Blocking and Failures. *Stochastic Models*, Vol. YY, pp. UU.

Hillier, F.S., and R.W. Boling, 1967, Finite Queues in Series with Exponential on Erlang Service Times — A Numerical Approach. *Operations Research*, Vol. 15, pp. 286–303.

Ho, Y.C., M. A. Eyler, and T. T. Chien, 1979, A Gradient Technique for General Buffer Storage Design in a Production Line Design. *Intl. J. Production Research*, Vol. 17, pp. 557–580.

Hong, Y., C. R. Glassey, and D. Seong, 1992, The Analysis of a Production Line with Unreliable Machines and Random Processing Times. *IIE Transactions*, Vol. 24, pp. 77–83.

Hooke, R., and T. A. Jeeves, 1966, Direct Search of Numerical and Statistical Problems. *ACM*, Vol. 8, pp. 212–229.

Hunt, G.C., 1956, Sequential Arrays of Waiting Lines. *Operations Research*, Vol. 4, pp. 674–683.

Jafari, M.A., and J.G. Shanthikumar, 1987, Exact and Approximate Solutions to Two-Stage Transfer Lines with General Uptime and Downtime Distributions. *IIE Transactions*, Vol. 19, pp. 412–420.

Jafari, M.A., and J.G. Shanthikumar, 1989, Determination of Optimal Buffer Storage Capacities and Optimal Allocation in Multistage Automatic Transfer Lines. *IIE Transactions*, Vol. 21, pp. 130–135.

Jensen, P.A., R. Pakath, and J.R. Wilson, 1988, Optimal Buffer Inventories for Multistage Production Systems with Failures, SMS 88-1, School of Industrial Engineering, Purdue University.

Jun, K., and H.G. Perros, 1987, An Approximate Analysis of Open Tandem Queueing Networks with Blocking and General Service Times. *Computer Science*, NCSU, Raleigh, NC.

Kerbache, L., J. M. Smith, 1987, The Generalized Expansion Method for Open Finite Queueing Networks, European J. Operations Research, Vol. 32, pp. 448-461.

Kerbache, L., and J. M. Smith, 1988, Asymptotic Behavior of The Expansion Method for Open Finite Queueing Networks. *Computers and Operations Research*, Vol. 15, pp. 157–169.

Kraemer, W., and M. Langenbach-Belz, 1976, Approximate Formulae for the Delay in the Queueing System G/G/1. *Proc. 8th International Teletraffic Congress*, Melbourne.

Liu, X., and J.A. Buzacott, 1989, A Balanced Local Flow Technique for Queueing Networks with Blocking. *Proc. 1st Intl. Workshop on Queueing Networks with Blocking*, Raleigh, NC.

Melamed, B., 1986, A Note on the Reversibility and Duality of Some Tandem Blocking Queueing Systems. *Management Science*, Vol. 32, pp. 1648–1650.

Mitra, D., and J. Mitrani, 1990, Analysis of a Novel Discipline for Cell Coordination in Production Lines, I. *Management Science*, Vol. 36, pp. 1548–1566.

Murphy, R.A., 1978, Estimating the Output of a Series Production Systems. *AIIE Transactions*, Vol. 10, pp. 139–148.

Muth, E.J., 1973, The Production Rate of a Series of Work Stations with Variable Service Times. *Intl. J. Production Research*, Vol. 11, pp. 155–169.

Muth, E.J., 1979, The Reversibility of Production Lines. *Management Science*, Vol. 25, pp. 152–158.

Onvural, R.O., H.G. Perros, and T. Altiok, 1987, On the Complexity of the Matrix- Geometric Solution of Exponential Open Queueing Networks with Blocking. *Proc. Intl. Workshop on Mod eling Techniques and Performance Evaluation* (G. Pujolle, S. Fdida, and A. Horlait, eds.), pp. 3–12.

Perros, H. G., 1994, *Queueing Networks with Blocking.* Oxford University Press, Oxford.

Perros, H.G., and T. Altiok, 1986, Approximate Analysis of Open Networks of Queues with Blocking, Tandem Configurations. *IEEE Trans. Soft. Eng.*, SE-12, pp. 450–461.

Pollock, S.M., J.R. Birge, and J.M. Alden, 1985, Approximation Analysis for Open Tandem Queues with Blocking: Exponential and General Distributions. IOE Rept. 85-30, Univ. of Michigan, Ann Arbor, MI.

Powell, M. J. D., 1964, An Efficient Method for Finding the Minimum of a Function of Several Variables Without Calculating Derivatives. *Computer J.*, Vol. 7, pp. 155–162.

Rosenbrock, H. H., 1960, An Automated Method for Finding the Greatest or Least Value of a Function. *Computer J.*, Vol. 3, pp. 175–184.

Sarker, B.R., 1984, Some Comparative and Design Aspects of Series Production Systems. *IIE Transactions*, Vol. 16, pp. 229–239.

Shanthikumar, J.G., and C.C. Tien, 1983, An Algorithmic Solution to Two-Stage Transfer Lines with Possible Scrapping of Units. *Management Science*, Vol. 29, pp. 1069–1086.

Smith, J. M., and S. Daskalaki, 1988, Buffer Space Allocation in Automated Assembly Lines. *Operations Research*, Vol. 36, pp. 343–358.

Smith, J. M., and N. Chikhale, 1994, Buffer Allocation for a Class of Nonlinear Stochastic Knapsack Problems. Dept. of Ind. Eng. and Oper. Res., University of Massachusetts, Amherst.

Stewart, W. J., 1978, Comparison of Numerical Techniques in Markov Modeling. *Comm. of the ACM*, Vol. 21, pp. 144–152.

Takahashi, Y., 1989, Aggregate Approximation for Acyclic Queueing Networks with Communication Blocking. , *Proc. 1st Intl. Workshop on Queueing Networks with Blocking*, Raleigh, NC..

Takahaski, Y., H. Miyahara, and T. Hasegawa, 1980, An Approximation Method for Open Restricted Queueing Network. *Operations Research*, Vol. 28, pp. 594–602.

Van Dijk, N., and B. Lamond, 1988, Bounds for the Call Congestion of Finite Single- Server Exponential Tandem Queues. *Operations Research*, Vol. 36, pp. 470–477.

Van Dijk, N, and J. van der Wal, 1989, Simple Bounds and Monotonicity Results for Finite Multi-Server Exponential Tandem Queues. *Queueing Systems,* Vol 4, pp. 1–16.

Varga, R. S., 1963, *Matrix Iterative Analysis*, Prentice Hall, Englewood Cliffs, N.J.

Wijngaard, J., 1979, The Effect of Interstage Buffer Storage on the Output of Two Unreliable Production Units in Series with Different Production Rates. *AIIE Transactions*, Vol. 11, pp. 42–47.

Yamashita, H. and T. Altiok, 1994, *Buffer Capacity Allocation for a Desired Throughput in Production Line.*, Dept. of Ind. Eng., Rutgers University, Piscataway, N.J.

Yamazaki, G., and H. Sakasegawa, 1975, Properties of Duality in Tandem Queueing Systems. *Annals of the Inst. of Statistical Math.*, Vol. 27, pp. 201–212.

Yamashita, H., and S. Suzuki, 1987, An Approximate Solution Method for Optimal Buffer Allocation in Serial n-Stage Automatic Production Lines. *Trans. Japan Soc. Mech. Eng.*, 53-C, pp. 807–817.

Yamazaki, G., T. Kawashima, and H. Sakasegawa, 1985, Reversibility of Tandem Blocking Systems. *Management Science*, Vol. 31, pp. 78–83.

Young, H. H., 1967, Optimization Models for Production Lines. *Intl. J. Production Research*, Vol. 18, pp. 70–78.

Problems

4.1. A two-stage flow line has two-stage Erlang processing times at both stations with a mean of 4.0 time units each. The intermediate buffer capacity is 5 units, including the position in the second station. Obtain the average work-in-process level in the line and the line throughput using the matrix-recursive procedure of Section 4.3.2.

4.2. Consider a two-stage production line with arbitrary processing times. The processing time at station 1 has a mean of 2.0 time units and a variance of 3 time units. The processing time at station 2 has a mean of 3.0 time units and a variance of 0.5 time units. The first station is always busy except when blocked. The intermediate buffer has a capacity of 4 units. Using the approximate phase-type representations of the processing times, obtain the steady-state output rate and the average number of units in the buffer waiting to be processed. Test your results by computing the average throughput in different ways.

4.3. Implement the recursive procedure of Section 4.3.2 in a three-station flow line with exponentially distributed processing times with rates μ_1, μ_2, and μ_3, respectively. Let the buffer capacities be N_2 and N_3. Describe the procedure in detail. Verify your results with Hunt's model.

4.4. A production system consists of two stations subject to failures. The processing times, the time to failures, and the repair times are all exponentially distributed. The processing rates are 2 and 3, the failure rates are 0.15 and 0.30, and the repair rates are 2 and 3, respectively. The failures are operation-dependent. The intermediate buffer has a capacity of 4 units. Obtain the exact values of the steady- state output rate, the average number of units waiting in the buffer, the station utilizations, and the down time probabilities. *Hint*: Construct a phase-type representation for the completion time that explicitly shows the down phase. Hence, the probability of being in this phase is equivalent to the downtime probability.

4.5. Consider the following two-station line with phase-type processing times. The intermediate buffer has a capacity of 3 units. BAP-type blocking is in effect.

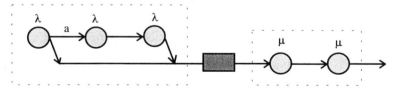

Assume that $\lambda = 1$, $a = 0.25$, and $\mu = 1$. Modify the matrix-recursive procedure of Section 4.3.2, and obtain the steady-state output rate, the average number of units in the buffer, and the station utilizations.

4.6. A flow line consists of three work stations with exponentially distributed processing times with rates μ_i for station i. There are no intermediate buffers, and the first station is always busy except when blocked. Assuming BAP-type blocking and $\mu_1 = 2$, $\mu_2 = 1$, and $\mu_3 = 1.5$:

(a) Obtain the exact throughput.
(b) Develop a two-node decomposition approximation to compute the steady-state distribution of the station status for each station. Compare the accuracy of the approximation with the exact result of (a).
(c) Repeat (b) by replacing the phase-type processing-time distributions by their approximate exponential counterparts with the same means. Compare the results with those of (a) and (b).

4.7. A flow line consists of three stations with phase-type processing times as shown below.

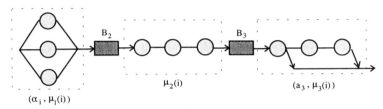

(a) In a two-node decomposition approach, characterize the subsystems pictorially. Also, develop the ($\boldsymbol{\alpha}$, **T**) representations of the effective processing times in the subsystems.
(b) Simplify the effective processing times by replacing the phase-type distributions by exponential distributions with the same means.
(c) Carry out the approximation algorithm for $\alpha_1(i) = (0.1, 0.3, 0.6)$, $\mu_1(i) = (2, 4, 6)$, $\mu_2(i) = (4, 2, 5)$, and $a_3 = 0.2$ and $\mu_3(i) = (3, 5, 6)$. Obtain the steady-state measures of the production line such as the output rate, the average number of units in each buffer, and the station utilizations. Also, obtain the average time a job spends in the entire system. Show details, including the approximate average idle times each station experiences once they become idle and the expected time each station remains blocked once they become blocked.

4.8. Consider a three-station flow line with an MGE-2 processing time at station 1 with parameters $(a_1, \mu_1(1), \mu_1(2))$, a three-phase Erlang processing time with a phase rate of μ_2 at station 2, and a two-phase hyperexponential processing time with parameters $(a_3(1), a_3(2), \mu_3(1), \mu_3(2))$ at station 3. The first station has an infinite-capacity buffer. The job interarrival times are also MGE-2 distributed with parameters $(a_0, \mu_0(1), \mu_0(2))$. How would you modify the approximate decomposition algorithm to accommodate this problem? Show the pictorial representations and describe the details.

4.9. A flow line consists of 5 work stations each with a deterministic processing time. The first station has an infinite capacity and jobs arrive at this station according to a Poisson process with rate 0.15 per unit time. The processing times are $D_1 = 2$, $D_2 = 3$, $D_3 = 1$, $D_4 = 5$, and $D_5 = 4$ time units. The capacities of the intermediate buffers are $\underline{N} = (4, 5, 2, 1)$. Blocking is of BAP type. Obtain the expected time a job spends in the entire system and the expected total number of jobs in the system. What would the results be if the job interarrival times had a mean of 10 time units and a squared coefficient of variation of 0.50? If we had $N_1 = 3$, which makes it a loss system, what would the range of the throughput be?

4.10. Consider a two-stage production system with Erlang-3 processing times with a mean of 1 and 1.5 at stations 1 and 2, respectively. The intermediate buffer has a capacity of 5 units. Generate the transition rate matrix of the Markov model in compact form, and obtain the steady-state probabilities of the number of jobs in the system using the power method in compact form. That is, the matrix products should be carried out in compact form. For both BAP- and BBP-type policies, relate the steady-state output rate and the distribution of the number of units in the buffer to the ones of the reversed line that has the same buffer capacity and the stations in the reverse order.

4.11. Consider a 4-station flow line with exponential processing times with rates 2, 4, 1 and 2, at stations 1 to 4, respectively. The capacities of the intermediate buffers are 3, 5, and 2 units.

(a) Obtain Muth's bounds for the throughput.
(b) Obtain Kerbache-Smith's, and Bell's bounds on the throughput.
(c) Develop your own nontrivial bounds for the throughput and compare with (a) and (b). The computation of the bounds should be simpler than the analysis of the system itself.

4.12. Consider a flow line with 5 work stations. Station 1 has an exponentially distributed processing time with rate 4. Stations 2 to 5 have fixed processing times with means 3, 2, 5 and 3 time units, respectively. The capacities of the intermediate buffers are 3, 4, 2, and 4. Blocking is of BAP type. Develop lower and upper bounds for the throughput of the system.

4.13. Hillier and Boling (1966) empirically showed that an unbalanced work-load allocation in production lines with exponential processing times improved the output rate. It is called the "bowl phenomenon," which assigns less work to the middle stations. For a total work content of 25 time units (sum of processing times), verify the bowl phenomenon in a 3-station production line with no buffers using Hunt's formula for the output rate.

4.14. Consider a 3-station flow line with mean processing times of 1.0 time unit each and a variance of 0.5, 1.5, and 0.5 at stations 1 to 3, respectively. Blocking is of BAP type. Using an approximate model, show the impact of the capacity of buffers B_2 and B_3 on the steady-state average inventory accumulation at each buffer and the output rate of the system. You may run your approximation program for a reasonable range of the buffer capacities (i.e., 2 to 10). Present your results in the form of graphs. Repeat your results for the variance of 0.5 at station 2, and compare the results.

4.15. Consider a two-station flow line presented in Section 4.3.1. The processing times are exponentially distributed with rates μ_1 and μ_2. Develop an explicit profit-maximizing objective function involving the revenue per unit produced and the unit holding cost per unit time. Also, consider the special case of $\mu_1 = \mu_2$. Knowing that both the output rate and the average number of jobs in the system are monotone nondecreasing functions of the buffer capacity N, suggest a procedure to find the profit-maximizing N.

4.16. Design problem: Consider a 3-station production line where the processing times, times to failure, and the repair times are exponentially distributed. The first station never starves and the blocking policy is BAP type. The failure, repair, and the processing rates are given below.

	Stat. 1	Stat. 2	Stat. 3
Processing rate	0.25	0.20	0.30
Failure rate	0.01	0.02	0.04
Repair rate	0.10	0.30	0.50

For a unit selling price of $30, a unit variable production cost of $10, and a unit holding cost of $0.20 /unit-unit time:

(a) Obtain the profit-maximizing values of N_2 and N_3.
(b) Redistribute $N_2^* + N_3^*$ to maximize the output rate only.
(c) Reallocate the buffer capacities so that the total buffer capacity is minimized subject to the constraint that the output rate is greater than the one obtained in (a).
(d) Interpret and compare (a)–(c).

4.17. Integrating design and performance: It is possible to devise a profit-maximizing buffer allocation algorithm in which the objective function is decomposed into a set of single-buffer objective functions, each corresponding to a subsystem in the decomposition algorithm. The objective functions of the subsystems are related to each other. Once the decomposition algorithm converges for an initial buffer configuration, each objective function can be partially improved (to a certain extent) by augmenting its buffer variable and performing a few backward and forward passes until the procedure converges again. This procedure may continue until no further improvement is obtained in each objective function. Develop a formal procedure along the above guidelines and implement it to a simpler case of a three- station flow line with exponential processing times. Choose your own values of the objective function parameters.

4.18. Consider a four-station flow line with exponential processing times. The first station has an infinite-capacity buffer. The processing time vector is $\underline{\mu} = (3, 4, 2, 5)$. Jobs arrive according to a Poisson process with rate $\lambda = 2$. At each station, a certain percentage of jobs, say d_i for station i, turns out to be defective and discarded. Let $\mathbf{d} = (0.1, 0.13, 0.06, 0.1)$. What is the system throughput? Implement the decomposition algorithm to obtain the average number of jobs at each buffer.

4.19. A three-station flow line has deterministic processing times of 3, 4, and 2.5 at stations 1 to 3, respectively. The line experiences failures. The first station, which has sufficient raw material not to experience starvation, is prone to failures upon which the job has to be discarded. The second station experiences failures upon which the process has to restart on the same job from scratch. The third station also fails, and the process has to continue from the point of interruption. All the failures are operation-dependent. The time to failure at each station is exponentially distributed with rate vector $\underline{\delta} = (0.10, 0.15, 0.20)$. Also, the repair times are all exponentially distributed with rates $\underline{\gamma} = (1, 1, 1)$. The buffer capacity vector is $\underline{N} = (3, 3)$. Implement the decomposition approach to obtain the line throughput as well as the average WIP level in the system.

5

Analysis of Transfer Lines

5.1 Introduction

Transfer lines are common in industries such as food, automotive, electronics, and pharmaceutical, among many others. A transfer line is a **synchronous** production line consisting of several work stations in series integrated into one system by a common transfer mechanism and a control system. Each station is a stopping point at which operations (machining, inspection, etc.) are performed on the work pieces. The main feature of transfer lines is that the machines are connected through conveyors and the units on the entire conveyor system move one slot at a time synchronously. The transfer points in time are determined by the maximum of the unit processing times at all the stations, which is referred to as the **cycle time**. Typical examples of these systems are filling and packaging lines in the pharmaceutical and food industries, and the assembly lines in various other industries. Transfer lines operate like a single machine when all the machines are up and producing one unit per cycle. However, they experience failures like any other production line. Different manufacturing practices may necessitate different line designs. Also, the procedures as to what to do when a failure occurs may vary from one manufacturing environment to another. For instance, the entire system may stop, or parts of the system may remain unaffected by a failure and may continue to operate utilizing the conveyor segments and the buffer storages. The units on failed machines may be discarded, reworked, or processed further for the remaining unfinished work after the repair is completed. In food and pharmaceutical manufacturing facilities, depending on the cost and the perishability concerns, the unit on a failed machine may be removed and discarded.

Buffers are used in low- to medium-speed (10–150 units per minute) transfer lines to absorb the impact of machine and conveyor stoppages or failures. Placing buffers in high-speed lines (500 or more units per minute) can be either impossible or impractical, since they are filled quickly, causing the system to stop in a short period of time. Therefore, high-speed lines are generally stopped upon a failure at any one of the machines. Completion of the remaining unfinished work on a failed machine is not a common practice in synchronous transfer lines since it jeopardizes synchronization. For this reason, in synchronous systems, the parts on

5. Analysis of Transfer Lines

failed machines have to be either discarded or reworked from scratch. Occasionally, depending on the jobs and the processes, it may be possible to single out the operation interrupted by a failure and complete the unfinished work. It may also be possible to consider the unfinished work on the interrupted unit as part of the repair process. This is experienced more in assembly lines such as those in the automotive industry, where a failure in a station stops the whole line. The transfer mechanism moves a unit from each station to the next, and the transfer is performed simultaneously on all parts in the entire conveyor system. The work stations may have either the same or different deterministic processing times. The maximum of the processing times determines the cycle time. Again, as in production lines, we are interested in designing and analyzing the performance of transfer lines by considering the standard measures of interest such as the throughput, the average inventory levels, and the percentage of machine down times.

Let us begin with the simplest possible transfer line to study, that is, the one stopped upon every failure at any one of the machines. Let us assume that it consists of K machines, as shown in Figure 5.1.

The other assumptions for this particular transfer line are as follows:

- There is an unlimited supply of work pieces in front of the first machine.
- The up and down times are mutually independent, and independent among themselves.
- Work pieces are not scrapped upon failure. The line ceases operation at the end of the cycle in which a failure occurs. Machines, including the failed machine, hold their finished units (or the interrupted unit). After the machine is repaired, at the next cycle beginning, the conveyor transports all the units on the machines to their next stops. Notice that the interrupted unit is considered as a good unit in this scenario. Another interpretation is that the unfinished work is completed during the repair time.
- No more than one machine may fail in a cycle.
- The line stops upon a failure at any machine since there are no intermediate buffer storages (other than the conveyor).

Let X_i and Y_i be the up and down times (in terms of the number of cycles) at machine i, denoted by M_i, respectively. Then, the isolated efficiency of machine M_i is given by $e_i = E[X_i]/\{E[X_i]+E[Y_i]\}$. Consequently, the up-time probability (or the efficiency) of the transfer line with no intermediate buffers, abbreviated by

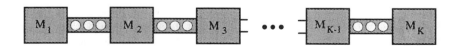

Figure 5.1. A transfer line with K machines.

$P_{\ell,o}(u)$, is given by

$$P_{\ell,o}(u) = \prod_{j=1}^{K} e_j \quad \text{for operation-independent failures,} \tag{5.1}$$

and by

$$P_{\ell,o}(u) = \frac{1}{1 + \sum_{k=1}^{K} \frac{E[Y_k]}{E[X_k]}} \quad \text{for operation-dependent failures.} \tag{5.2}$$

Equation (5.1) is well known in reliability theory as the availability of a system with K independent components connected in series. We should mention that (5.1) allows simultaneous machine failures. Equation (5.2) can be shown as follows: The average rate at which the system fails is $\sum_{i=1}^{K} \frac{1}{E[X_i]}$. When the system is down, the probability that it is due to the kth machine is

$$\Pr(\text{the failure is due to machine } k \mid \text{line is down}) = \frac{\frac{1}{E[X_k]}}{\sum_{i=1}^{K} \frac{1}{E[X_i]}}. \tag{5.3}$$

Since the expected down time of machine k is $E[Y_k]$, the average time that the line is down is given by

$$E[\text{line down time}] = \sum_{k=1}^{K} E[Y_k] \cdot \frac{\frac{1}{E[X_k]}}{\sum_{i=1}^{K} 1/E[X_i]} = \sum_{k=1}^{K} \frac{E[Y_k]}{E[X_k]} \bigg/ \sum_{i=1}^{K} \frac{1}{E[X_i]}. \tag{5.4}$$

Then, the ratio of the expected up time to the sum of the expected up and down times gives us the steady-state probability that the line is up at any arbitrary point in time, that is,

$$P_{\ell,o}(u) = \frac{[\sum_{i=1}^{K} \frac{1}{E[X_i]}]^{-1}}{\sum_{k=1}^{K} \frac{E[Y_k]}{E[X_k]} \cdot [\sum_{i=1}^{K} \frac{1}{E[X_i]}]^{-1} + [\sum_{i=1}^{K} \frac{1}{E[X_i]}]^{-1}}$$

$$= \frac{1}{1 + \sum_{k=1}^{K} \frac{E[Y_k]}{E[X_k]}}. \tag{5.5}$$

Notice that (5.5) relies only on the up/down cycle arguments of a failed machine and is independent of the distributions of X_i and Y_i. Moreover, (5.1) to (5.5) are also valid for continuous-time (as opposed to counting time in terms of cycles) repair time and time-to-failure distributions.

EXAMPLE 5.1. Consider a four-machine transfer line where the conveyor is also prone to failures. Table 5.1 has the individual mean time to repairs (MTTR), the failure rates, and the isolated efficiencies of the components in the line. The conveyor can simply be viewed as another machine in the line.

TABLE 5.1. Input parameters for Example 5.1

	M_1	M_2	M_3	M_4	Conveyor
Isolated efficiencies	0.9	0.95	0.92	0.98	0.9
MTTR (in cycles)	5	3	6	5	3
Failure rate	0.022	0.018	0.0157	0.004	0.037

Clearly, the system failure rate is the sum of the failure rates of all the machines and the conveyor, that is 0.096 failures/cycle. Remember that only one component may fail at a time. The probabilities that a failure is due to machine k, $k = 1, 4$, or the conveyor (using (5.3)) are 0.229, 0.188, 0.156, 0.042, and 0.385, respectively. The expected line down time (using (5.4)) is 4.01 cycles. Finally, the system up-time probability is 0.72.

In transfer lines, in principle, the up or down status of a machine may change only at transfer points in time, that is, the beginning or the ending points of a cycle. Earlier, we have assumed that when a machine fails, the line continues to operate until the end of the cycle time. Therefore, in the long run, the percentage of the time the line is up, $P_{\ell,o}(u)$, is the ratio of the cycles at the end of which a unit departs from the line to the total number of cycles observed. Consequently, $P_{\ell,o}(u)$ becomes the time-average probability that a unit departs from the line at the end of a given cycle, and it is also the system throughput per cycle, that is, $\bar{o}_{\ell,o} = P_{\ell,o}(u)$. Then, the throughput per T cycles for any T is given by $T\, P_{\ell,o}(u)$. This argument is valid only when we do not scrap the jobs on failed machines. Otherwise, $P_{\ell,o}(u)$ will still be the up-time probability but will no longer be the throughput. The throughput will be smaller in the case of scrapping.

5.2 A Two-Station Line with No Intermediate Buffer: A Regenerative Approach

Let us next look at a two-machine line with no buffer in between, with the policy of scrapping the job on the machine upon a failure (operation-dependent) and with the possibility of both stations failing during the same cycle. Clearly, one can study the simple Markov chain representing the behavior of this special transfer line. However, we will leave the Markovian approach to later sections for the analysis of systems with more general assumptions. Here, we choose to focus on a regeneration argument to obtain the throughput of the transfer line. Let us define the failure and repair processes as follows: Assume that time takes discrete values, that is $t = 0, 1, 2, 3, \ldots$, and these epochs are the beginning points of cycles in the line. Then, let $p_i(t)$ be the probability that machine i is under repair during the cycle starting at epoch t given that it was working during the cycle starting at epoch $t - 1$. Conventionally, it is assumed that $\lim_{t \to \infty} p_i(t) = p_i$, which is the time-independent probability that machine i experiences a failure during a

5.2. A Two-Station Line with No Intermediate Buffer: A Regenerative Approach

cycle. Consequently, the number of cycles starting from the one where machine i becomes operational to the end of the cycle during which the next failure occurs, denoted by X_i, is a geometrically distributed random variable with mean $1/p_i$. Keep in mind that the number of cycles from any up cycle to the cycle of the next failure is also a geometrically distributed random variable with the same parameter p_i due to the memoryless property of the geometric distribution. Similarly, $r_i(t)$ is the probability that machine i is operational during the cycle starting at epoch t given that it was under repair in the previous cycle. When repaired at epoch t, a machine starts operating at epoch $t+1$ if a unit is available for processing. Again, over a long period of operation, we have $\lim_{t \to \infty} r_i(t) = r_i$, and the number of cycles until the end of the cycle during which the repair activity is completed, denoted by Y_i, is geometrically distributed with mean $1/r_i$. Also, by convention, let us assume that failures and repairs become effective at the end of the cycles during which they occur. In fact, this assumption will remain effective in the rest of this chapter. Notice that the assumption of geometric time to failure and the repair time bring the independence of what may happen in one cycle from the history of the previous cycles through the memoryless property of the geometric distribution. First of all, both the time to failure and the repair time are continuous random variables by nature. That is, one has to compute p and r for every machine from the actual failure and repair data. However, only the exponential failure and repair times yield fixed ps and rs that do not change from one cycle to another, such as the ones we have introduced above. Therefore, if the true failure and repair times do not possess the exponentiality assumption, models utilizing geometric up and down times will provide only approximate results.

Let us be specific about the operating policy of the two-machine line:

- Upon a failure, the system ceases operation, the part being processed on the failed machine is discarded at the end of the cycle, and the repair activity is started at the beginning of the next cycle. If M_1 is the failed machine, M_2 ejects the finished job and becomes idle. If M_2 fails, M_1 becomes blocked, keeping the unit on the machine ready to be ejected to M_2 as soon as it is repaired.
- Upon a failure recovery during a cycle, M_1 starts processing in the next cycle, and M_2 starts processing only after a unit leaves M_1. Notice that M_1 may go through a series of up and down cycles until a unit departs from M_1. Both machines start processing in the next cycle after a failure is recovered in M_2. A simultaneous recovery on both machines is equivalent to a recovery at M_1.

Let us develop the following regeneration argument: Consider those epochs in time where either of the machines is recovered from a failure (or both are recovered), and both of the machines are ready for a fresh start to process work pieces in the next cycle. Let us denote this particular epoch by s_1. Clearly, the system regenerates a new sample path for its behavior at every entry into s_1. Consequently, the number of cycles in the transfer line between any two consecutive entries into s_1 forms a regeneration period. Based on this regeneration argument, the throughput of the transfer line is given by

$$\bar{o}_{\ell,o} = \frac{E[\text{the number of jobs produced during a regeneration period}]}{E[\text{the length of the regeneration period}]}. \quad (5.6)$$

Once the system is in state s_1, it will remain in this state until either M_1 or M_2 fails or both fail. Therefore, there are three cases to be studied to obtain the expected number of jobs produced in a regeneration period and the expected length of this period.

Case 1: M_1 fails first with probability

$$\Pr(X_1 < X_2) = 1 - \frac{p_2}{1 - (1 - p_1)(1 - p_2)}, \quad (5.7)$$

which can be shown by conditioning on X_2 and then removing the condition:

$$\Pr(X_1 < X_2) = \sum_{k=1}^{\infty} \Pr(X_1 < k) \Pr(X_2 = k),$$

where

$$\Pr(X_1 < k) = \sum_{j=1}^{k-1} p_1(1 - p_1)^{j-1} = 1 - (1 - p_1)^{k-1}.$$

In this case, the system is in state s_1 until M_1 fails. The expected time until this event occurs given that it occurs before M_2 fails is given by

$$E[X_1 \mid X_1 < X_2] = \frac{p_1(1 - p_2)}{\Pr(X_1 < X_2)[1 - (1 - p_1)(1 - p_2)]^2}, \quad (5.8)$$

which can be obtained using

$$E[X_1 \mid X_1 < X_2] = \sum_{x_1=1}^{\infty} x_1 \Pr(X_1 = x_1 \mid X_1 < X_2),$$

where

$$\Pr(X_1 = x_1 \mid X_1 < X_2) = \frac{p_1(1 - p_1)^{x_1-1}(1 - p_2)^{x_1}}{\Pr(X_1 < X_2)}.$$

Notice that $E[X_1 \mid X_1 < X_2]$ is also the expected number of units produced during a regenerative period.

For the expected cycle length, let us discuss the following: When M_1 fails, M_2 becomes idle (starving), and the system will not enter s_1 until M_1 transfers a job

5.2. A Two-Station Line with No Intermediate Buffer: A Regenerative Approach

to M_2. Let us denote this time by Z, which is depicted in Figure 5.2. Note that, in the first period just after its repair, M_1 may fail with probability p_1 and the job may be discarded. This process repeats itself until M_1 does process a unit without experiencing a failure. Then, Z is a geometrically distributed random variable in the sense that there are geometric terms in its probability distribution.[1]

$$\Pr(Z = k) = \frac{r_1(1 - p_1)}{c} \left[a^{k-1} - b^{k-1} \right], k = 1, 2, \ldots \quad (5.9)$$

where

$$c = \sqrt{(1 - r_1)^2 + 4r_1 p_1}, \quad a = \frac{2p_1 r_1}{c - 1 + r_1} \quad \text{and} \quad b = \frac{2p_1 r_1}{-c - 1 + r_1}.$$

Note that $P(Z = 1) = 0$, for $0 < a < 1$, and $-1 < b < 0$. It has an expected value of

$$E[Z] = \sum_{y=1}^{\infty} \sum_{k=1}^{\infty} k(y+1)(1-p_1)p_1^{k-1} \Pr(Y_1 = y)$$

$$= \frac{E[Y_1] + 1}{1 - p_1} = \frac{r_1 + 1}{r_1(1 - p_1)}. \quad (5.10)$$

Then, the expected length of the regeneration period in this case is

$$E[C_1] = E[X_1 \mid X_1 < X_2] + E[Z], \quad (5.11)$$

Figure 5.2. A pictorial representation of the random variable Z.

---

[1] Equation (5.9) can be obtained by inverting its generating function. First of all, we have $\Pr(Y_1 + 1 = k) = r_1(1 - r_1)^{k-2}, k = 2, 3, \ldots$. Also, the number of times $Y_t + 1$ is repeated is geometrically distributed with parameter p_1. Consequently, the generating function of (5.9) can be shown to be

$$\Phi(v) = \frac{(1 - p_1)r_1 v^2}{1 - (1 - r_1)v - p_1 r_1 v^2}.$$

192   5. Analysis of Transfer Lines

and the expected number of jobs produced during this period is given by

$$E[L_1] = E[X_1 \mid X_1 < X_2]. \tag{5.12}$$

Notice that (5.12) compensates the very first scrapped unit by the one produced by M_1 at the end of the regenerative cycle. Possible events in case 1 are shown in Figure 5.3.

Case 2: M_2 fails first with probability

$$\Pr(X_1 > X_2) = \frac{(1 - p_1)p_2}{1 - (1 - p_1)(1 - p_2)}. \tag{5.13}$$

The expected time until M_2 fails given that M_2 fails before M_1 is given by

$$E[X_2 \mid X_1 > X_2] = \frac{1}{1 - (1 - p_1)(1 - p_2)}, \tag{5.14}$$

Hence, the length of the regeneration period will be the sum of the time until M_2 fails and its repair time. At the end of the unit cycle in which M_2 is repaired, it receives the already processed job from M_1 and the system enters state s_1 again.

$$E[C_2] = E[X_2 \mid X_1 > X_2] + E[Y_2] \tag{5.15}$$

The expected number of jobs produced during this period will simply be

$$E[L_2] = E[X_2 \mid X_1 > X_2] - 1, \tag{5.16}$$

due to the fact that the job on M_2 is lost.

Case 3: M_1 and M_2 fail simultaneously with probability

$$\Pr(X_1 = X_2) = \frac{p_1 p_2}{1 - (1 - p_1)(1 - p_2)}. \tag{5.17}$$

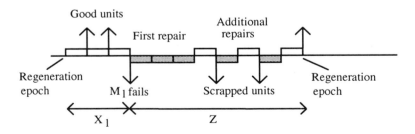

Figure 5.3. Regenerative period in case 1.

5.2. A Two-Station Line with No Intermediate Buffer: A Regenerative Approach

That is, the system remains in state s_1 until both machines fail during the same cycle. The expected time until this event happens is

$$E[X_1 \mid X_1 = X_2] = E[X_2 \mid X_1 = X_2] = \frac{1}{1 - (1-p_1)(1-p_2)}. \quad (5.18)$$

When both machines fail, there are two scenarios to be considered:

(i) M_1 becomes operational before M_2 and produces one unit to be transferred to M_2; (that is, $Z < Y_2$). In this case, we have

$$E[C_{31}] = E[X_1 \mid X_1 = X_2] + E[Z \mid Z < Y_2] + E[Y_2] \quad (5.19)$$

and

$$E[L_{31}] = E[X_1 \mid X_1 = X_2] - 1. \quad (5.20)$$

Notice that, at the point of recovery at M_1, the time until M_2 is repaired is again geometrically distributed with parameter p_2 due to the memoryless property.

(ii) M_2 becomes operational first (that is, $Z \geq Y_2$), yielding

$$E[C_{32}] = E[X_1 \mid X_1 = X_2] + E[Z \mid Z \geq Y_2] \quad (5.21)$$

and

$$E[L_{32}] = E[X_1 \mid X_1 = X_2] - 1. \quad (5.22)$$

Since (ii) includes the case in which both machines are repaired simultaneously, (i) and (ii) can be combined as follows:

$$\begin{aligned} E[C_3] &= E[X_1 \mid X_1 = X_2] + \{E[Z \mid Z < Y_2] + E[Y_2]\} \Pr(Z < Y_2) \\ &\quad + E[Z \mid Z \geq Y_2] \Pr(Z \geq Y_2) \\ &= E[X_1 \mid X_1 = X_2] + E[Y_2] \Pr(Z < Y_2) + E[Z], \end{aligned} \quad (5.23)$$

and

$$E[L_3] = E[X_1 \mid X_1 = X_2] - 1, \quad (5.24)$$

where $\Pr(Z < Y_2)$ can be obtained using (5.9), as we demonstrate in Example 5.2 ahead. Thus, the expected length of the regenerative period for the two-machine transfer line becomes

$$E[C] = \Pr(X_1 < X_2) E[C_1] + \Pr(X_1 < X_2) E[C_2] + \Pr(X_1 = X_2) E[C_3], \quad (5.25)$$

Similarly, the expected number of jobs produced during this period is

$$E[L] = \Pr(X_1 < X_2)E[L_1] + \Pr(X_1 < X_2)E[L_2] + \Pr(X_1 = X_2)E[L_3]. \tag{5.26}$$

Consequently, the expected throughput of the line can be obtained using (5.6), (5.25), and (5.26):

$$\bar{o}_{\ell,o} = \frac{E[L]}{E[C]}. \tag{5.27}$$

Notice that $\bar{o}_{\ell,o}$ is not the same as the probability that the line is up at any point in time, that is, $P_{\ell,o}(u)$, due to scrapping. Remember that a machine is still considered to be up in those cycles where scrapping occurs. Clearly, $P_{\ell,o}(u)$ can also be obtained by going through the same analysis and counting the number of up cycles in a regeneration period. However, a much simpler approach would be to use $P_{\ell,o}(u) = \bar{o}_{\ell,o}/(1 - p_2)$, on which we will elaborate in the following sections.

EXAMPLE 5.2. Consider the following two-machine symmetrical transfer line with no internal storage. Machine parameters are assumed to be $p_1 = p_2 = 0.1$ and $r_1 = r_2 = 0.7$. Let us evaluate the throughput by using the above regenerative approach.

Case 1: $\{X_1 < X_2\}$

Using (5.7), (5.13), and (5.17), we have

$$\Pr(X_1 < X_2) = 0.4737,$$
$$\Pr(X_2 < X_1) = 0.4737,$$
$$\Pr(X_1 = X_2) = 0.0526,$$
$$E[X_1 \mid X_1 < X_2] = 5.2630,$$
$$E[Z] = 2.6984.$$

Using (5.9) and (5.10),

$$E[C_1] = 7.9614 \quad \text{and} \quad E[L_1] = 5.2630.$$

Case 2: $\{X_2 < X_1\}$

Using (5.14) to (5.16), we obtain

$$E[X_2 \mid X_2 < X_1] = 5.2630,$$
$$E[C_2] = 6.6916 \quad \text{and} \quad E[L_2] = 4.2630.$$

Case 3: $\{X_1 = X_2\}$

Using (5.9), we can write

$$\Pr(Z = k) = 0.7180 \left[0.2424^{k-1} - (-0.1189)^{k-1} \right], k = 1, 2, \ldots,$$

and derive

$$\Pr(Z < Y_2) = \frac{r_1(1-p_1)}{c(1-a)} \left\{ 1 - \frac{r_2}{1-a(1-r_2)} \right\}$$
$$- \frac{r_1(1-p_1)}{c(1-b)} \left\{ 1 - \frac{r_2}{1-b(1-r_2)} \right\} = 0.0243.$$

Using (5.23) and (5.24),

$$E[C_3] = 7.9995 \quad \text{and} \quad E[L_3] = 4.2663.$$

Finally, using (5.25) to (5.27), we obtain $E[C] = 7.3619$, $E[L] = 4.7369$, and the throughput

$$\overline{o}_\ell = \frac{E[L]}{E[C]} = 0.6434.$$

It is clear from Chapter 4 that placing buffers in a transfer line improves its throughput. The throughput of a two-machine transfer line with a buffer storage in between can be expressed as a function of the throughput of the line with no buffer as follows:

$$\overline{o}_{\ell,N} = \overline{o}_{\ell,o} + h, \qquad (5.28)$$

where h is the percentage of the time that the second machine is able to operate due to the existence of the jobs in the buffer when the first machine is down. That is, h is the contribution of having a buffer in the system to the percentage of the system up time. Clearly, (5.28) is not an expression to compute $\overline{o}_{\ell,N}$ but rather to show that $\overline{o}_{\ell,N} > \overline{o}_{\ell,o}$. We usually obtain $\overline{o}_{\ell,N}$ by studying the stochastic process underlying the behavior of the transfer line, as we will do in the next section.

5.3 Transfer Lines with Work-in-Process Buffers

Let us consider a transfer line consisting of K machines with processing times of similar magnitudes such that each process starts and finishes almost simultaneously. The conveyor moves a fixed distance at every process completion, starting a new cycle for all the machines. Failures are operation-dependent, scrapping occurs, and the time to failure and the repair time are geometrically distributed as introduced in Section 5.2. Also, more than one event may occur during a cycle. That is, two or more machines may fail simultaneously, or a machine may fail and another may be repaired during the same cycle.

Secondary buffers (as opposed to the primary buffer, which is the conveyor) with finite capacities are placed between the machines to absorb the impact of failures, as shown in Figure 5.4. Let N_i be the capacity of the ith secondary buffer, denoted by B_i, between machines M_i and M_{i+1}. The conventional measures of performance for such a transfer line can be listed as follows:

$P_i(d)$: probability that M_i is down at any epoch;
$P_i(u)$: probability that M_i starts processing a job at any epoch;
$P_i(b)$: probability that M_i is blocked at any epoch;
$P_i(id)$: probability that M_i is idle (starving) at any epoch;
$\bar{o}_i$: average throughput of M_i, for $i = 1, 2, \ldots, K$;
$\bar{o}_\ell$: average line throughput;
$P_i(n_i)$: probability that there are n_i jobs in secondary buffer B_i at any epoch, for $n_i = 0, 1, 2, \ldots, N_i$;
$\overline{N}_i$: average inventory level in B_i, for $i = 1, 2, \ldots, K - 1$.

In principle, the exact analysis of a K-machine transfer line can be carried out through the analysis of the underlying Markov chain imbedded at cycle beginnings (or epochs). For instance, consider the stochastic process

$$\{(\alpha_1(t), \alpha_2(t), \ldots, \alpha_K(t), N_1(t), N_2(t), \ldots, N_{K-1}(t)), t = 1, 2, \ldots\},$$

where $\alpha_i(t)$ is the status of machine i, and $N_i(t)$, $N_i(t) = 0, 1, 2, \ldots, N_i$, is the number of jobs in B_i at epoch t, which is the beginning of the tth cycle. It is assumed that each cycle has a unit length. In the above representation, $\alpha_i(t) = 0$ if M_i is under repair; $\alpha_i(t) = 1$ if it starts processing a job; $\alpha_i(t) = i$ if M_i is idle (starving); and $\alpha_i(t) = b$ if M_i is blocked at epoch t. Due to the geometric up and down times, $\{\alpha_1(t), \alpha_2(t), \ldots, \alpha_K(t), N_1(t), N_2(t), \ldots, N_{K-1}(t)\}$ gives rise to a discrete-time Markov chain with a finite number of states. Let

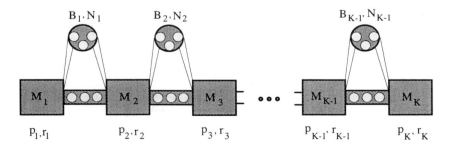

Figure 5.4. A transfer line with primary and secondary buffers.

$(\alpha_1, \alpha_2, \ldots, \alpha_K, n_1, n_2, \ldots, n_{K-1})$ be a particular state of the Markov chain at any epoch and $P(\alpha_1, \alpha_2, \ldots, \alpha_K, n_1, n_2, \ldots, n_{K-1})$ be its steady-state probability. In general, one is interested in obtaining $P(\bullet)$ from which most of the measures of performance can be obtained. Unfortunately, it is not possible to deal with the above stochastic process analytically for any practical values of the system parameters due to the magnitude of its state space. Therefore, one may resort to numerical solutions of these equations using direct methods (as termed in the numerical analysis literature, such as Gaussian elimination) or iterative methods (such as the power method or Jacobi iteration), again depending on the size of the statespace. Another important approach to analyze transfer lines has been the approximation of the line behavior by various decomposition principles. In the remainder of this chapter, we will study exact methods and approximations to evaluate the above measures of performance in various transfer lines and focus on some design issues.

5.4 Buffer Clearing Policies in Transfer Lines

There are a number of different policies regarding the operation of a transfer line, such as reworking or scrapping the part on a failed machine, resuming the conveyor movement, or clearing the upstream buffer upon the repair of a machine, and others. Here, let us consider two buffer clearing policies, namely **buffer clearing policy 1 (BCP1)** and **buffer clearing policy 2 (BCP2)**. The operation under **BCP1** is as follows: When machine i fails during a cycle, the adjacent conveyor segments cease operation. Machine M_{i-1} starts placing its output into buffer B_{i-1}, and machine M_{i+1} starts removing jobs (if any) from buffer B_i. At the end of a cycle, M_{i-1} is blocked when M_i is down and B_{i-1} is full of work pieces. The unit that is processed just before blocking occurs resides on M_{i-1} until M_i is repaired. Note that the blocking mechanism in use is the blocking-after-processing (BAP) type. At the epoch when M_{i-1} is ready for processing, the conveyor segment between machines M_{i-1} and M_i moves by one slot, removing the unit from M_{i-1} and placing a new one on M_i. On the other hand, M_{i+1} remains idle when M_i is down and there are no jobs in B_i. M_{i+1} starves until M_i recovers from the failure and processes a unit, in which case the conveyor segment between M_i and M_{i+1} moves by one slot, placing a new job on M_{i+1}. The number of jobs in B_i for any i remains unchanged during the cycles when M_i and M_{i+1} are both operating, since the primary buffer between these machines is active during this period. The secondary buffers are cleared only when their upstream machines are down and their downstream machines are up. Consequently, under BCP1, the same group of jobs may reside in a secondary buffer for a long time. BCP1 has been the common operating policy in the analysis of transfer lines in the literature.

Depending on the nature of the jobs and the processes, upon a failure recovery at M_i, it may be necessary to let M_i clear the secondary buffer B_{i-1} and keep M_{i-1} idle (or blocked) until the buffer is cleared. This policy, which we refer to as the buffer clearing policy 2 (**BCP2**), results in a smaller throughput when compared to BCP1, since it keeps a machine idle (blocked) while that machine is operational. Clearing buffers upon repair completions is a common practice in the food and pharmaceutical industries due to concerns of perishability, batch mixing, and product contamination.

The critical features of BCP2 are as follows: Upon the failure of M_i, the output of M_{i-1} is switched into buffer B_{i-1} until either B_{i-1} has N_{i-1} jobs, or M_{i-1} itself fails, or M_i becomes operational. There are several scenarios that may occur during the time M_i is under repair. The first one is associated with the failure of M_{i-1} before the repair completion of M_i. During the period M_i is under repair, if M_{i-1} fails, the job that is on M_{i-1} is discarded, which is a common assumption to both policies. The jobs that are in B_{i-1} upon the failure of M_{i-1} are to be cleared by M_i when it is operational. The cycle in which M_{i-1} becomes operational may be before or after (or on) the cycle M_i becomes operational. In either case, once repaired, M_{i-1} gets blocked if there are jobs in B_{i-1}, and it remains so until all the jobs in B_{i-1} are cleared by M_i. It is also possible that the repair of M_{i-1} is completed after B_{i-1} is cleared, in which case M_{i-1} starts processing provided that there are jobs to process.

If M_{i-1} does not experience a failure during the repair of M_i, M_i becomes operational either before or after M_{i-1} fills B_{i-1}. In the former case, starting at the end of the cycle in which it becomes operational, M_i processes the jobs in B_{i-1}. Once M_i starts processing, M_{i-1} stops and gets blocked until M_i clears all the jobs in B_{i-1}. In the latter case, B_{i-1} becomes full, blocking M_{i-1} before M_i becomes operational. That is, at the beginning of a cycle, if B_{i-1} has N_{i-1} jobs and M_{i-1} is operational, then M_{i-1} stops operating and is blocked until M_i clears all the jobs in B_{i-1} after it becomes operational.

Under BCP2, unlike BCP1, the machine does not have an already-processed job during the period it is blocked. That is, BCP1 assumes blocking after processing (BAP), while BCP2 assumes a modified version of blocking-before-processing (BBP) mechanisms. Also, under BCP2, there are no jobs in B_i for any i during the cycles M_i and M_{i+1} are both busy. Under both operating policies, our implicit assumption is that M_1 never starves and M_K never gets blocked. Also, any machine M_i, for $i = 2, 3, \ldots, K$, becomes idle at the beginning of a cycle if it is operational, M_{i-1} is under repair (for at least one cycle), and B_{i-1} is empty.

5.5 Analysis of a Two-Machine Synchronous Line Under BCP1

The two-machine synchronous transfer line shown in Figure 5.5 can be analyzed via a scaled-down version of the stochastic process introduced in Section 5.4, that is, $\{\alpha_1(t), \alpha_2(t), N(t)\}$. Let (α_1, α_2, n) be a particular state of the Markov chain, and let $P(\alpha_1, \alpha_2, n)$ be its steady-state probability. There are $4N + 6$ states in the state space of the Markov chain. The boundary states are $(0, i, 0)$, $(1, i, 0)$, $(b, 0, N)$. State $(0, i, 0)$ is visited when M_1 fails and no parts are in the secondary buffer, thus making M_2 idle. When M_1 is repaired, it starts processing a job in

5.5. Analysis of a Two-Machine Synchronous Line Under BCP1

the next cycle due to the assumption of an infinite supply of jobs for M_1. At the beginning of this cycle, if there are no jobs in B and M_1 is idle, the process visits state $(1, i, 0)$. If M_2 has been under repair for a long time, the secondary buffer may become full and M_1, though operational, becomes blocked, and consequently the process visits state $(b, 0, N)$. During blocking of M_1, the conveyor-operating policy assumes that there is a finished job on M_1, namely the blocking job. This job will be released through the movement of the conveyor at the end of the cycle in which M_2 becomes operational. The system state at the termination of blocking will be $(1, 1, N)$. For convenience, let us introduce the following probability vectors:

$$\tilde{\underline{P}}(0) = \begin{pmatrix} P(1, i, 0) \\ P(0, i, 0) \\ P(1, 0, 0) \\ P(0, 0, 0) \\ P(1, 1, 0) \\ P(0, 1, 0) \end{pmatrix}, \quad \tilde{\underline{P}}(n) = \begin{pmatrix} P(1, 0, n) \\ P(0, 0, n) \\ P(1, 1, n) \\ P(0, 1, n) \end{pmatrix}, \quad n = 1, 2, \ldots, N-1$$

and

$$\tilde{\underline{P}}(N) = \begin{pmatrix} P(1, 0, N) \\ P(0, 0, N) \\ P(1, 1, N) \\ P(b, 0, N) \end{pmatrix}.$$

Then, the steady-state flow-balance equations of $\{\alpha_1(t), \alpha_2(t), N(t)\}$ can be written in the matrix form as shown below:

$$\mathbf{G}\tilde{\underline{P}}(0) + \mathbf{H}\tilde{\underline{P}}(1) = \underline{0}, \tag{5.29}$$

$$\mathbf{D}\tilde{\underline{P}}(n) + \mathbf{C}\tilde{\underline{P}}(n+1) = \underline{0}, \quad n = 1, 2, \ldots, N-2, \tag{5.30}$$

$$\mathbf{D}\tilde{\underline{P}}(N-1) + \mathbf{E}\tilde{\underline{P}}(N) = \underline{0}, \tag{5.31}$$

$$\mathbf{F}\tilde{\underline{P}}(N) = \underline{0}, \tag{5.32}$$

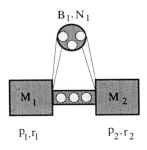

Figure 5.5. Two-machine line.

where $\underline{0}$ is a vector of zeroes with an appropriate dimension. The coefficient matrices are given below:

$$\mathbf{C} = \begin{bmatrix} 0 & 0 & 0 & 0 \\ 0 & r_1 r_2 & 0 & r_1(1-p_2) \\ p_1 r_2 & (1-r_1)r_2 & p_1(1-p_2) & (1-r_1)(1-p_2) \\ -1 & r_1(1-r_2) & 0 & r_1 r_2 \end{bmatrix},$$

$$\mathbf{D} = \begin{bmatrix} p_1(1-r_2) & r_1 r_2 - r_1 - r_2 & p_1 p_2 & (1-r_1)p_2 \\ (1-p_1)r_2 & 0 & p_1 p_2 - p_1 - p_2 & 0 \\ 0 & 0 & 0 & -1 \\ (1-p_1)(1-r_2) & 0 & (1-p_1)p_2 & 0 \end{bmatrix},$$

$$\mathbf{E} = \begin{bmatrix} 0 & 0 & 0 & 0 \\ 0 & r_1 r_2 & 0 & 0 \\ p_1 r_2 & (1-r_1)r_2 & p_1(1-p_2) & 0 \\ -1 & r_1(1-r_2) & 0 & 0 \end{bmatrix},$$

$$\mathbf{F} = \begin{bmatrix} p_1(1-r_2) & r_1 r_2 - r_1 - r_2 & p_1 p_2 & 0 \\ (1-p_1)r_2 & 0 & p_1 p_2 - p_1 - p_2 & r_2 \\ (1-p_1)(1-r_2) & 0 & (1-p_1)p_2 & -r_2 \end{bmatrix},$$

$$\mathbf{G} = \begin{bmatrix} -1 & r_1 & 0 & r_1 r_2 & 0 & r_1(1-p_2) \\ p_1 & -r_1 & p_1 r_2 & (1-r_1)r_2 & p_1(1-p_2) & (1-r_1)(1-p_2) \\ 0 & 0 & -1 & r_1(1-r_2) & 0 & r_1 p_2 \\ 0 & 0 & p_1(1-r_2) & r_1 r_2 - r_1 - r_2 & p_1 p_2 & (1-r_1)p_2 \\ (1-p_1) & 0 & (1-p_1)r_2 & 0 & p_1 p_2 - p_1 - p_2 & 0 \\ 0 & 0 & 0 & 0 & 0 & -1 \end{bmatrix},$$

$$\mathbf{H} = \begin{bmatrix} 0 & 0 & 0 & 0 \\ 0 & 0 & 0 & 0 \\ 0 & 0 & 0 & 0 \\ 0 & 0 & 0 & 0 \\ 0 & r_1 r_2 & 0 & r_1(1-p_2) \\ p_1 r_2 & (1-r_1)r_2 & p_1(1-p_2) & (1-r_1)(1-r_2) \end{bmatrix}.$$

Let us consider one of the equations to be familiar with the flow-balance argument. For state $(0, 1, n)$, with $n < N$, that is, M_1 is down, M_2 is up, and there are n units in the buffer, we have

$$P(0, 1, n) = P(1, 1, n+1)p_1(1-p_2) + P(1, 0, n+1)p_1 r_2$$
$$+ P(0, 0, n+1)(1-r_1)r_2 + P(0, 1, n+1)(1-r_1)(1-p_2), \quad (5.33)$$

which is written by considering the possible ways the process can get into state $(0, 1, n)$.

The bi-diagonal structure of the transition probability matrix allows us to develop a matrix-geometric algorithm to obtain the steady-state probabilities of the Markov

5.5. Analysis of a Two-Machine Synchronous Line Under BCP1

chain. Using (5.32), that is $F\tilde{\underline{P}}(N) = \underline{0}$, with $P(b, 0, N) = 1$, we obtain

$$\tilde{\underline{P}}(N) = \begin{bmatrix} p_1 r_2 / [(1 - p_1)(p_1 + p_2 - p_1 p_2 - p_1 r_2)] \\ p_1 r_2 / [(1 - p_1)(r_1 + r_2 - r_1 r_2)] \\ r_2 / [(p_1 + p_2 - p_1 p_2 - p_1 r_2)] \\ 1 \end{bmatrix}.$$

The special structure of (5.29)–(5.32) allows us to obtain $\tilde{\underline{P}}(\cdot)$ using the following matrix-geometric expressions: From (5.31), we have

$$\tilde{\underline{P}}(N - 1) = -\mathbf{D}^{-1}\mathbf{E}\tilde{\underline{P}}(N),$$

and from (5.30),

$$\tilde{\underline{P}}(N - 2) = -\mathbf{D}^{-1}\mathbf{C}\tilde{\underline{P}}(N - 1) = (-\mathbf{D}^{-1}\mathbf{C})(\mathbf{D}^{-1}\mathbf{E})\tilde{\underline{P}}(N).$$

Similarly,

$$\tilde{\underline{P}}(N - 3) = (-\mathbf{D}^{-1}\mathbf{C})^2 (\mathbf{D}^{-1}\mathbf{E})\tilde{\underline{P}}(N),$$

resulting in the following matrix-geometric expression:

$$\tilde{\underline{P}}(N - j) = -\mathbf{W}^{j-1}\mathbf{Y}\tilde{\underline{P}}(N), \qquad j = 1, \ldots, N - 1 \tag{5.34}$$

where

$$\mathbf{W} = -\mathbf{D}^{-1}\mathbf{C} \quad \text{and} \quad \mathbf{Y} = \mathbf{D}^{-1}\mathbf{E}.$$

Finally, using (5.29) and (5.34), we obtain $\tilde{\underline{P}}(0)$ in terms of $\tilde{\underline{P}}(N)$:

$$\tilde{\underline{P}}(0) = \mathbf{Z}\tilde{\underline{P}}(1) = -\mathbf{Z}\mathbf{W}^{N-2}\mathbf{Y}\tilde{\underline{P}}(N), \tag{5.35}$$

where $\mathbf{Z} = -\mathbf{G}^{-1}\mathbf{H}$. $\mathbf{D}$ and $\mathbf{G}$ are invertible matrices. Using (5.34) and (5.35), $\tilde{\underline{P}}(0), \ldots, \tilde{\underline{P}}(N - 1)$ are obtained as functions of $\tilde{\underline{P}}(N)$. Finally, the steady-state probabilities $P(\alpha_1, \alpha_2, n)$ for all α_i and n are obtained through normalization.

5.5.1 Measures of Performance

As usual, we are interested in obtaining the steady-state probabilities of the machine status, the output rate, and the average number of jobs in the buffer. The output rate, $\bar{o}_\ell$, is the output rate of M_2, that is,

$$\bar{o}_\ell = \bar{o}_2 = (1 - p_2) \sum_{\alpha_1, n} P(\alpha_1, 1, n). \tag{5.36}$$

TABLE 5.2. The steady-state probabilities of machine status

$P_1(u)$	$\sum_{\alpha_2, n} P(1, \alpha_2, n)$
$P_1(d)$	$\sum_{\alpha_2, n} P(0, \alpha_2, n)$
$P_1(b)$	$P(b, 0, N)$
$P_1(id)$	0
$P_2(u)$	$\sum_{\alpha_1, n} P(\alpha_1, 1, n)$
$P_2(d)$	$\sum_{\alpha_1, n} P(\alpha_1, 0, n)$
$P_2(b)$	0
$P_2(id)$	$P(1, i, 0) + P(0, i, 0)$

The summation term in (5.36) is the steady-state probability that machine M_2 is processing a job at the beginning of a cycle, which is $P_2(u)$ (see Table 5.2). Therefore, the output rate of a two-machine synchronous production line with a secondary buffer and with job scrapping upon failure is

$$\bar{o}_\ell = (1 - p_2) P_2(u). \tag{5.37}$$

Let us keep in mind that $P_1(u) = e_1[1 - P_1(b)]$ and $P_2(u) = e_2[1 - P_2(i)]$, with e_i being the isolated efficiency of machine M_i. The output rate of the system can also be obtained from the output rate of M_1, using the following proposition.

Proposition. *The steady-state output rate of the system can also be expressed as*

$$\bar{o}_\ell = (1 - p_1)(1 - p_2) P_1(u). \tag{5.38}$$

PROOF. In the long run, $\bar{o}_1$ represents the probability that M_2 is processing jobs regardless that some may be discarded, which is $P_2(u)$. That is,

$$\bar{o}_1 = P_2(u), \tag{5.39}$$

which can also be shown by using the flow-balance equations of the Markov chain. On the other hand, by definition we have

$$(1 - p_1) P_1(u) = (1 - p_1) \sum_{\alpha_2, n} P(1, \alpha_2, n),$$

$$= (1 - p_1) \sum_{\alpha_2, n \neq N} P(1, \alpha_2, n) + P(1, 1, N),$$

5.5. Analysis of a Two-Machine Synchronous Line Under BCP1

and replacing $P(1, 1, N)$ by its equivalent in the flow-balance equations, we obtain

$$(1 - p_1)P_1(u) = (1 - p_1) \sum_{\alpha_2, n \neq N} P(1, \alpha_2, n) + P(1, 0, N)(1 - p_1)r_2$$
$$+ P(1, 1, N)(1 - p_1)(1 - p_2) + P(b, 0, N)r_2 = \bar{o}_1,$$

resulting in

$$\bar{o}_1 = (1 - p_1)P_1(u). \tag{5.40}$$

Using (5.37), (5.39), and 5.40) we obtain

$$\bar{o}_\ell = \bar{o}_2 = (1 - p_1)(1 - p_2)P_1(u). \tag{5.41}$$

$\square$

The average number of jobs in the secondary buffer is given by

$$\overline{N} = \sum_{\alpha_1, \alpha_2, n} n Q(\alpha_1, \alpha_2, n). \tag{5.42}$$

The matrix-geometric algorithm is highly efficient and stable. Only two matrices with negligible dimensions are inverted at the beginning of the algorithm. Table 5.3 shows the output rate of the system, with parameters $p_1 = 0.2$, $p_2 = 0.1$, $r_1 = 0.3$, and $r_2 = 0.3$, demonstrating the stability of the algorithm for large buffer capacities. Table 5.3 gives us an opportunity to discuss what happens as the buffer capacity gets larger. The input rate to the buffer is $\bar{o}_{in} = e_1[1 - P_1(b)](1 - p_1)$, and the rate of units coming out of the buffer is $\bar{o}_{out} = e_2[1 - P_2(i)]$. As N approaches infinity, we expect a reasonable accumulation in the buffer if $\bar{o}_{in} < \bar{o}_{out}$. In this case, $P_1(b)$ approaches zero and $\bar{o}_\ell$ approaches $e_1(1 - p_1)(1 - p_2)$, which is the

TABLE 5.3. The system output rate for increasing values of N (under BCP1)

N	$\bar{o}$
3	0.4132
5	0.4221
10	0.4299
25	0.4319
50	0.4319
100	0.4320
250	0.4320

case in Table 5.3. In the opposite case, that is, if $\bar{o}_{in} \geq \bar{o}_{out}$, then we expect a large accumulation in the buffer, causing instability in the system. In this case, as N approaches infinity, $P_2(i)$ approaches zero, resulting in $\bar{o}_\ell = e_2(1 - p_2)$. Hence, the failure and repair rates and the scrap probabilities play a crucial role in the stability of the system for large buffer capacities.

EXAMPLE 5.3. Consider a two-machine transfer line operating under the BCP1 rule with the following parameters: $p_1 = p_2 = 0.2$, $r_1 = r_2 = 0.3$, and $N = 2$. Let us calculate the performance measures of the system using the matrix-geometric algorithm. The state space of the underlying stochastic process is $\{(1, i, 0), (0, i, 0), (1, 0, 0), (0, 0, 0), (1, 1, 0), (0, 1, 0), (1, 0, 1), (0, 0, 1),$ $(1, 1, 1), (0, 1, 1), (1, 0, 2), (0, 0, 2), (1, 1, 2), (b, 0, 2)\}$, having $(4)(2) + 6 = 14$ states. The probability vectors of $\tilde{\underline{P}}(0)$, $\tilde{\underline{P}}(1)$, and $\tilde{\underline{P}}(2)$ are as follows:

$$\tilde{\underline{P}}(0) = \begin{pmatrix} P(1, i, 0) \\ P(0, i, 0) \\ P(1, 0, 0) \\ P(0, 0, 0) \\ P(1, 1, 0) \\ P(0, 1, 0) \end{pmatrix}, \quad \tilde{\underline{P}}(1) = \begin{pmatrix} P(1, 0, 1) \\ P(0, 0, 1) \\ P(1, 1, 1) \\ P(0, 1, 1) \end{pmatrix}, \quad \tilde{\underline{P}}(2) = \begin{pmatrix} P(1, 0, 2) \\ P(0, 0, 2) \\ P(1, 1, 2) \\ P(b, 0, 2) \end{pmatrix}.$$

The corresponding steady-state flow-balance equations are

$$\mathbf{G}\tilde{\underline{P}}(0) + \mathbf{H}\tilde{\underline{P}}(1) = \underline{0},$$
$$\mathbf{D}\tilde{\underline{P}}(1) + \mathbf{E}\tilde{\underline{P}}(2) = \underline{0},$$
$$\mathbf{F}\tilde{\underline{P}}(2) = \underline{0},$$

where the matrices $\mathbf{G}$, $\mathbf{H}$, $\mathbf{D}$, and $\mathbf{E}$ are as follows:

$$\mathbf{D} = \begin{bmatrix} .14 & -.51 & .04 & .14 \\ .24 & 0 & -.36 & 0 \\ 0 & 0 & 0 & -1 \\ .56 & 0 & .16 & 0 \end{bmatrix}, \quad \mathbf{E} = \begin{bmatrix} 0 & 0 & 0 & 0 \\ 0 & .09 & 0 & 0 \\ .06 & .21 & .16 & 0 \\ -1 & .21 & 0 & 0 \end{bmatrix},$$

$$\mathbf{D}^{-1} = \begin{bmatrix} 0 & .6667 & 0 & 1.5 \\ -1.9608 & 0 & -0.2745 & 0.4902 \\ 0 & -2.333 & 0 & 1 \\ 0 & 0 & -1 & 0 \end{bmatrix},$$

$$\mathbf{G} = \begin{bmatrix} -1 & .3 & 0 & .09 & 0 & .24 \\ .2 & -.3 & .06 & .21 & .16 & .56 \\ 0 & 0 & -1 & .21 & 0 & .06 \\ 0 & 0 & .14 & -.51 & .04 & .14 \\ .8 & 0 & .24 & 0 & -.36 & 0 \\ 0 & 0 & 0 & 0 & 0 & -1 \end{bmatrix},$$

5.5. Analysis of a Two-Machine Synchronous Line Under BCP1

$$\mathbf{H} = \begin{bmatrix} 0 & 0 & 0 & 0 \\ 0 & 0 & 0 & 0 \\ 0 & 0 & 0 & 0 \\ 0 & 0 & 0 & 0 \\ 0 & .09 & 0 & .24 \\ .06 & .21 & .16 & .49 \end{bmatrix}.$$

$$\mathbf{G}^{-1} = \begin{bmatrix} -2.6195 & -2.6195 & -0.7445 & -1.8474 & -1.3695 & -2.3989 \\ -5.2512 & -8.5846 & -2.3346 & -5.4228 & -4.4179 & -6.9669 \\ -0.1029 & -0.1029 & -1.1029 & -0.5147 & -0.1029 & -0.2206 \\ -0.4902 & -0.4902 & -0.4902 & -2.4510 & -0.4902 & -0.7647 \\ -5.8897 & -5.8897 & -2.3897 & -4.4485 & -5.8897 & -5.4779 \\ 0 & 0 & 0 & 0 & 0 & -1 \end{bmatrix}.$$

Using (5.34) and (5.35), we obtain

$$\tilde{\underline{\mathbf{P}}}(2) = \begin{pmatrix} 0.25 \\ 0.147 \\ 1 \\ 1 \end{pmatrix}, \quad \text{(before normalization)}.$$

We have $\mathbf{W} = -\mathbf{D}^{-1}\mathbf{C}$ and $\mathbf{Y} = \mathbf{D}^{-1}\mathbf{E}$. Then, $\tilde{\underline{\mathbf{P}}}(1) = -\mathbf{W}^0\mathbf{Y}\tilde{\underline{\mathbf{P}}}(2) = -\mathbf{D}^{-1}\mathbf{E}\tilde{\underline{\mathbf{P}}}(2)$, that is,

$$\tilde{\underline{\mathbf{P}}}(1) = \begin{pmatrix} .0320 \\ .1639 \\ .2500 \\ .2059 \end{pmatrix}, \quad \text{(before normalization)}.$$

Finally, $\tilde{\underline{\mathbf{P}}}(0) = -\mathbf{Z}\mathbf{W}^{N-2}\mathbf{Y}\tilde{\underline{\mathbf{P}}}(N) = \mathbf{G}^{-1}\mathbf{H}\mathbf{D}^{-1}\mathbf{E}\tilde{\underline{\mathbf{P}}}(2)$, that is,

$$\tilde{\underline{\mathbf{P}}}(0) = \begin{pmatrix} .5545 \\ 1.6386 \\ .0496 \\ .1802 \\ 1.4434 \\ .1945 \end{pmatrix}, \quad \text{(before normalization)}.$$

The normalized steady-state probabilities and the performance measures of the system are as follows:

$$\underline{\tilde{P}}(0) = \begin{pmatrix} 0.0749 \\ 0.2216 \\ 0.0067 \\ 0.0244 \\ 0.1951 \\ 0.0263 \end{pmatrix}, \quad \underline{\tilde{P}}(1) = \begin{pmatrix} 0.0432 \\ 0.0221 \\ 0.0338 \\ 0.0278 \end{pmatrix}, \quad \underline{\tilde{P}}(2) = \begin{pmatrix} 0.0338 \\ 0.0199 \\ 0.1352 \\ 0.1352 \end{pmatrix},$$

and $\bar{o}_\ell = 0.3346$, $\overline{N} = 0.7751$, $P_1(d) = 0.3421$, and $P_2(d) = 0.2853$.

5.5.2 A Special Case: $N = 0$

In the case of two-machine lines with no intermediate buffer, one can study the reduced Markov chain representing the behavior of the two machines. Its state space is $S = \{(1, i), (0, i), (1, 0), (0, 0), (1, 1), (b, 0)\}$. Due to scrapping, $(0,1)$ is not a valid state in this line. The state in which both machines are processing a work piece at the beginning of a cycle is $(1,1)$. If M_1 fails during that cycle, the work piece being processed on M_1 is discarded at the beginning of the next cycle, and the system enters into state $(0, i)$. Also, state $(1,0)$ may be visited from state $(0,0)$ when M_1 is repaired. The transition probability matrix, $\tilde{Q}$, of the Markov chain is

$$\tilde{Q} = \begin{bmatrix} 0 & p_1 & 0 & 0 & 1-p_1 & 0 \\ r_1 & 1-r_1 & 0 & 0 & 0 & 0 \\ 0 & p_1 r_2 & 0 & p_1(1-r_2) & (1-p_1)r_2 & (1-p_1)(1-r_2) \\ r_1 r_2 & (1-r_1)r_2 & r_1(1-r_2) & (1-r_1)(1-r_2) & 0 & 0 \\ 0 & p_1(1-p_2) & 0 & p_1 p_2 & (1-p_1)(1-p_2) & (1-p_1)p_2 \\ 0 & 0 & 0 & 0 & r_2 & 1-r_2 \end{bmatrix}.$$

The steady-state probabilities, $P(\alpha_1, \alpha_2)$, can easily be found using the following set of equations:

$$(\tilde{Q} - I)^T \underline{P} = \underline{0},$$

and $\sum P(\alpha_1, \alpha_2) = 1$.

Clearly, the output rate of the machine is $\bar{o}_\ell = P(1, 1)(1 - p_2)$. This should coincide with the results of Section 5.2. It can be verified using Example 5.2.

5.6 Operation Under BCP2

Under this policy, the number of jobs in the secondary buffer B is zero during the cycles when M_1 and M_2 are both processing jobs, which corresponds to state

5.6. Operation Under BCP2

(1,1,0). Upon a failure of M_2, the conveyor stops and M_1 places units into B, until either B has N jobs, or M_2 becomes operational, or M_1 itself fails. In the first case, the system finally enters state (b, 0, N). In the second case, the system enters state (b, 1, n), where $n < N$. In either of these cases, M_1 remains blocked until M_2 clears all the jobs in B, after which it becomes operational, and the system returns to state (1,1,0) again. In the third case, if M_1 fails before M_2 is repaired, the system first enters a state (0, 0, n), where $n < N$. The next transition depends on the repair times of the machines. If M_1 becomes operational first, it does not start to process jobs immediately (given $n > 0$); hence the system enters state (b, 0, n). M_1 remains blocked until M_2 becomes operational and clears all the jobs in B. On the other hand, if M_2 becomes operational first, the system enters state (0, 1, $n-1$), given $n > 0$. In this case, M_2 starts to clear the jobs in B immediately. If M_1 becomes operational before M_2 clears B, it will be blocked again until all the jobs in B are cleared. If M_1 remains under repair for a long time, M_2 may clear all the jobs in B and become idle, and the system enters state (0, i, 0). Upon the repair of M_1, the system will then enter state (1, i, 0), and if M_1 does not fail during the cycle of this state, the two-machine line will finally enter state (1,1,0) at the beginning of the next cycle. The above arguments correspond to the basic sample path realizations under the BCP2 operating policy. There are $5N+1$ states in the state space of the Markov chain $\{\alpha_1(t), \alpha_2(t), N(t)\}$ under BCP2.

Let us consider the following probability vectors, representing the steady-state probabilities of the Markov chain $\{(\alpha_1(t), \alpha_2(t), N(t)), t = 1, 2, \ldots\}$:

$$\underline{\tilde{P}}(0) = \begin{pmatrix} P(1, i, 0) \\ P(0, i, 0) \\ P(1, 0, 0) \\ P(0, 0, 0) \\ P(1, 1, 0) \\ P(0, 1, 0) \end{pmatrix},$$

$$\underline{\tilde{P}}(n) = \begin{pmatrix} P(1, 0, n) \\ P(b, 0, n) \\ P(0, 0, n) \\ P(b, 1, n) \\ P(0, 1, n) \end{pmatrix}, \quad n = 1, 2, \ldots, N-2,$$

$$\underline{\tilde{P}}(N-1) = \begin{pmatrix} P(1, 0, N-1) \\ P(b, 0, N-1) \\ P(0, 0, N-1) \\ P(b, 1, N-1) \\ P(b, 0, N) \end{pmatrix}.$$

Notice that these probability vectors are slightly different than those under BCP1.

The flow-balance equations of $\{(\alpha_1(t), \alpha_2(t), N(t))\}$ in the matrix form are of the same type as Eqs. (5.29)–(5.32) except that the contents and the dimensions of matrices **C, D, E, F, G**, and **H** are different in this case. The flow-balance equations for the line under the BCP2 policy are

$$\mathbf{G}\tilde{\underline{P}}(0) + \mathbf{H}\tilde{\underline{P}}(1) = \underline{0} \qquad (5.43)$$

$$\mathbf{D}\tilde{\underline{P}}(n) + \mathbf{C}\tilde{\underline{P}}(n+1) = \underline{0}, \qquad n = 1, 2, \ldots, N-2, \qquad (5.44)$$

$$\mathbf{D}\tilde{\underline{P}}(N-2) + \mathbf{E}\tilde{\underline{P}}(N-1) = \underline{0}, \qquad (5.45)$$

$$\mathbf{F}\tilde{\underline{P}}(N-1) = \underline{0}, \qquad (5.46)$$

where matrices **C**, **D**, **E**, **F**, **G**, and **H** are as follows:

$$\mathbf{C} = \begin{bmatrix} 0 & 0 & 0 & 0 & 0 \\ 0 & 0 & 0 & 0 & 0 \\ 0 & r_2 & r_1 r_2 & 1-p_2 & r_1(1-p_2) \\ p_1 r_2 & 0 & (1-r_1)r_2 & 0 & (1-r_1)(1-p_2) \\ -1 & 0 & 0 & 0 & 0 \end{bmatrix},$$

$$\mathbf{D} = \begin{bmatrix} 0 & -r_2 & r_1(1-r_2) & p_2 & r_1 p_2 \\ p_1(1-r_2) & 0 & r_1 r_2 - r_1 - r_2 & 0 & (1-r_1)p_2 \\ (1-p_1)r_2 & 0 & 0 & -1 & 0 \\ 0 & 0 & 0 & 0 & -1 \\ (1-p_1)(1-r_2) & 0 & 0 & 0 & 0 \end{bmatrix},$$

$$\mathbf{E} = \begin{bmatrix} 0 & 0 & 0 & 0 & 0 \\ 0 & 0 & 0 & 0 & 0 \\ 0 & r_2 & r_1 r_2 & 1-p_2 & 0 \\ p_1 r_2 & 0 & (1-r_1)r_2 & 0 & 0 \\ -1 & 0 & 0 & 0 & 0 \end{bmatrix},$$

$$\mathbf{F} = \begin{bmatrix} 0 & -r_2 & r_1(1-r_2) & p_2 & 0 \\ p_1(1-r_2) & 0 & r_1 r_2 - r_1 - r_2 & 0 & 0 \\ (1-p_1)r_2 & 0 & 0 & -1 & r_2 \\ (1-p_1)(1-r_2) & 0 & 0 & 0 & -r_2 \end{bmatrix},$$

$$\mathbf{G} = \begin{bmatrix} -1 & r_1 & 0 & r_1 r_2 & 0 & r_1(1-p_2) \\ p_1 & -r_1 & p_1 r_2 & (1-r_1)r_2 & p_1(1-p_2) & (1-r_1)(1-p_2) \\ 0 & 0 & -1 & r_1(1-r_2) & 0 & r_1 p_2 \\ 0 & 0 & p_1(1-r_2) & r_1 r_2 - r_1 - r_2 & p_1 p_2 & (1-r_1)p_2 \\ (1-p_1) & 0 & (1-p_1)r_2 & 0 & p_1 p_2 - p_1 - p_2 & 0 \\ 0 & 0 & 0 & 0 & 0 & -1 \end{bmatrix},$$

$$\mathbf{H} = \begin{bmatrix} 0 & 0 & 0 & 0 & 0 \\ 0 & 0 & 0 & 0 & 0 \\ 0 & 0 & 0 & 0 & 0 \\ 0 & 0 & 0 & 0 & 0 \\ 0 & r_2 & r_1 r_2 & 1-p_2 & r_1(1-p_2) \\ p_1 r_2 & 0 & (1-r_1)r_2 & 0 & (1-r_1)(1-p_2) \end{bmatrix}.$$

5.6. Operation Under BCP2

Again due to the bi-diagonal structure of the transition probability matrix, the matrix-geometric approach of Section 5.5 can also be used in this case. Using (5.46), namely $\mathbf{F}\tilde{\mathbf{P}}(N-1) = \mathbf{0}$, we have

$$\tilde{\mathbf{P}}(N-1) = \begin{pmatrix} r_2/[(1-p_1)(1-r_2)] \\ r_1 p_1 (1-r_2)/[(1-p_1)(r_1+r_2-r_1 r_2)] + p_2/(1-r_2) \\ p_1 r_2/[(1-p_1)(r_1+r_2-r_1 r_2)] \\ r_2/(1-r_2) \\ 1 \end{pmatrix}.$$

Given that matrices $\mathbf{W}$, $\mathbf{Y}$, and $\mathbf{Z}$ are defined as before, (5.44) and (5.45) yield the following expression:

$$\tilde{\mathbf{P}}(N-j) = -\mathbf{W}^{j-2}\mathbf{Y}\tilde{\mathbf{P}}(N-1), \qquad j=2,\ldots,N-1. \tag{5.47}$$

Using (5.44) and (5.47) for $j = N-1$, we obtain

$$\tilde{\mathbf{P}}(0) = \mathbf{Z}\tilde{\mathbf{P}}(1) = -\mathbf{Z}\mathbf{W}^{N-3}\mathbf{Y}\tilde{\mathbf{P}}(N-1). \tag{5.48}$$

Hence, $\tilde{\mathbf{P}}(0), \ldots, \tilde{\mathbf{P}}(N-2)$ are all expressed in terms of $\tilde{\mathbf{P}}(N-1)$. Again, the values of the steady-state probabilities, $P(\alpha_1, \alpha_2, n)$, are obtained through normalization. The system throughput, $\bar{o}_\ell$, and the average inventory level in the secondary buffer, $\overline{N}$, are obtained in the same way as we have done in Section 5.5.

We should note that the expressions representing the steady-state machine status probabilities in Table 5.2 remain unchanged under BCP2, except $P_1(b)$. In this case, $P_1(b) = \sum_{\alpha_2, n} P(b, \alpha_2, n)$. Also, the Eqs. (5.37) and (5.38) are still valid for BCP2.

The throughput of the line operating under BCP2 is less than that of the line operating under BCP1, since we stop M_1, though it is operational, during those periods when M_2 clears the secondary buffer under BCP2. The difference may be significant if M_2 fails often and remains down for long periods. In this case, most of the time there will be some jobs in the secondary buffer and during those periods M_2 is up, it will be processing jobs in the buffer, blocking M_1 even though M_1 is operational. Illustrating the stability of the matrix-geometric algorithm and the decrease in the output rate under BCP2, Table 5.4 shows the results of the system studied in Example 5.3, operating under BCP2, for the corresponding buffer capacities of Table 5.3.

Table 5.4 shows that the matrix-geometric approach is also stable for the computation of a large number of probabilities in the two-machine line operating under the BCP2 policy. The difference in the throughputs of the two buffer clearing policies shown in Tables 5.3 and 5.4 is significant, and it remains significant in the entire range of the buffer capacities. Notice that as N increases, $\bar{o}_\ell$ approaches an asymptotic value, and it does so faster under BCP2 than under BCP1. Intuitively, this is due to the fact that buffer contents do not get too large under BCP2, and therefore it does not have a significant impact on the throughput. Unlike that under

TABLE 5.4. The system output rate for increasing values of N (under BCP2)

N	$\bar{o}_\ell$
3	0.36343
5	0.37158
10	0.37498
25	0.37518
50	0.37518
100	0.37518
250	0.37518

BCP1, the asymptotic throughput under BCP2 can no longer be expressed using simple relations. No matter what the value of N is, the machines will experience some idleness and blocking, and the probabilities of these events must be obtained to evaluate $\bar{o}_\ell$. However, since $P_1(b), P_2(i) > 0$, the system will remain stable under BCP2 as N approaches infinity. The relative increase in $\bar{o}_\ell$ when N increases from 3 to 250 is 4.6 percent under BCP1 and 3.2 percent under BCP2. Although not significant, these percentages are indicative of the relative impact of the buffer contents on the throughput. We will examine the behavior of a two-machine line as a function of the machine parameters in the next section.

5.7 Line Behavior and Design Issues

The performance of a transfer line is significantly affected by the failures and the repair activities that the machines go through. Clearly, if the line does not experience stoppages, buffers will remain empty and the throughput will always be one unit per cycle. However, stoppages do occur and mostly in the form of failures. The frequency of failures and how quickly the repairs are completed affect the throughput and the average buffer levels. Furthermore, different machines may have a different impact on the performance of the system. For instance, the throughput is a decreasing function of the failure probabilities (that is, p_i for machine i), and an increasing function of the repair probabilities (r_i), independent of the machine location (that is, the subscript i) in the line. Also, a frequently failing machine causes an inventory buildup in the upstream buffers, having the biggest impact on the immediate upstream buffer. Similarly, a frequently failing machine causes starvation in the downstream stations, resulting in empty buffers. The impact of the repair rate on the average buffer contents is opposite that of failure rates. Naturally, higher machine repair probabilities tend to reduce the content of the upstream buffers and cause a buildup in the downstream buffers. These arguments are also valid for the flow lines discussed in Chapter 4.

5.7.1 Throughput and Average Buffer Contents Versus Machine Parameters

Let us observe the behavior of a two-machine transfer line operating under BCP1 and verify the impact of p_1, p_2, r_1, and r_2 on the throughput and the average buffer level. Consider a symmetric two-machine system with $N = 5$, $p_1 = p_2 = 0.02$, and $r_1 = r_2 = 0.3$. Figures 5.6, 5.7, and 5.8 show $\bar{o}_\ell$ and $\overline{N}$ as functions of p and r. In each function, while the specific parameter p or r is allowed to vary in (0,1), the rest of the parameters are kept at the above given values. The output rate is affected by p_1 and p_2 in almost exactly the same way. Figure 5.6 shows $\bar{o}_\ell$ as a function of p_2, and as expected, $\bar{o}_\ell$ decreases as p_2 increases (w.r.t. in Figure 5.6 stands for "with respect to". It decreases with a slower rate for r_1, or for $r_2 > 0.30$.

The average buffer level $\overline{N}$ approaches zero as p_1 increases, as shown in Figure 5.7, and it is practically zero at $p_1 = 0.5$. Clearly, $\overline{N}$ (as a function of p_1) should approach zero with a slower rate for $r_1 > 0.30$, which will provide a faster recovery from failures. $\overline{N}$ increases as the repair probability r_1 increases, since the failure probability p_2 of the second machine remains unchanged. The accumulation in B as a function of r_1 will be higher for $p_2 > 0.02$.

The effects of p_2 and r_2 on $\overline{N}$ are completely opposite to those of p_1 and r_1, as shown in Figure 5.8. That is, higher values of p_2 cause accumulation in the buffer and make M_2 the bottleneck machine. Clearly, higher values of r_1 cause an even faster and larger buildup in B. A faster repair process in M_2 helps reduce the accumulation in the buffer. $\overline{N}$ as a function of r_2 will be higher for higher values of r_1 and lower for higher values of p_1.

Let us next analyze the relative impact of M_1 and M_2 on the system throughput. The right way of measuring the throughput sensitivity to the machine parameters is to develop derivative expressions. However, since this is not possible in the absence of a throughput expression as a function of p and r, let us rather observe Figure 5.9, which shows $\bar{o}_\ell$ as a function of p_1, p_2, r_1, and r_2, respectively. While each of the p and r values varies in (0,1), the rest of the parameters are kept at the

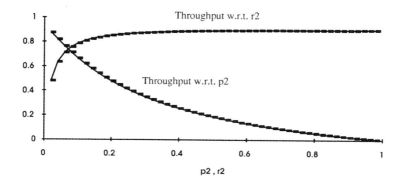

Figure 5.6. Throughput as a function of p_2 and r_2.

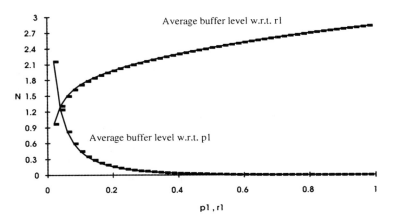

Figure 5.7. The average buffer content as a function of p_1 and r_1.

value of 0.5. It is a symmetric system, which is why all the functions cross each other at 0.50 on the parameter axis, resulting in the same throughput of 0.12437. The impact of p_1 on $\bar{o}_\ell$ ranges from almost zero at $p_1 = 0.02$ to a slightly more significant value at $p_1 = 0.35$, and it reduces thereafter. However, the impact of p_2 is significantly higher than that of p_1 until $p_2 = 0.35$, and it remains comparable thereafter.

Thus, for practical purposes, one can say that the failure process of the second machine is more effective on $\bar{o}_\ell$ than the failure process of the first machine if the repair processes on both machines are comparable. On the other hand, the repair

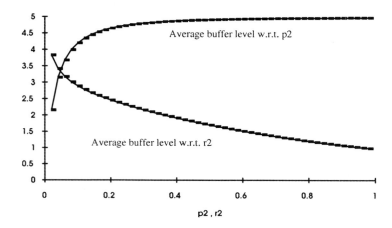

Figure 5.8. The average buffer content as a function of p_2 and r_2.

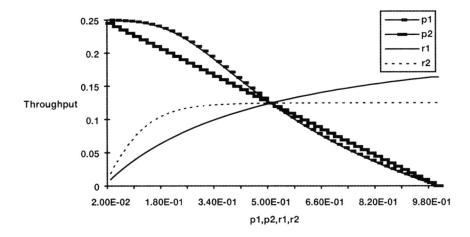

Figure 5.9. Throughput versus machine parameters.

probability of the second machine has a higher impact on $\bar{o}_\ell$ until $r_2 = 0.15$, and it reduces significantly thereafter. The repair process of the first machine has a significant impact on $\bar{o}_\ell$ for $r_1 > 0.15$. Clearly, these arguments are not conclusive; however, one may conjecture a rule of thumb that the failure processes of the machines toward the downstream end and the repair processes of the machines toward the upstream end of the line have a higher impact on the system throughput if the failure and repair processes of the machines are comparable to each other. This conjecture makes sense because the impact of failures from the upstream machines are absorbed by the downstream buffers to a certain extent, and this is not the case for the failures of the downstream machines. That is, the output is more likely to be affected by a failure occurring toward the downstream of the line rather than a failure occurring at an upstream machine. However, since the upstream end of the line is the source of the flow of units into the system, the recovery rate from the stoppages at the upstream end of the system becomes a critical issue. Again, note that these arguments are valid for symmetric or nearly symmetric lines. In cases where some of the machines have significantly different failure and/or repair probabilities, the allocation of the repair effort to machines becomes a design problem.

5.7.2 Related Design Problems

Let us next consider some of the **design problems** in transfer lines pertaining to how the best performance of the line can be achieved. The best performance of a transfer line implies the maximum throughput. As shown in Tables 5.3 and 5.4, for fixed values of machine parameters, the throughput is a nondecreasing function of the buffer capacities. This leads to the optimization problem of finding the buffer allocation, with a fixed total buffer capacity N_T (known) that maximizes the throughput:

$$\max_{N_1,\ldots,N_K} \bar{o}_\ell(N_1,\ldots,N_K) \quad \text{such that} \quad N_1 + \cdots + N_K = N_T. \tag{5.49}$$

One may also want to find the buffer allocation that minimizes the total buffer capacity with the restriction of achieving a desired throughput $\bar{o}_d$:

$$\min_{N_1,\ldots,N_K} N_1 + \ldots + N_K \quad \text{such that} \quad \bar{o}_\ell \le \bar{o}_d. \tag{5.50}$$

While both of the above optimization problems are legitimate design problems for transfer lines, allocating buffers may not be the most viable solution to make the line continue to produce in the event of stoppages. This is simply due to the fact that transfer lines usually produce with significantly high throughput rates, and buffers are filled quickly when downstream machines fail. Therefore, an effective solution to maximizing the throughput may come from the speedy recoveries from the stoppages as suggested by the conjecture made earlier in this section. The repair probability r_i for machine i is a direct consequence of the facility's emphasis on the maintenance activities. Thus, in principle, rs can be used as decision variables, and there should be a restriction on the sum of the repair probabilities due to available funding for maintenance in the facility. Then, we have

$$\max_{r_1,\ldots,r_K} \bar{o}_\ell(r_1,\ldots,r_K) \quad \text{such that} \quad r_1 + \cdots + r_K \le c. \tag{5.51}$$

Based on our earlier discussion, in the case of equal failure probabilities, (5.51) will allocate higher repair probabilities toward the upstream end of the line. In the case of unequal failure probabilities, it will allocate higher repair probabilities to those machines with higher failure probabilities.

It is conceivable that one may also want to control the repair effort to reduce the irregularities in the output of the system, which leads to the minimization of the variance of the throughput, which is another legitimate design problem. We have

$$\min_{r_1,\ldots,r_K} \bar{v}_\ell(r_1,\ldots,r_K) \quad \text{such that} \quad r_1 + \cdots + r_K \le c, \tag{5.52}$$

where $\bar{v}_\ell$ is the variance of the throughput per cycle for a given set of repair rates. Note that the failure probabilities are fixed at their given values.

The variance of the throughput can be perceived as follows: Let us assume that each unit that machine K produces yields a reward of one unit. Any of those states where M_K is operational at the beginning of a cycle will produce a unit of reward in that cycle with probability $1 - p_K$. Then, the variance of the throughput becomes the variance of the reward we obtain over a period of h cycles, averaged over h for large h. Using the concept of the variance of reward in Markov chains (see

Kemeny and Snell (1983)), one may obtain the variance of the throughput:

$$\bar{v} = \sum_{i,j \in S_p} c_{ij}(1 - p_2)^2 + \sum_{i \in S_P} \pi_i (1 - p_2) p_2, \qquad (5.53)$$

where S_p contains all the states that the line is producing a unit with probability $1 - p_2$. That is, $S_P = \{i \mid i = (\alpha_1, 1, n)\}$, and π_i is the steady-state probability of being in state i. Furthermore,

$$c_{ij} = \pi_i z_{ij} + \pi_j z_{ji} - \pi_i d_{ij} - \pi_i \pi_j,$$

where

$$d_{ij} = \begin{cases} 1 & \text{if } i = j \\ 0 & \text{otherwise,} \end{cases}$$

and z_{ij} is the (i, j)th element of the fundamental matrix[1]

$$\mathbf{Z} = (\mathbf{I} - \tilde{\mathbf{Q}} + \mathbf{A})^{-1}.$$

$\tilde{\mathbf{Q}}$ is the one-step transition probability matrix of the Markov chain, $\mathbf{A}$ is the matrix with each row being its steady-state probabilities, and $\mathbf{I}$ is the identity matrix.

Figure 5.10 shows the behavior of $\bar{\sigma}_\ell$, $\bar{v}_\ell$, and the squared coefficient of variation of the throughput with respect to r_1 while $p_1 = p_2 = 0.1$, $N = 4$, and $r_1 + r_2 = 1$ in the two-machine line. First of all, with the restriction on the total repair probabilities, the squared coefficient of variation and the variance of the throughput behave almost the same. Therefore, let us focus our attention on the relative behaviors of the variance and the throughput. In the symmetric case shown in Figure 5.10, the corresponding values of r_1 for the maximum throughput and the minimum variance are very close to each other. The same behavior is observed in the asymmetric cases of $p_1 < p_2$ and $p_1 > p_2$, as Figures 5.11 and 5.12 show. It is quite likely that this observation is also valid for a K-machine transfer line, yet still with the restriction on the total repair probabilities. Then, one can conclude here that the maximization of the throughput with respect to the repair rates also nearly minimizes the irregularities in the output in the existence of a limitation on the total repair effort. This result is also intuitive since the maximum throughput is one unit per cycle when the variance is zero.

---

[1] The sum of the differences between the n-step transition probabilities and the steady-state probabilities constitutes the fundamental matrix of the Markov chain, that is

$$\mathbf{Z} = \sum_{n=0}^{\infty} (\tilde{\mathbf{Q}}^n - \mathbf{A}) = (\mathbf{I} - \tilde{\mathbf{Q}} + \mathbf{A})^{-1}$$

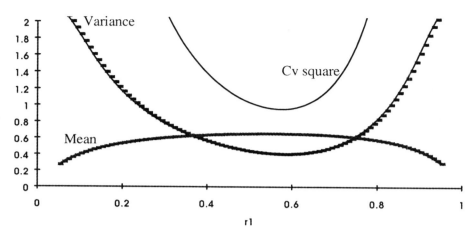

Figure 5.10. Mean, variance and Cv^2 of the throughput versus r_1 when $p_1 = p_2 = 0.1$, $N = 4$.

The ultimate design problem would probably entail the inclusion of both the buffer capacities and the repair rates as decision variables for an objective function such as profit maximization:

$$\max_{N_i, r_i, \forall i} \text{Profit} \equiv \bar{o}_\ell \cdot (\text{unit revenue}) - \sum_{i=1}^{K} \overline{N}_i h_i \quad (5.54)$$

$$\text{such that} \quad r_1 + \cdots + r_K \leq c$$

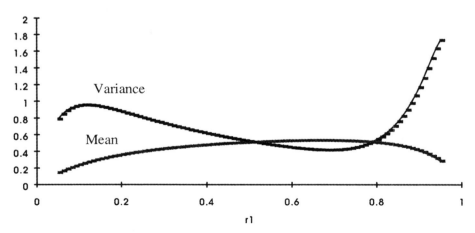

Figure 5.11. Mean and variance of the throughput when $p_1 > p_2$, $N = 4$, $p_1 = 0.2$, $p_2 = 0.1$, $r_2 = 1 - r_1$.

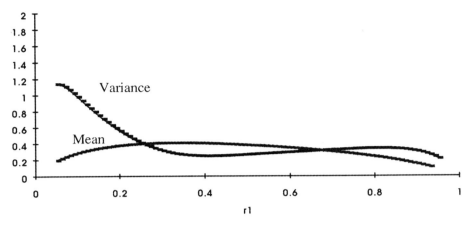

Figure 5.12. Mean and variance of the throughput when $p_1 < p_2$, $N = 10$, $p_1 = 0.1$, $p_2 = 0.35$, $r_2 = 1 - r_1$.

where $\overline{N}_i$ and h_i are the average buffer contents and the associated unit holding cost per unit time for B_i, respectively. The problem has both integer and continuous decision variables. The lack of explicit expressions for $\overline{o}_\ell$ and $\overline{N}_i$ as functions of N_i and r_i makes the problem difficult. A possible approach to generate acceptable (but not necessarily optimal) solutions is to create two problems, one strictly focusing on the buffer allocation and the other on the repair effort allocation problem. An iterative procedure solving one problem and then the other in a back-and-forth manner may prove useful in obtaining acceptable solutions. However, one still has to develop a measure for the acceptability of a solution.

5.8 Analysis of Larger Systems

As mentioned in Section 5.3, it is practically impossible to deal with the state space of the underlying Markov chain when there are more than two or three machines in the transfer line. Therefore, in this section we will attempt to introduce an approximation procedure to analyze systems with a larger number of machines. Based on our experience in Chapter 4, it is conceivable that an approximation procedure similar to the one used to analyze flow lines can be developed for transfer lines also. One such approach would be to decompose the K-machine synchronous line into a set of two-machine subsystems as shown in Figure 5.13 and analyze them in isolation. There exists a subsystem for every buffer in the transfer line, and the behavior of the buffer contents in a subsystem is expected to replicate the behavior of its counterpart in the original system. That is, for each buffer B_j, there exists a subsystem denoted by $\Omega(j)$, $j = 1, \ldots, K - 1$, where the first machine replicates the impact of the upstream machines (starvation effect), and the second machine replicates that of the downstream machines (blocking effect) on the behavior of the contents of B_j. An iterative scheme may relate these subsystems to each other by formulating the effects of blocking and starvation on each machine and may achieve consistency among the subsystems by forcing their throughputs to follow a relationship that takes scrapping into account. Next, we focus on how subsystem $\Omega(i)$ is characterized.

5.8.1 Characterization of $\Omega(i)$

Let us focus on $\Omega(i)$ and denote its machines by M_{i1} and M_{i2} and its buffer by B_i, which has the same capacity N_i as the corresponding buffer in the transfer line. Let us define $\bar{o}_{i,1}$ as the output rate of the first machine in $\Omega(i)$, and $\bar{o}_{i,2}$ as that of the second machine. We will use the following notation in this section:

$P_i(\alpha_1, \alpha_2, n)$ = steady-state probability distribution of the Markov chain representing $\Omega(i)$, as described in Section 5.5;
$P_{ij}(\text{u})$ = steady-state probability that the jth machine is up at the beginning of a cycle in $\Omega(i)$, $j = 1, 2$;
$P_{ij}(\text{d})$ = steady-state probability that the jth machine is down at the beginning of a cycle in $\Omega(i)$, $j = 1, 2$;
$P_{ij}(\text{id})$ = steady-state probability that the jth machine is idle at the beginning of a cycle in $\Omega(i)$, $j = 2$;
$P_{ij}(\text{b})$ = steady-state probability that the jth machine is blocked at the beginning of a cycle in $\Omega(i)$, $j = 1$;
$\bar{N}_j$ = the expected number of jobs in buffer j.

M_{i1} and M_{i2} also experience failures and go through their associated repair processes. Since there are no upstream machines to machine M_{i1} in $\Omega(i)$, the failure of M_{i1} represents either the actual failure of M_i or the idleness of M_i due to a lack of jobs in B_{i-1} in the original system. Consequently, the repair of M_{i1} is either the repair of an actual failure of M_i or the termination of the idleness of M_i. Let us denote the parameters of M_{i1} by $p'_i(t)$ and $r'_i(t)$ and the parameters of M_{i2} by $p''_i(t)$ and $r''_i(t)$, respectively. Note that $p'_i(t)$ is the probability that M_{i1} fails during the cycle starting at epoch t. Then, we have

$$p'_i(t) = \Pr(M_{i1} \text{ is down at epoch } t+1 \mid \text{It was processing a job at epoch } t)$$
$$= \Pr(M_i \text{ is down at epoch } t+1 \mid \text{It was processing a job at epoch } t)$$
$$+ \Pr(M_i \text{ is idle at epoch } t+1 \mid \text{It was processing a job at epoch } t).$$

Notice that $M_{i-1,2}$ and M_{i1} both represent machine M_i in the transfer line. Therefore, if M_i is under repair, $M_{i-1,2}$ should not be idle (starving), or vice versa. Hence, the events that M_i is under repair at epoch t and $M_{i-1,2}$ is idle at epoch t are mutually exclusive. Also, $M_{i-1,2}$ and M_{i1} should process jobs whenever

5.8. Analysis of Larger Systems

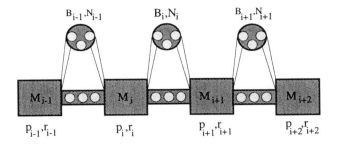

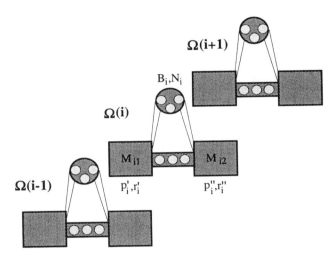

Figure 5.13. Decomposition of a transfer line.

M_i processes a job. These arguments show that the above expression for p'_i is equivalent to the following:

$p'_i = \Pr(M_i$ is under repair at epoch $t+1 \mid M_i$ was processing a job at epoch $t)$
$+ \Pr(M_{i-1,2}$ is idle at epoch $t+1 \mid M_{i-1,2}$ was processing
a job at epoch $t)$.
(5.55)

The first term in the above expression is simply p_i of M_i, and the second term can be obtained from the analysis of $\Omega(i-1)$. Then, in the long run, under both buffer clearing policies (BCP1 and BCP2), we can write

$$p'_i = p_i + \frac{[P_{i-1}(1,1,0)p'_{i-1}(1-p''_{i-1}) + P_{i-1}(0,1,0)(1-p''_{i-1})]}{\sum_{\alpha_1,n} P_{i-1}(\alpha_1,1,n)}. \quad (5.56)$$

220   5. Analysis of Transfer Lines

Notice that the only states where $M_{i-1,2}$ becomes idle at $t+1$ while processing at t are $(1,1,0)$ and $(0,1,0)$. In the second term in the numerator of (5.56), we are not concerned with whether or not the first machine is repaired. This is because, once in $(0,1,0)$, $M_{i-1,2}$ will be idle regardless of the state of $M_{i-1,1}$ at the next epoch.

Using a similar argument, the repair probability r'_i of M_{i1} can be expressed as follows:

$$r'_i = \Pr(M_{i1} \text{ is up at epoch } t+1 \mid M_{i1} \text{ was not operational at epoch } t).$$

M_{i1} becomes up when M_i becomes operational after a repair or recovers from starvation. One way of achieving consistency among the throughputs of $\Omega(i-1)$ and $\Omega(i)$ is to incorporate them into the approximation method. First of all, one can extend the proposition of Section 5.5.1 to an M-station line and write

$$\bar{o}_\ell = \bar{o}_1 (1-p_2)(1-p_3) \cdots (1-p_M). \tag{5.57}$$

which is merely the flow-conservation principle. The job on machine M_i is discarded at the end of a cycle with probability p_i, yielding $\bar{o}_i = \bar{o}_{i-1}(1-p_i)$, which leads to (5.57). Since $\bar{o}_i$ is approximated by $\bar{o}_{i,1}$, using the above flow conservation, we can write

$$\bar{o}_i \cong \bar{o}_{i-1,1}(1-p_i). \tag{5.58}$$

On the other hand, the proposition in Section 5.5.1 provides $\bar{o}_{i,1} = P_{i1}(u)(1-p'_i)$, resulting in

$$\bar{o}_i \cong P_{i1}(u)(1-p'_i). \tag{5.59}$$

From (5.58) and (5.59), we have

$$P_{i1}(u) = \frac{\bar{o}_{i-1,1}(1-p_i)}{(1-p'_i)}. \tag{5.60}$$

Also, for M_{i1}, we have

$$P_{i1}(d) = 1 - P_{i1}(u) - P_{i1}(b). \tag{5.61}$$

Notice that we can find $P_{i1}(u)$ using (5.60) from the analysis of subsystem $\Omega(i-1)$. Then, $P_{i1}(u)$ and the most recent value for $P_{i1}(b)$ can be used to obtain $P_{i1}(d)$. Given these two probabilities, we have the following relationship for M_{i1} (to be used as a means to obtain r'_i):

$$\Pr(M_{i1} \text{ is up} \mid \text{It is not blocked}) = \frac{P_{i1}(u)}{P_{i1}(u) + P_{i1}(d)} = \frac{r'_i}{r'_i + p'_i},$$

which results in

$$r_i' = \frac{P_{i1}(u)}{P_{i1}(d)} p_i'. \tag{5.62}$$

It is fortunate that we are able to incorporate the impact of the flow-conservation principle, that is, the impact of the throughput through (5.60), into the preceding simple formula for r_i'. Machine M_{i2} of $\Omega(i)$ has the parameters p_i'' and r_i'' to be defined. The failure of M_{i2} represents either an actual failure of M_{i+1} or the blocking of it. Since M_{i2} of $\Omega(i)$ does not experience blocking, the periods during which machine M_{i+1} in the original line is under repair or is blocked are the down periods of M_{i2}. It leaves its down state when M_{i+1} recovers from a failure or from the blocked state in the original line. Then, an expression similar to (5.56) can be written for p_i'':

$$p_i'' = \Pr(M_{i+1} \text{ is under repair at epoch } t + 1 \mid \text{It was processing}$$
$$\text{a job at epoch } t)$$
$$+ \Pr(M_{i+1,1} \text{ is blocked at epoch } t + 1 \mid \text{It was processing}$$
$$\text{a job at epoch } t). \tag{5.63}$$

The first term in the above expression is simply p_{i+1} of M_{i+1}, and the second term can be obtained from the analysis of $\Omega(i + 1)$. Notice that the blocking of $M_{i+1,1}$ in $\Omega(i + 1)$ represents the blocking effect of the downstream machines on M_{i+1}. Then, in the long run, under the two different buffer clearing policies, the expressions for p_i'' are as follows: Under BCP1,

$$p_i'' = p_{i+1}$$
$$+ \frac{[P_{i+1}(1, 0, N_{i+1})(1 - p_{i+1}')(1 - r_{i+1}'') + P_{i+1}(1, 1, N_{i+1})(1 - p_{i+1}')p_{i+1}'']}{\sum_{\alpha_2, n} P_{i+1}(1, \alpha_2, n)}. \tag{5.64}$$

Under BCP2, we have

$$p_i'' = p_{i+1}$$
$$+ \frac{[(1 - p_{i+1}')r_{i+1}'' \sum_{n=1}^{N_{i+1}-2} P_{i+1}(1, 0, n) + (1 - p_{i+1}')P_{i+1}(1, 0, N_{i+1} - 1)]}{\sum_{\alpha_2;n} P_{i+1}(1, \alpha_2, n)}. \tag{5.65}$$

In both of the preceding expressions, the first term is the failure probability of M_{i+1} within a cycle and the second set of terms is the probability of M_{i+1} being blocked in a cycle given it was operating in the previous cycle. Since $M_{i+1,1}$ approximately replicates the behavior of M_{i+1}, the second set of terms in (5.64) and (5.65) is obtained from the analysis of $\Omega(i + 1)$.

222   5. Analysis of Transfer Lines

The repair probability r_i'' of $M_{i,2}$ includes the probability of M_{i+1} being actually repaired or recovering from blocking. That is,

$$r_i'' = \Pr(M_{i,2} \text{ is operational at epoch } t+1 \mid \text{It was under repair at epoch } t),$$

which can be rewritten as

$$r_i'' = \Pr[M_{i+1} \text{ is processing a work piece at epoch } t+1 \mid ((M_{i+1} \text{ was under repair at epoch } t) \cup (M_{i+1} \text{ was blocked at epoch } t))].$$

Since the blocking of M_{i+1} is represented by the blocking of $M_{i+1,1}$ of $\Omega(i+1)$, and the events that M_{i+1} is under repair and M_{i+1} is blocked cannot happen simultaneously at a given epoch, we have

$$r_i'' = \frac{\{\Pr(M_{i+1} \text{ is processing at } t+1 \mid M_{i+1} \text{ is under repair at } t) \cdot \Pr(M_{i+1} \text{ is under repair at } t) + \Pr(M_{i+1} \text{ is processing at } t+1 \mid M_{i+1} \text{ is blocked at } t) \cdot \Pr(M_{i+1} \text{ is blocked at } t)\}}{\Pr(M_{i+1} \text{ is under repair at } t) + \Pr(M_{i+1} \text{ is blocked at } t)}$$

$$= \{r_{i+1} \Pr(M_{i+1} \text{ is under repair at an epoch}) + \Pr(\text{status of } M_{i+1,1} \text{ changes from blocked (b) to operating (1) in a cycle}) \cdot \Pr(M_{i+1,1} \text{ is blocked at an epoch})\} /$$

$$\{\Pr(M_{i+1} \text{ is under repair at an epoch}) + \Pr(M_{i+1,1} \text{ is blocked at an epoch}).$$
(5.66)

Here, the probability $\Pr(M_{i+1}$ is under repair at an epoch) is simply $P_{i+1}(d)$. Since the blocking of M_{i+1} is approximately represented by the blocking of $M_{i+1,1}$ in $\Omega(i+1)$, we have

$$r_i'' = \frac{\{r_{i+1} P_{i+1}(d) + \Pr(\text{Status of } M_{i+1,1} \text{ changes from blocked (b) to processing (1))} \cdot \Pr(M_{i+1,1} \text{ is blocked at an epoch})\}}{P_{i+1}(d) + \Pr(M_{i+1,1} \text{ is blocked at an epoch})}.$$

Then, the expressions for r_i'' under the two operating policies are as follows: Under BCP1,

$$r_i'' = \frac{r_{i+1} P_{i+1}(d) + r_{i+1}'' P_{i+1}(b, 0, N_{i+1})}{P_{i+1}(d) + P_{i+1}(b, 0, N_{i+1})}. \tag{5.67}$$

Under BCP2,

$$r_i'' = \frac{r_{i+1} P_{i+1}(d) + r_{i+1}'' P_{i+1}(b, 0, 1) + (1 - p_{i+1}'') P_{i+1}(b, 1, 1)}{P_{i+1}(d) + \sum_{\alpha_2, n} P_{i+1}(b, \alpha_2, n)}. \tag{5.68}$$

In general, p_i' and r_i' are obtained using p_i and r_i, respectively, together with the required steady-state probabilities of $\Omega(i-1)$. Also, p_i'' and r_i'' are obtained by making use of the parameters p_i and r_i together with the steady-state probabilities of $\Omega(i+1)$. Notice that $p'_1 = p_1, r'_1 = r_1$ and $p''_{K-1} = p_K, r''_{K-1} = r_K$.

The preceding expressions for the parameters of $\Omega(i)$ give rise to a fixed-point iteration algorithm that analyzes Ωs in isolation and relates them to each other by passing information to obtain the relevant system parameters. The algorithm analyzes $\Omega(1), \ldots, \Omega(K-2)$ in each forward pass and obtains improved values for p'_i and r'_i for $i = 2, \ldots, K-1$. It also executes a backward pass from $\Omega(K-1)$ to $\Omega(2)$ to obtain the improved values of p''_i and r''_i for $i = K-2, \ldots, 1$. The algorithm executes a series of forward and backward passes until a flow-conservation criteria that takes into account the number of jobs discarded is satisfied. The down-time probability, $P_i(d)$, for each i, is obtained using the following expression:

$$P_i(d) = 1 - P_i(id) - P_i(b) - P_i(u). \tag{5.69}$$

5.8.2 The Decomposition Algorithm

For a K-machine transfer line with parameters p_i and r_i, $i = 1, \ldots, K$, and N_j, $j = 1, \ldots, K-1$, the fixed-point iteration algorithm is summarized in Table 5.5. Notice that in step (iv) of the forward pass, we use the most recent value of $P_{i+1}(b)$ to obtain $P_{i+1}(d)$. Since the idleness of M_{i+1} is approximated by the idleness of M_{i2}, we have $P_{i+1}(id) = P_i(1, i, 0) + P_i(0, i, 0)$ from the analysis of $\Omega(i)$. By the proposition in Section 5.5.1, we also have $P_{i2}(u) = \bar{o}_{i,1}$, which in turn approximates $P_{i+1}(u)$. As a result, we have the probabilities $P_{i+1}(id)$ and $P_{i+1}(u)$ for M_{i+1} from the analysis of $\Omega(i)$. In each of the forward and backward passes, we obtain the blocking probability of M_i, $P_i(b)$, for $i = 1, 2, \ldots, K-1$. Thus, in step (iv) of the forward pass, we obtain $P_{i+1}(d)$ using the information from $\Omega(i)$ and the most recent value of $P_{i+1}(b)$.

The down-time probabilities can be obtained in the following manner:

$$P_1(d) \cong \sum_{\alpha_2, n} P_1(0, \alpha_2, n),$$

$$P_K(d) \cong \sum_{\alpha_1, n} P_{K-1}(\alpha_1, 0, n), \quad \text{and}$$

$$P_{i+1}(d) \cong 1 - \{P_i(1, i, 0) + P_i(0, i, 0)\} - P_{i+1}(b, 0, N_{i+1}) - \bar{o}_{i,1}$$

(under BCP1),

$$P_{i+1}(d) \cong 1 - \{P_i(1, i, 0) + P_i(0, i, 0)\} - \sum_n P_{i+1}(b, 0, n) - \bar{o}_{i,1}$$

(under BCP2), for $i = 2, 3, \ldots, M - 1$.

Finally, the probability distribution of the buffer contents is approximated by

$$P_i(n_i) \cong \sum_{\alpha_1, \alpha_2} P_i(\alpha_1, \alpha_2, n_i).$$

TABLE 5.5. The fixed-point iteration algorithm for a K-machine transfer line

INITIALIZATION
Set $k = 1$, $p'_i = p_i$, $r'_i = r_i$, and $p''_i = p_{i+1}$, $r''_i = r_{i+1}$, $i = 1, \ldots, K-1$.
Set $P_i(d) = 0$, $P_i(b) = 0$, $i = 1, 2, \ldots, K$.

ITERATION k

FORWARD PASS: $i = 1, \ldots, K-2$
- (i) Analyze $\Omega(i)$ to obtain $P_i(\alpha_1, \alpha_2, n)$, $\bar{o}_{i,1}$.
- (ii) If $i = 1$ then $P_1(d) \cong \sum_{\alpha_2, n} P_1(0, \alpha_2, n)$, else go to (iii).
- (iii) $P_i(b) \cong P_i(b, 0, N_i)$ (under BCP1), $P_i(b) = \sum_n P_i(b, 0, n)$ (under BCP2).
- (iv) $P_{i+1}(d) \cong 1 - \{P_i(1, i, 0) + P_i(0, i, 0)\} - P_{i+1}(b) - \bar{o}_{i,1}$
- (v) Obtain p'_{i+1} using (5.56).
- (vi) $P_{i+1,1}(u) = \bar{o}_{i,1}(1 - p_{i+1})/(1 - p'_{i+1})$, $P_{i+1,1}(d) = 1 - P_{i+1,1}(u) - P_{i+1}(b)$.
- (vii) Obtain r'_{i+1} using (5.62).

BACKWARD PASS: $i = K-1, \ldots, 2$
- (i) Analyze $\Omega(i)$ to obtain $P_i(\alpha_1, \alpha_2, n)$, $\bar{o}_{i,1}$ and $\bar{o}_{i,2}$.
- (ii) If $i = K-1$, then $P_M(d) \cong \sum_{\alpha_1, n} P_{M-1}(\alpha_1, 0, n)$, else go to (iii).
- (iii) $P_i(b) \cong P_i(b, 0, N_i)$ (under BCP1), $P_i(b) \cong \sum_n P_i(b, 0, n)$ (under BCP2)
- (iv) Obtain p''_{i-1} and r''_{i-1}, using (5.64) and (5.67) under BCP1, or (5.65) and (5.68) under BCP2.
- (v) Let $\bar{o}_\ell^{(k)} = \bar{o}_{M-1,2}$.

STOPPING CRITERIA: If $|\bar{o}_\ell^{(k)} - \bar{o}_\ell^{(k-1)}| < \varepsilon$, Stop.
Otherwise, $k \leftarrow k+1$, and execute iteration k.

EXAMPLE 5.4. Let us focus on the behavior of the algorithm in a three-machine transfer line operating under the BCP1 policy. There are two buffers and consequently two subsystems, namely Ω_1 and Ω_2. Assume the following machine parameters: $p_1 = 0.1$, $p_2 = 0.2$, $p_3 = 0.2$, $r_1 = r_2 = r_3 = 0.3$, and $N_1 = N_2 = 4$. Figure 5.14 shows the parameters of the subsystems, and Table 5.6 shows the values of p''_1, r''_1, p'_2, and r'_2 (parameters related to M_2 in Ω_1 and Ω_2) in the first five iterations of the algorithm, where an iteration is equivalent to a forward and then a backward pass.

Let us observe the behavior of the parameters of machine 2 in Ω_1 and Ω_2. First of all, the convergence is fast. We normally expect p'_i and p''_i to be greater than p_i and r'_i and r''_i to be less than r_i. In case we have $r_{i+1} = r''_{i+1}$ for any i, then we have $r''_i = r_{i+1}$ due to (5.56), as is the case for r''_1 in this example. The approximate steady-state probabilities of idleness and being blocked for machine 2 are 0.0971 and

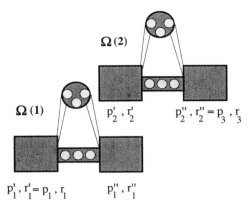

Figure 5.14. Decomposition of the three-machine line.

0.0617, respectively. Since these two probabilities are not significantly different, p_1'' and p_2' are quite close to each other.

Let us change the values of the parameter set to $p_1 = 0.05$, $p_2 = 0.1$, $p_3 = 0.15$, and $r_1 = 0.2$, $r_2 = 0.25$, $r_3 = 0.3$, which will yield a higher blocking probability (0.1026) and a slightly lower probability of idleness (0.0884). The immediate consequence of this is the increased absolute difference between p_1'' and p_2', the two parameters representing the failure probability for the same machine in the two subsystems. The resulting parameter values for Ω_1 and Ω_2 are given in Table 5.7. The relative difference between p_1'' and p_2' has increased due to the higher blocking experienced by M_2. On the other hand, r_1'' and r_2 are quite close to each other, since $P_2(\text{id})$ is insignificant. In the next few examples, we look at the accuracy of the approximation.

TABLE 5.6. Convergence of the decomposition method in a three-machine line with $p_1 = 0.1$, $p_2 = 0.2$, $p_3 = 0.2$, $r_1 = r_2 = r_3 = 0.3$, and $N_1 = N_2 = 4$

Iteration	Ω_1		Ω_2	
	p_1''	r_1''	p_2'	r_2'
0	0.2	0.3		
1	0.233379	0.3	0.238876	0.286977
2	0.235494	0.3	0.233359	0.290137
3	0.235574	0.3	0.233103	0.290270
4	0.235577	0.3	0.233091	0.290277
5	0.235577	0.3	0.233091	0.290277

TABLE 5.7. Convergence of the decomposition method in a three-machine line with $p_1 = 0.05$, $p_2 = 0.1$, $p_3 = 0.15$, and $r_1 = 0.2$, $r_2 = 0.25$, $r_3 = 0.3$

	Ω_1		Ω_2	
Iteration	p_1''	r_1''	p_2'	r_2'
0	0.1	0.25	0.1	0.25
1	0.149455	0.261312	0.128471	0.227878
2	0.152040	0.262460	0.123278	0.235871
3	0.152147	0.262506	0.123060	0.236130
4	0.152151	0.262510	0.123051	0.236140
5	0.152151	0.262510	0.123051	0.236140

The purpose of numerical examples in a study involving approximations is to identify the impact of important system parameters on the accuracy of the approximation. This effort requires testing the approximation against exact (if available) numerical or simulation results in a number of examples. Let us consider the following examples for illustrative purposes.

EXAMPLE 5.5. Consider the three-machine line studied in Example 5.4. Again, the parameters of the line are $p_1 = 0.1$, $p_2 = 0.2$, $p_3 = 0.2$, $r_1 = r_2 = r_3 = 0.3$, and $N_1 = N_2 = 4$. The approximate and the simulation values of $\bar{o}_\ell$, $\overline{N}_k$, and $P_k(d)$ for $k = 1, 2, 3$ are given in Table 5.8. The simulation results are based on one million job departures from the line. The approximate results are highly accurate, especially when the machine parameters are comparable to each other. The throughput and the down-time probabilities are in general highly accurate.

EXAMPLE 5.6. Consider a five-machine transfer line with the following parameters:

TABLE 5.8. Approximation versus simulation results for Example 5.5

	Approximation	Simulation
$\bar{o}_\ell$	0.32876	0.32708(±0.0013)
$\overline{N}_1$	2.71070	2.71056(±0.0004)
$\overline{N}_2$	1.15653	1.15578(±0.0005)
$P_1(d)$	0.19056	0.19142(±0.0015)
$P_2(d)$	0.34230	0.34189(±0.0022)
$P_3(d)$	0.26607	0.27342(±0.0025)

5.9. Transfer Lines with No Scrapping

	Machines				
	1	2	3	4	5
p_i	0.05	0.03	0.01	0.005	0.001
r_i	0.2	0.2	0.2	0.2	0.2
N_i	4	4	4	4	4

The approximation and simulation results for both the BCP1 and BCP2 policies are given in Table 5.9. Again, the simulation results are based on one million job departures from the line. Observe the downward shift in both the throughput and the average buffer levels from BCP1 to BCP2. The throughput results are again highly accurate (less than 2 percent relative error). However, in this type of approximation, it is possible to have relative error figures larger than 10 percent in the average buffer contents, as in the case of $\overline{N}_3$ in this example under BCP1. The reason is that, in general, the throughput is less sensitive to the approximating assumptions used in the decomposition approximations than the average buffer contents. The accuracy tends to decrease in systems with significant differences in ps and rs from one machine to another.

EXAMPLE 5.7. Finally, consider the following two lines:
(1) an 8-machine line with $p_i = 0.01$, $r_i = 0.2$, and buffer capacities of 10, 15, 15, 10, 10, 15, 15;
(2) a 50-machine line with $p_i = 0.01$, $r_i = 0.3$, and $N_i = 5$, for all i.

A comparison of the simulated and the approximate throughputs is given in Table 5.10 for both BCP1 and BCP2. Notice the very low throughput in the 50-machine transfer line. The throughput reaches an asymptotic value as the number of machines increases.

TABLE 5.9. Results of Example 5.6

BCP1	$\overline{o}$	$\overline{N}_1$	$\overline{N}_2$	$\overline{N}_3$	$\overline{N}_4$
Approximation	0.67831	1.38613	0.42053	0.18374	0.03415
Simulation	0.67696	1.38568	0.43500	0.20779	0.03630
(95% CI)	(+0.0037)	(+0.0229)	(+0.0101)	(+0.0114)	(+0.0034)

BCP2	$\overline{o}$	$\overline{N}_1$	$\overline{N}_2$	$\overline{N}_3$	$\overline{N}_4$
Approximation	0.63933	0.44360	0.14517	0.05834	0.01021
Simulation	0.63658	0.46293	0.15290	0.05969	0.01028
(95% CI)	(+0.0009)	(+0.0073)	(+0.0054)	(+0.0042)	(+0.0015)

TABLE 5.10. Approximation and simulation values for the throughput in Example 5.7

	Throughput			
	BCP1		BCP2	
	Simulation (95% CI)	Approximation	Simulation (95% CI)	Approximation
8-m/c line	0.84727 (+0.0017)	0.85132 (+0.0016)	0.72133	0.74329
50-m/c line	0.00339 (+0.0001)	0.00349 (+0.0001)	0.00273	0.00297

5.9 Transfer Lines with No Scrapping

Let us next consider the case where the interrupted unit is not scrapped but rather is kept on the machine during repair and departs from the machine as a good unit at the end of the repair. Let us assume that failures are operation-dependent and that no more than one event may occur in a cycle. The blocking mechanism is still the BAP type. Let us observe the line at the end of each cycle and just before the transfer of units takes place (Buzacott's observations; see Buzacott and Shanthikumar (1993); as opposed to observing after the transfer of units, as we have done in Section 5.5). For a two-machine line, $\{\alpha_1(t), \alpha_2(t), N(t)\}$ again constitutes a Markov chain, which is slightly different than what we have in Section 5.5. Here, $\alpha_i(t)$ is again the state of machine i and $N(t)$ is the number of jobs in the buffer. However, $N(t)$ includes the job on M_2. A typical flow-balance equation for the underlying Markov chain is

$$P(1, 1, n) = (1 - p_1)(1 - p_2)P(1, 1, n) + (1 - p_1)r_2 P(1, 0, n - 1) + r_1(1 - p_2)P(0, 1, n + 1). \tag{5.70}$$

The throughput of the line is given by

$$\overline{o}_\ell = \begin{cases} \dfrac{1 - u\sigma^N}{(1 + x_1) - (1 + x_2)u\sigma^N} & \text{if } u \neq 1, \\[2mm] \dfrac{r_1[1 + q - q(p_1 + r_1)] + Nq(p_1 + r_1)}{(2p_1 + r_1)[1 + q - q(p_1 + r_1)] + Nq(p_1 + r_1)^2} & \text{if } u = 1, \end{cases} \tag{5.71}$$

where $x_i = p_i/r_i$, $u = x_2/x_1$, $q = p_2/p_1$, and σ is

$$\sigma = \frac{(p_1 + p_2)(r_1 + r_2) - p_1 r_2 (p_1 + p_2 + r_1 + r_2)}{(p_1 + p_2)(r_1 + r_2) - p_2 r_1 (p_1 + p_2 + r_1 + r_2)}.$$

If $r_1 = r_2 = r$ and $p_1 = p_2 = p$, then the throughput simplifies to

$$\bar{o}_\ell = \frac{2r + (p+r)(N-r)}{(2p+r)(2-r-p) + N(p+r)^2}. \tag{5.72}$$

If more than one event occurs in a cycle, the throughput for the two-machine line (identical machines) is given by

$$\bar{o}_\ell = \frac{2 - r + N(r+p)}{2(1 + 2p/r) + r(N-1)(1 + p/r)^2}, \tag{5.73}$$

for nonidentical stages, the throughput of a two-machine line is

$$\bar{o}_\ell = \begin{cases} \dfrac{1 - u\Phi\Lambda^N}{1 + x_1 - (1 + x_2)u\Phi\Lambda^N} & \text{if } u \neq 1, \\[1em] \dfrac{1 + q - r_2 + Nr_2(1 + x)}{(1 + q - r_2)(1 + 2x) - r_2 x^2 + Nr_2(1 + x)^2}, & \text{if } u = 1, \end{cases} \tag{5.74}$$

where

$$\Phi = \frac{r_1 + r_2 - r_1 r_2 - p_1 r_2}{r_1 + r_2 - r_1 r_2 - p_2 r_1} \quad \text{and} \quad \Lambda = \Phi \frac{p_1 + p_2 - p_1 p_2 - p_1 r_2}{p_1 + p_2 - p_1 p_2 - p_2 r_1}.$$

5.9.1 Explicit Probabilities for Two-Machine Lines

Under certain assumptions, it is also possible to obtain explicit expressions for the steady-state probabilities of the Markov chain. For instance, let us assume that the interrupted unit is transferred back to the immediate upstream buffer at the end of the cycle during which the failure occurs. The unit will be reprocessed after the machine is repaired. Also assume that the blocking mechanism is the BBP (blocking before processing) type. The immediate consequence of the above assumptions is that $(0, 0, 0)$, $(1, 0, 0)$, $(1, 1, 0)$, $(1, 0, 1)$, $(0, 1, N-1)$, $(0, 0, N)$, $(0, 1, N)$, and $(1, 1, N)$ are transient states since they cannot be reached from any other state. The buffer becomes full only in state $(1, 0, N)$, during which the first machine is blocked. The following are two typical flow-balance equations of the underlying Markov chain:

$$P(1, 1, n) = r_1 r_2 P(0, 0, n) + r_1(1 - p_2)P(0, 1, n) + (1 - p_1)r_2 P(1, 0, n)$$
$$+ (1 - p_1)(1 - p_2)P(1, 1, n) \tag{5.75}$$

and

$$P(0, 1, n) = (1 - r_1)r_2 P(0, 0, n+1) + (1 - r_1)(1 - p_2)P(0, 1, n+1)$$
$$+ p_1 r_2 P(1, 0, n+1) + p_1(1 - p_2)P(1, 1, n) \tag{5.76}$$

Note that it is possible to have more than one event to occur simultaneously. The following equations are the explicit steady-state probabilities obtained by guessing a type of solution to the set of flow-balance equations of the Markov chain (Gershwin and Schick (1983)):

$$P(0, 1, 0) = cx \frac{r_1 + r_2 - r_1 r_2 - r_1 p_2}{r_1 p_2},$$

$$P(0, 0, 1) = cx,$$

$$P(0, 1, 1) = cxy_2,$$

$$P(1, 1, 1) = \frac{cx}{p_2} \frac{r_1 + r_2 - r_1 r_2 - r_1 p_2}{p_1 + p_2 - p_1 p_2 - r_1 p_2},$$

$$P(\alpha_1, \alpha_2, n) = cx^n y_1^{\alpha_1} y_2^{\alpha_2}, \quad n = 2, \ldots, N-2, \quad \alpha_i \in [0, 1], \quad (5.77)$$

$$P(0, 0, N-1) = cx^{N-1},$$

$$P(1, 0, N-1) = cx^{N-1} y_1,$$

$$P(1, 1, N-1) = \frac{cx^{N-1}}{p_1} \frac{r_1 + r_2 - r_1 r_2 - r_2 p_1}{p_1 + p_2 - p_1 p_2 - r_2 p_1},$$

$$P(1, 0, N) = cx^{N-1} \frac{r_1 + r_2 - r_1 r_2 - r_2 p_1}{r_2 p_1},$$

where

$$x = \frac{y_2}{y_1}, \quad \text{with} \quad y_1 = \frac{r_1 + r_2 - r_1 r_2 - r_1 p_2}{p_1 + p_2 - p_1 p_2 - r_2 p_1}$$

$$\text{and} \quad y_2 = \frac{r_1 + r_2 - r_1 r_2 - r_2 p_1}{p_1 + p_2 - p_1 p_2 - r_1 p_2}.$$

The throughput of the above two-machine line is given by

$$\overline{o}_\ell = \sum_{\alpha_1} \sum_{n>0} P(\alpha_1, 1, n) = \overline{o}_{GS}(N).$$

The throughput of the above two-machine line approaches $\min(e_1, e_2)$ as $N \to \infty$. For a finite N, its throughput multiplied by $(1 - p_1)(1 - p_2)$ will approximate the throughput of the two-machine line with a buffer capacity of $N - 1$, with scrapping and operated under the BCP1 rule, if $e_1 < e_2$. This is not true for $e_1 > e_2$ since the line studied earlier is not reversible due to scrapping. Table 5.11 illustrates this argument by comparing the throughputs of the two lines. The equality of the throughput in the original and the reversed lines for the no-scrap case is apparent from Examples 1 and 2, 4 and 5, and 6 and 7. Also, observe that in those cases where $e_2 \geq e_1$, $\overline{o}_{GS}(N)(1 - p_1)(1 - p_2)$ gives quite close results to $\overline{o}_{BCP1}(N - 1)$. This is not true for $e_1 > e_2$, since $\overline{o}_{BCP1}(N - 1)$ increases for

TABLE 5.11. The use of GS model for the case with scrapping

	p_1, p_2, r_1, r_2, N	e_1	e_2	$\bar{o}_{GS}(N)$	$\bar{o}_{GS}(N)(1-p_1)(1-p_2)$	$\bar{o}_{BCP1}(N-1)$
1	0.05,.02,.2,.2,5	0.8000	0.90910	0.76440	0.71170	0.7190
2	0.02,.05,.2,.2,5	0.9091	0.80000	0.76440	0.71170	0.7219
3	0.001,.001,.1,.1,4	0.9901	0.99010	0.98130	0.97940	0.9798
4	0.001,.001,.2,.1,4	0.9950	0.99010	0.98610	0.98410	0.9846
5	0.001,.001,.1,.2,4	0.9901	0.99500	0.98610	0.98410	0.9846
6	0.2,.1,.3,.3,∞	0.6000	0.75000	0.60000	0.43200	0.4320
7	0.1,.2,.3,.3,∞	0.7500	0.60000	0.60000	0.43200	0.4800
8	0.2,.1,.3,.3,11	0.6000	0.75000	0.58930	0.42430	0.4299
9	0.2,.1,.3,.3,6	0.6000	0.75000	0.56850	0.40930	0.4221
10	0.1,.1,.2,.2,125	0.6667	0.66670	0.65710	0.53220	0.5399

increased e_1. This also indicates that the transfer line with scrapping policy is not reversible. That is, it does not preserve its throughput when reversed.

5.9.2 Decomposition of Lines with Multiple Machines and No Scrapping

Let us go back to the decomposition scheme sketched in Figure 5.13. First of all, one can modify the algorithm presented in Section 5.8.2 to approximate the performance measures of the transfer lines with no scrapping. However, for the purpose of introducing other arguments about transfer lines, let us focus on a slightly different procedure to decompose systems with no scrapping. We will adhere to our earlier notation in Figure 5.13 with the addition of $\bar{o}(i)$ representing the throughput of $\Omega(i)$. We assume operation-dependent failures, the possibility of multiple events in a cycle, and BBP-type blocking. In systems where units are not scrapped upon failures, we have

$$\bar{o}_1 = \bar{o}_2 = \cdots = \bar{o}_K = \bar{o}_\ell, \tag{5.78}$$

for machines $i = 1, \ldots, K$. Also, when the decomposition method converges, we have

$$\bar{o}(1) \cong \bar{o}(2) \cong \cdots \cong \bar{o}(K-1) \cong \bar{o}_\ell, \tag{5.79}$$

for subsystems $\Omega(1), \ldots, \Omega(K-1)$. Using the isolated efficiency of machine i, that is, $e_i = r_i/(r_i + p_i)$, we can write

$$\bar{o}_i = e_i[1 - P_{i-1}(0) - P_i(N_i)], \tag{5.80}$$

or from $\Omega(i-1)$, we have

$$\bar{o}_i \cong \bar{o}(i-1) = \frac{r''_{i-1}}{r''_{i-1} + p''_{i-1}}[1 - P_{i-1}(0)]. \tag{5.81}$$

Also, from $\Omega(i)$,

$$\bar{o}_i \cong \bar{o}(i) = \frac{r'_i}{r'_i + p'_i}[1 - P_i(N)]. \tag{5.82}$$

Combining (5.78)–(5.82), we obtain

$$\frac{p''_{i-1}}{r''_{i-1}} + \frac{p'_i}{r'_i} = \frac{1}{\bar{o}_i} + \frac{1}{e_i} - 2, \quad i = 2, \ldots, K-1. \tag{5.83}$$

Let us next revisit the definition of r'_i. M_{i1} becomes up when M_i is repaired or when its state changes from idle to busy. Then, we may write the following:

$$r'_i = \Delta_{i-1} r'_{i-1} (1 - \Delta_{i-1}) r_i, \tag{5.84}$$

where

$$\Delta_{i-1} = \Pr(n_{i-1} = 0, n_i < N \text{ in cycle } t \mid n_{i-1} = 0,$$
$$\text{or } \alpha_i = 0, n_i < N, \text{ in cycle } t)$$
$$= \frac{\Pr(n_{i-1} = 0, n_i < N \text{ in cycle } t)}{\Pr(n_{i-1} = 0 \text{ or } \alpha_i = 0, n_i < N \text{ in cycle } t)},$$

where the numerator can be approximated by $P_{i-1}(0)$ from $\Omega(i-1)$, and the denominator can be evaluated as follows:

$$r'_i \Pr(n_{i-1} = 0 \text{ or } \alpha_i = 0, n_i < N \text{ in cycle } t)$$
$$= p'_i \Pr(n_{i-1} > 0 \text{ or } \alpha_i = 1, n_i < N \text{ in cycle } t) = p'_i \bar{o}_i \tag{5.85}$$

which implies that the rate of recovery from a failure on machine i is equal to the rate into a failure on the same machine. Consequently, we have

$$\Delta_{i-1} = \frac{P_{i-1}(0) r'_i}{\bar{o}_i p'_i}. \tag{5.86}$$

Using similar arguments, we can also write

$$r''_i = \Pi_{i+1} r''_{i+1} + (1 - \Pi_{i+1}) r_{i+1}, \tag{5.87}$$

where

$$\Pi_{i+1} = \frac{P_{i+1}(N_{i+1})r''_{i+1}}{p''_{i+1}\overline{o}_{i+1}}. \tag{5.88}$$

Thus, we can construct set of $4(K-1)$ equations, including (5.78), (5.83), (5.84), and (5.87), and the boundary equations $r'_1 = r_1$, $p'_1 = p_1$, $r''_{K-1} = r_K$, and $p''_{K-1} = p_K$, with a total of $4(K-1)$ unknowns, namely r'_i, p'_i, r''_i, and p''_i, $i = 1, \ldots, K-1$. Clearly, due to the dependence among the subsystems, again we have to resort to an iterative solution procedure. With this in mind, let us rewrite (5.83) to make it convenient for a fixed-point iteration algorithm, that is,

$$\frac{p'_i}{r'_i} = \frac{1}{\overline{o}(i-1)} - \frac{p''_{i-1}}{r''_{i-1}} + \frac{1}{e_i} - 2, \tag{5.89}$$

as well as

$$\frac{p''_{i-1}}{r''_{i-1}} = \frac{1}{\overline{o}(i)} - \frac{p'_i}{r'_i} + \frac{1}{e_i} - 2. \tag{5.90}$$

Notice the change of notation in the throughputs. Equations (5.89) and (5.90) are valid since all the throughputs (of machines as well as of subsystems) are either the same or approximately equal upon convergence of the iterative procedure. Again, $P_i(n_i)$ is approximated by

$$P_i(n_i) \cong \sum_{\alpha_1,\alpha_2} P_i(\alpha_1, \alpha_2, n_i).$$

Now, for a K-machine transfer line, we can construct an algorithm that is similar in nature to what is presented in Section 5.8.2. It consists of forward and backward passes, each yielding a new value for an unknown parameter in each subsystem. It can be implemented as shown in Table 5.12.

The algorithm is efficient and highly accurate. Although the two decomposition algorithms presented in this chapter are highly similar in nature, they differ slightly in the equations through which the unknown parameters are obtained. The equations in the algorithm in Table 5.5 emerge from the parameter definitions such as (5.55), whereas they come from properties of the transfer line such as (5.80)–(5.82) in the above algorithm. None of the algorithms experiences convergence problems.

Related Bibliography

There is a significant amount of literature on synchronous transfer lines focusing on the effect of station breakdowns using exact (when possible) or approximate

TABLE 5.12. The decomposition algorithm for transfer lines with no scrapping

INITIALIZATION: Set $k = 1$, $p'_i = p_i, r'_i = r_i, p''_i = p_{i+1}, r''_i = r_{i+1}$, $i = 1, \ldots, K - 1$

ITERATION k

STEP 1: FORWARD PASS ($i = 1, \ldots, K - 2$)
Analyze $\Omega(i)$, obtain r'_{i+1} using (5.84) and p'_{i+1} using (5.89) to be used in the analysis of $\Omega(i + 1)$.

STEP 2: BACKWARD PASS ($i = K - 1, \ldots, 2$)
Analyze $\Omega(i)$, obtain r''_{i-1} using (5.87) and p''_{i-1} using (5.90) to be used in the analysis of $\Omega(i - 1)$.
Obtain $\overline{o}_\ell^{(k)} = e_K[1 - P_{K-1}(0)]$.
Stop if $|\overline{o}_\ell^{(k)} - \overline{o}_\ell^{(k-1)}| < \varepsilon$;
else let $k \leftarrow k + 1$ and go to iteration k.

methods to compute the measures of performance. According to Buzacott (1967a), the earliest studies were by Vladzievskii (1953) and Koenigsberg (1959) and were mainly descriptive. Also, Sevast'Yanov (1962) studied two-machine systems exactly and proposed approximations for larger systems. Sevast'Yanov extended the results of Vladzievskii for identical stages to nonidentical stages where each machine may have different failure and/or repair processes. The basic idea of decomposing a larger system into a set of smaller subsystems was first suggested by Zimmern (1956). Freeman (1964) gave guidelines for buffer-capacity allocation using simulation. Although Buzacott's papers (1967a, 1968) were not the first to focus on transfer lines, they did construct the basic mathematical model for a variety of assumptions to analyze the effect of station breakdowns on the production rate of the system. Buzacott showed the impact of the number of machines, failures, and the buffer capacities on the efficiency of the system. Consequently, synchronous geometric-repair-time, geometric-failure-time transfer-line models are referred to as **Buzacott-type models** (as rightly suggested by Dallery and Gershwin (1992)). The impact of the intermediate buffer storage, through Eq. (5.28), was studied by Buzacott, and Eqs. (5.71)-(5.73) were also due to Buzacott (1967a). Buzacott (1982) verified that a policy such as BCP2 yields a smaller throughput than BCP1. In the basic Buzacott model, there are two machines, M_1 and M_2, having the same unit cycle time, separated by a buffer of capacity N. Each machine undergoes geometric up and down times due to operation-dependent failures. The two main assumptions of the Buzacott model that differ from the two-node transfer line operating under the BCP1 policy introduced in Section 5.5 are that no scrapping is allowed and that only one event (either failure or repair) can take place during a cycle. In two-machine lines, Buzacott (1967a) obtained product-form solutions for the steady-state probabilities of the underlying Markov chain in the general form of

$$P(\alpha_1, \alpha_2, n) = \sum_j D_j X_j^n \Phi_j(\alpha_1, \alpha_2)$$

for some scalars D_j and X_j and for some scalar functions Φ_j (see Dallery and Gershwin (1992)).

Artamanov (1977) solved the basic model allowing more than one event occurrence in a cycle. Also, Buzacott and Hanifin (1978) studied the same problem. Equation (5.74) is due to Buzacott and Hanifin. Artamanov allowed the addition of partly processed work pieces into the system. That is, upon a failure of a machine, the part being processed on it is assumed to be complete and transferred to the downstream buffer at the end of the cycle. Later, Dudick (1979) and Gershwin and Schick (1983) obtained a product-form solution for essentially the same model of Artamanov, without allowing partly processed work pieces in the line. However, Dudick did not allow simultaneous repairs of the machines. Buzacott's basic model differs in three aspects from that of Gershwin and Schick (1983):

(1) The order of events are reversed in the Gershwin-Schick model;
(2) More than one event occurrence in a cycle is allowed in the Gershwin-Schick model;
(3) Blocking is of before-processing type in the Gershwin-Schick model.

Gershwin and Schick showed that (for the two-machine case) if the first machine is slower than the second, the average buffer level approaches a limit as the buffer size increases. Jafari and Shanthikumar (1987b) extended the two-stage model to accommodate general up and down times. Dallery and Gershwin (1992) further indicated that the same approach can be used to show that if the first machine is faster than the second, the average buffer level is unbounded as the buffer size increases with no limit. Also, if the two machines are identical (having the same repair and failure rates), then the average buffer level is equal to half of the buffer capacity. Equation (5.77) is cleverly obtained by Gershwin and Schick (1983).

Basically, all the earlier work done in synchronous production lines used the same model with different operating policies and assumptions and focused on the effect of the buffer capacities on the output rate of the system. The operating policies in the studies reviewed up to this point differ only in whether to allow the partly processed work pieces in the line, or to perform rework on them. However, in both policies, a part will definitely leave the line after processing (either partly or completely). In this case, the transfer line is reversible for both blocking-after processing (Muth (1979)) as well as blocking-before processing (Gordon and Newell (1967)) mechanisms. However, the reversibility property no longer remains valid if scrapping of work pieces takes place upon machine failures.

The steady-state flow-balance equations of a two-stage synchronous line with geometric-failure, geometric-repair times can be put in the form of a set of matrix equations, which can be solved recursively as we have done in Section 5.5. Let S_i

be the set of states for which the buffer level is equal to i, $i = 1, 2, \ldots, N$. States S_0 and S_N are called **boundary states**. Then, if $\mathbf{P}$ is the transition probability matrix of the Markov chain, the matrix $(P - I)^T$ has the tridiagonal structure shown below:

$$(\mathbf{P} - \mathbf{I})^T = \begin{bmatrix} B_0 & A_1 & & & & & \\ K_0 & B & A & & & & \\ & K & B & A & & & \\ & & & \ddots & & & \\ & & & & K & B & A & \\ & & & & & K & B & A_N \\ & & & & & & K & B_N \end{bmatrix}.$$

Let $\underline{\mathbf{b}} = (b_0, b_1, \ldots, b_N)$ be such that $\mathbf{b}_i$ is the subvector containing the nonnormalized steady-state probabilities of the states for which there are i units in the buffer. Then, the solution to $(\tilde{\mathbf{Q}} - I)^T \underline{b}^T = 0$ can be obtained recursively. That is, there is a sequence of submatrices $\mathbf{Q}_0, \ldots, \mathbf{Q}_N$ (which can be found in terms of the submatrices of $(\tilde{\mathbf{Q}} - I)^T$) such that the vector $\underline{b}$ is found by a set of equations of type

$$\mathbf{Q}_0 \underline{b}_0^T = \underline{0},$$
$$\underline{b}_i^T - \mathbf{Q}_i \underline{b}_{i-1}^T = \underline{0}, \qquad i = 1, \ldots, N.$$

$(\tilde{\mathbf{Q}} - I)^T$ has the matrix-geometric property if $\mathbf{Q}_2 = \mathbf{Q}_3 = \cdots = \mathbf{Q}_{N-2} = \mathbf{Q}$, and the corresponding solution is called the **matrix-geometric solution**. Yeralan and Muth (1987) showed that $\mathbf{Q}_2 = \mathbf{Q}_3 = \cdots = \mathbf{Q}_{N-2} = \mathbf{Q}$ if at least one of the matrices $\mathbf{K}$ or $\mathbf{A}$ has rank 1 and the submatrix $(\mathbf{B} + \mathbf{I})$ is nonsingular. Yeralan and Muth (1987) gave a general review of the two-stage synchronous transfer lines. They developed conditions of a general matrix-geometric solution for the steady-state probabilities of the underlying Markov chain. They compared the previous work with various assumptions and established a unifying view of the existing models. The matrix-geometric model presented in Section 5.5 is due to Dogan and Altiok (1995a) and is a special case of Yeralan-Muth model.

Sheskin (1976) studied Buzacott-type synchronous models with the assumption of $p_i + r_i = 1$. Through an empirical analysis, Sheskin showed that larger buffers should be allocated around machines with less reliability. Okamura and Yamashina (1977) studied the Gershwin-Schick model (with the assumption that the part on a failed machine is discarded from the system) by a matrix-geometric approach. Through an extensive set of numerical experiments, they showed that the output rate is a convergent function of the buffer size, and the shape of the average buffer level versus the buffer-size curve depends on the relative values of the failure and repair probabilities. Soyster et al. (1979) considered the optimal allocation of buffer capacities to maximize the output rate and provided guidelines to allocate buffers for better throughput. Soyster et al. also used the same assumption of $p_i + r_i = 1$.

With this restriction, they were able to study the balance equations more easily for systems with multiple machines. They also obtained bounds on the output rate of the line. Shanthikumar and Tien (1983) also used the matrix-geometric approach to solve the same model with the assumption that the part on a failed machine is scrapped with a given probability.

Buzacott (1967b) extended his two-stage equations to three-stage synchronous transfer lines with geometric failure and repair-time distributions (identical) (Dallery and Gershwin (1992)). Gershwin and Schick (1983) extended their two-stage model to exactly analyze three-stage systems.

For longer lines, Gershwin (1987) developed an approximation method that decomposes a K-stage synchronous line with finite buffers into $K - 1$ two-stage lines (subsystems), $\Omega(1)$, (2), ..., $\Omega(K - 1)$, one for each buffer in the original line. Later, Dallery et al. (1988) improved Gershwin's decomposition equations to make the algorithm faster and to avoid possible convergence problems. Their algorithm is presented in Table 5.12. Jafari and Shanthikumar (1987a) proposed an approximation method for synchronous transfer lines with operation-dependent failures and possible scrapping of parts. Their method is a combination of decomposition and aggregation methods. They decompose the line into $K - 1$ subsystems. The basic idea of aggregation is to replace a two-stage one-buffer subsystem, $\Omega(i)$, by a single equivalent stage, and let this equivalent stage be the first stage of $\Omega(i + 1)$. The algorithm in Table 5.5 was developed by Dogan and Altiok (1995a). It is highly similar in nature to Gershwin's (1987) and Dallery et al.'s (1988) algorithms. The regenerative analysis in Section 5.2, the BCP2 buffer clearing policy of Section 5.4, and design issues involving the repair probabilities as the decision variables in Section 5.7 are due to Dogan and Altiok (1995b). Variance of the throughput is an extension of the variance of the number of visits to a set of states in finite Markov chains (see Kemeny and Snell (1983)).

The design problem of allocating buffers in transfer lines has been studied. Soyster and Toof (1976) showed that a single buffer should be placed in the middle of the line when all the machines have the same failure probability. Yamashita and Suzuki (1988) and Jafari and Shanthikumar (1989) both applied a dynamic programming procedure to allocate buffers in transfer lines to maximize throughput for a given total buffer capacity.

The reader can also find extensive coverage of transfer lines in Buzacott and Shanthikumar (1993), Gershwin (1994), Papadopoulos et al. (1993), and Perros (1994) in the form of tandem queues with synchronous servers and blocking.

References

Artamanov, G. T., 1977, Productivity of a Two-Instrument Discrete Processing Line in the Presence of Failures (English translation) *Cybernetics*, Vol. 12, pp. 464–468.

Buzacott, J. A., 1967a, Automatic Transfer Lines with Buffer Stocks, *Intl. J. Production Research*, Vol. 5, pp. 183–200.

Buzacott, J. A., 1967b, Markov Chain Analysis of Automatic Transfer Lines with Buffer Stock. Ph.D. Thesis, Department of Engineering Production, University of Birmingham.

Buzacott, J. A., 1968, Prediction of the Efficiency Production Systems without Internal Storage, *Intl. J. Production Research*, Vol. 6, pp. 173–188.

Buzacott, J. A., 1982, Optimal Operating Rules for Automated Manufacturing Systems, *IEEE Transactions on Automatic Control*, Vol. 27, no. 1, pp. 80–85.

Buzacott, J. A. and L. E. Hanifin, 1978, Models of Automatic Transfer Lines with Inventory Banks-A Review and Comparison, *IIE Transactions*, Vol. 10, pp. 197-207.

Buzacott, J. A., and J. G.Shanthikumar, 1993, *Stochastic Models of Manufacturing Systems*, Prentice Hall, Englewood Cliffs, NJ.

Dallery, Y., R. David, and X. L. Xie, 1988, An Efficient Algorithm for Analysis of Transfer Lines with Unreliable Machines and Finite Buffers, *IIE Transactions*, Vol. 20, pp. 280–283.

Dallery, Y., S. B. Gershwin, 1992, Manufacturing Flow Line Systems: A Review of Models and Analytical Results, *Queueing Systems Theory and Applications*, Vol. 12, pp. 3–94.

Dogan, E., and T. Altiok, 1995a, Approximate Analysis of Transfer Lines with Finite Buffers, *Transactions on Operational Research*, Vol. 7, no. 1, pp. 13-44.

Dogan, E., and T. Altiok, 1995b, Planning Repair Effort in Transfer Lines, Submitted to *IIE Transactions*.

Dudick, A., 1979, Fixed Cycle Production Systems with In-Line Inventory and Limited Repair Capability, Ph.D. Thesis, Columbia University.

Freeman, M. C., 1964, The Effect of Breakdowns and Interstage Storage on Production Line Capacity, *J. of Industrial Engineering*, Vol. 15, pp. 194–200.

Gershwin, S. B., 1987, An Efficient Decomposition Method for the Approximate Evaluation of Tandem Queues with Finite Storage Space and Blocking, *Operations Research*, Vol. 35, pp. 291–305.

Gershwin, S. B., 1994, *Manufacturing Systems Engineering*, Prentice Hall, Englewood Cliffs, N.J.

Gershwin, S. B., I. C. Schick, 1983, Modeling and Analysis of Three-Stage Transfer Lines with Unreliable Machines and Finite Buffers, *Operations Research*, Vol. 31, pp. 354-380.

Gordon, W. K., and G. F. Newell, 1967, Cyclic Queueing Systems with Restricted Length Queues, *Operations Research*, Vol. 15, pp. 266-277.

Jafari, M. and J. G. Shanthikumar, 1987a, An Approximate Model of Multi-Stage Automatic Transfer Lines with Possible Scrapping of Work Pieces, *IIE Transactions*, Vol. 19, pp. 252–265.

Jafari, M. and J. G. Shanthikumar, 1987b, Exact and Approximate Solutions to Two-Stage Transfer Lines with General Uptime and Downtime Distributions, *IIE Transactions*, Vol. 19, pp. 412 -420.

Jafari, M. and J. G. Shanthikumar, 1989, Determination of Optimal Buffer Storage Capacities and Optimal Allocation in Multi-Stage Automatic Transfer Lines, *IEE Transactions*, Vol. 21, pp. 130–135.

Kemeny, J. G.and J. L. Snell, 1983, *Finite Markov Chains*, Springer-Verlag, New York.

Koenigsberg, E., 1959, Production Lines and Internal Storage: A Review, *Management Science*, Vol. 5, pp. 410-433.

Muth, E. J., 1979, The Reversibility Property of Production Lines, *Management Science*, Vol. 25, pp. 152–158.

Okamura, K. and H. Yamashina, 1977, Analysis of the Effect of Buffer Storage Capacity in Transfer Line Systems, *AIIE Transactions*, Vol. 9, pp. 127-135.

Papadopoulos, H.T., C. Heavey, and J. Browne, 1993, *Queueing Theory in Manufacturing Systems Analysis and Design*, Chapman & Hall, New York, New York.

Perros, G, 1994, *Queueing Networks with Blocking*, Oxford University Press, New York.

Sevast'Yanov, B. A., 1962, How Bunker Capacity Influences the Average Standstill Time of an Automatic Machine Tool Line, *Teoriya Veroyatnostey i ee Primeneniya*, Vol. 7, pp. 429–438, (English Translation: *Theory of Probability and Its Applications*).

Shanthikumar, J. G., and C. C. Tien, 1983, An Algorithmic Solution to Two-Stage Transfer Lines with Possible Scrapping of Units, *Management Science*, Vol. 29, pp. 1069-1086

Sheskin, T. J., 1976, Allocation of Interstage Storage Along an Automatic Production Line, *AIIE Transactions*, Vol. 8, pp. 146–152.

Soyster, A. L., J. W. Schmidt, and M. W. Rohrer, 1979, Allocation of Buffer for a Class of Fixed Cycle of Production Systems, *AIIE Transactions*, Vol. 11, pp. 140–146.

Soyster, A. L. and D. I. Toof, 1976, Some Comparative and Design Aspects of Fixed Cycle Production Systems, *Naval Res. Logistics Quart.*, Vol. 23, pp. 437–454.

Viladzievskii, A. P., 1953, Losses of Working Time and the Division of Automatic Lines into Sections, *Stankii Instrument*, Vol. 36, pp. 470–477.

Yamashita, H. and S. Suzuki, 1988, An Approximation Method for Line Production Rate of a Serial Production Line with a Common Buffer, *Computers and Operations Research*, Vol. 15, pp. 395–402.

Yeralan, S., and E. Muth, 1987, A General Model of a Production Line with Intermediate Buffer and Station Breakdowns, *IIE Transactions*, Vol. 19, pp. 130–139.

Zimmern, B., 1956, Etudes de la Propagation des Arrets Aleatoires dans les Chaines de Production, *Revue de Statistique Appliquee*, Vol. 4, pp. 85–104.

Problems

5.1. Consider the following single machine where the processing time X is geometrically distributed with parameter p, that is, $P(X = k) = (1 - p)p^{k-1}, k = 1, 2, \ldots$.

(a) At the end of the processing time, the process restarts from scratch with probability q. Using the moment-generating functions, find the distribution of the process completion time (the total time each job spends on the machine). Find its mean, the variance, and the squared coefficient of variation.

(b) If the process is adjusted for one cycle before it starts each time, what would be the mean and the coefficient of variation of the process completion time?

5.2. Consider a two-machine transfer line without internal storage. Machines do not experience failures. However, at the end of each cycle, the process on machine i is repeated with probability α_i. The first machine becomes blocked when the process is complete and the second machine is busy. Obtain expressions for the throughput and the blocking probability.

5.3. Obtain the probability distribution of the random variable Z that is depicted in Figure 5.2. You may choose to work with the moment-generating functions. Note that Z is also the random variable representing the process completion time in Problem 5.1.(b).

5.4. Consider a two-machine transfer line, subject to operation-dependent failures, with parameters p_1, r_1, p_2 and r_2 with no intermediate buffer. Both machines may fail during the same cycle. Assume that the interrupted unit is not scrapped but departs as a good unit at the end of the cycle during which the repair is completed.

240   5. Analysis of Transfer Lines

(a) Obtain the throughput as a function of the machine parameters using the regenerative approach of Section 5.2.
(b) Verify your results using the underlying Markov chain as well as (5.73)
(c) How is (a) affected if the part is reprocessed from scratch upon repair completion?
(d) Repeat (a) and (b) for the special case of fixed repair length of one cycle for each machine.

5.5. Consider the following two-machine transfer line with a secondary buffer with a capacity of 6 units. The machine parameters are $(p_1, r_1) = (0.2, 0.8)$ and $(p_2, r_2) = (0.1, 0.9)$. Failures are operation dependent, and the interrupted units are scrapped upon failures. Assuming operating under the BCP1 policy, obtain the steady-state marginal distributions of the buffer contents and the machine status at cycle beginnings. Calculate the average buffer contents and the throughput. Repeat your analysis for the operating policy of BCP2.

5.6. Extend the matrix-geometric algorithm of Section 5.5 to the case of operation-independent failures (under BCP1). Implement it in Problem 5.5 and compare the numerical results.

5.7. Extend the geometric solution of Gershwin and Schick given in Section 5.9 to the case with operation-dependent failures with scrapping, operation under BCP1, BBP blocking, and $r_1 = r_2 = r$ and $p_1 = p_2 = p$.

5.8. Obtain an explicit expression for the throughput in a two-machine transfer line operating under BCP1 with job scrapping. Failures are operation dependent, and simultaneous event occurrence is not possible (see Eq. (5.71)).

5.9. Consider a two-machine line with a secondary buffer with a capacity of one unit. Failures are operation dependent. Simultaneous failures are not allowed; however, the interrupted jobs are scrapped. The repair time is equal to one cycle on each machine. Assuming BAP blocking policy, develop an expression for the throughput and the probability that the buffer is full for operation under BCP1 and BCP2. Apply your results to cases with $(p_1, p_2) = (0.1, 0.3)$ and $(p_1, p_2) = (0.2, 0.2)$.

5.10. Design problem: Let us design the following two-machine line with BAP-type blocking: Machines experience operation-dependent failures. The interrupted jobs are scrapped, and the line operates under BCP1. The machine parameters are $(p_1, r_1) = (0.2, 0.4)$ and $(p_2, r_2) = (0.1, 0.6)$. Obtain the throughput in each of the following steps:

(a) Assume that there is no intermediate buffer
(b) Assume that $N = \max\lceil 1/0.4, 1/0.6 \rceil = 3$. Compare the throughput with that of $N = \infty$.
(c) At the end of every repair, restore the buffer to the capacity and resume operation with $N = 3$ units.
(d) Can you think of other ways of improving the throughput without changing the values of the machine parameters?
(e) If you are allowed to change the repair probabilities for better throughput, how would you change them while the sum remains the same?

5.11. Design problem: Consider a two-machine line with operation-dependent failures, operating under BCP1 policy with the interrupted jobs being scrapped. Blocking is of BAP type. Assume that $r_1 + r_2 = 1$, and $p_1 = p_2 = 0.2$. Develop a procedure of your own (exactly or approximately) to obtain values for N, r_1 and r_2 such that

$$\max_{N_i, r_i, \forall i} \text{Profit} \equiv \bar{o}_\ell (revenue/unit) - \overline{N}h, \qquad \text{such that} \quad r_1 + r_2 = 1.$$

The choice of the unit revenue and the unit holding cost (h) is left to the reader. You may wish to work with the increments of 0.1 for the repair probabilities.

5.12. Consider a symmetric three-machine line, operating under BCP1, experiencing operation-dependent failures with job scrapping and with BAP-type blocking. Let us assume that $p_i = p = 0.2$, $r_i = r = 0.8$, and $N_1 = N_2 = 3$. Implement the decomposition algorithm of Section 5.8 to obtain the approximate values of the throughput, the average buffer contents, as well as the machine status probabilities. If necessary, write a computer program that may be specific to this problem.

5.13. Modify the algorithm in Table 5.5 by using an equation similar to (5.83) to obtain p' and p'' in each subsystem. Resolve Problem 5.12 using the new algorithm.

5.14. Modify the algorithm in Table 5.12 for the case with no scrapping by using an equation similar to (5.62) to obtain r'_i. Implement the new algorithm on a three- machine line with parameters of your choice.

5.15. Modify the algorithm in Table 5.12 by using the basic definitions of r' and r'' from Section 5.8. Implement the new algorithm on a three-machine line with parameters of your choice.

5.16. Develop an approximation method to analyze the four-machine line shown below:

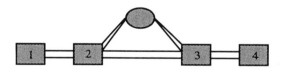

There is a single buffer that is placed between the second and the third machines. Failure on the first or second machines stops both machines 1 and 2, and a failure on the third or fourth machines stops both machines 3 and 4. Assume operation-dependent failures with the interrupted jobs being scrapped. The machine parameters are $N = 4$, $p_i = (0.05, 0.1, 0.15, 0.20)$, and $r_i = 0.90$ for all i. Assuming operation under BCP1, obtain the approximate values for $\bar{o}_\ell$ and $\overline{N}$.

5.17. Consider the following five-machine transfer line with operation-dependent failures and job scrapping:

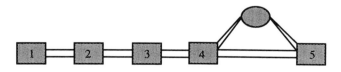

Assume operation under the BCP2 policy, job scrapping, and the BAP blocking mechanism. Develop an approximation algorithm to obtain the average throughput. Implement to the symmetric case with $N = 10$, $p = 0.1$, and $r = 0.85$.

6

Assembly Systems

6.1 Introduction

In manufacturing systems, the finished products are frequently produced by assembling a set of semifinished products or subassemblies. The fundamental characteristic of assembly systems is that a certain number of units from each input material are assembled to produce one unit of the finished product. The assembled units may include raw materials, subassemblies, semifinished units, or even finished products. The assembly activity may be a single operation that takes place in one work station or it may be a sequence of operations performed in a group of stations. For instance, the transfer lines covered in Chapter 5 may be considered as the special case of multistage assembly systems in which one or more units of raw material are assembled to the main assembly at every station. A manufacturing system may consist of a network of assembly, subassembly, and regular work stations, as shown in Figure 6.1. The immediate upstream buffers of an assembly station may be called "matching buffers" and may be assumed to be part of the assembly station. The stations that are feeding the matching buffers of an assembly station are referred to as the "subassembly stations."

Similar to production lines, the assembly systems may be synchronous or asynchronous. In synchronous systems, jobs in all the stations are processed or assembled and transferred to the downstream buffers simultaneously. The assembly times are less than or equal to the cycle time, and jobs are transferred to other stations at the end of the cycle time. Again, intermediate buffers are justified due to random disruptions such as failures. In asynchronous systems, the assembly times may be random and the stations are not synchronized with each other. Intermediate buffers have finite capacities resulting in blockages that are of blocking-after-processing (BAP) type. Later in this chapter, we will also consider the blocking-before-processing (BBP) type. We refer the reader to Section 4.1 for the definitions of different types of blocking mechanisms.

The objectives in studying assembly systems are again to understand the behavior of these systems and to be able to obtain the values of some performance measures that are useful in designing such systems. In addition to the measures such as station utilizations, average inventory levels, and the throughput, one is

6. Assembly Systems

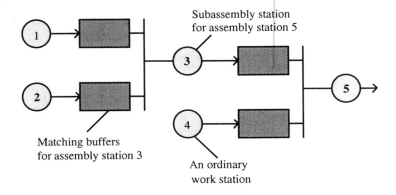

Figure 6.1. A network of assembly stations.

generally interested in the average number of pairs ready to be assembled and the average number of (excess) units waiting for their matching units to arrive at an assembly station. These measures will be obtained through either exact or approximation methods. We will start with the exact analysis of a simple asynchronous assembly system.

6.2 An Asynchronous Assembly System

Let us consider the assembly station shown in Figure 6.2, which assembles two units, one from each matching buffer, to make a unit of the finished product. The subassembly stations are always busy when not blocked. The capacities of the matching buffers B_1 and B_2 are N_1 and N_2, respectively. They both include the space on the machine at the assembly station. The system is asynchronous and the assembly times are exponentially distributed with rates μ_1, μ_2, and μ_3.

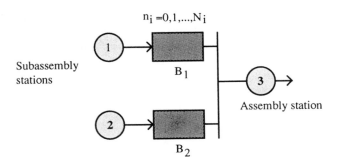

Figure 6.2. A two-node assembly system.

At the moment of an assembly completion, station 3 becomes idle if there are no units in either or both of the matching buffers. Otherwise, the next assembly starts immediately. The subassembly stations produce until their downstream buffers are full. Blocking occurs if a departing job finds the next buffer full. During blocking, the blocking unit resides on the machine at the subassembly station, keeping the station idle, that is, the BAP-type blocking. It will be over as soon as an assembly leaves the system. It is the assembly operation that makes the difference between the blocking phenomenon in production lines and the one in assembly systems. The job departure at a blocking station in a production line guarantees termination of blocking. However, in the system shown in Figure 6.2, a departure from station 3 while it is blocking station 1 cannot terminate blocking if buffer B_2 is empty at this point in time. In this case, it is the process completion at station 2 that terminates blocking at station 1.

6.2.1 A Recursive Method to Obtain the Steady-State Measures

To obtain the steady-state performance measures of the system in Figure 6.2, let us study the stochastic process $\{N_1(t), N_2(t), t \geq 0\}$ where $N_i(t) = 0, 1, \ldots, N_i + 1$ is the number of units in buffer B_i at time t. $N_i(t) = N_i + 1$ represents that station i is blocked, and the $(N+1)$st unit resides at station i. $N_1(t) = n_1$ implies that there is one unit on the machine at station 3, and there are $n_1 - 1$ units in B_1 waiting to be assembled. Station 3 starts the assembly operation at time t only when $N_1(t)$ and $N_2(t)$ are both greater than zero. Let (n_1, n_2) be a particular state of this process with $P(n_1, n_2)$ being its steady-state probability. At any point in time, the time until the state of the system changes is the minimum of the remaining processing times in all or some of the stations. Clearly, this time is exponentially distributed, and $N_1(t), N_2(t)$ constitutes a finite-state, irreducible Markov chain having a unique steady-state distribution.

The transition diagram of the above Markov chain is given in Figure 6.3. The vertical and horizontal transitions represent the process completions at stations 1 and 2, respectively. The upward diagonal transitions represent the assembly completions at station 3. Notice that the total number of states in the above Markov chain is $(N_1 + 2)(N_2 + 2)$. Let us define the following vectors of probabilities:

$$\underline{\tilde{P}}(i) = \begin{pmatrix} P(0, i) \\ P(1, i) \\ \vdots \\ P(N_1 + 1, i) \end{pmatrix}, \quad i = 0, \ldots, N_2 + 1.$$

The steady-state flow-balance equations for states $(0, N_2 + 1), \ldots, (N_1 + 1, N_2 + 1)$ can be written in matrix form as follows:

$$\mathbf{A}\underline{\tilde{P}}(N_2 + 1) = \mu_2 \underline{\tilde{P}}(N_2), \tag{6.1}$$

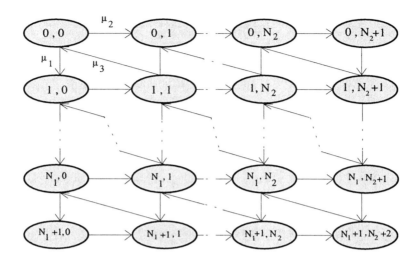

Figure 6.3. The transition diagram of $\{N_1(t), N_2(t)\}$.

where

$$\mathbf{A} = \begin{pmatrix} \mu_1 & 0 & \cdots & 0 & 0 \\ -\mu_1 & \mu_1 + \mu_3 & \cdots & 0 & 0 \\ 0 & -\mu_1 & \cdots & 0 & 0 \\ \vdots & \vdots & \ddots & \vdots & \vdots \\ 0 & 0 & \cdots & \mu_1 + \mu_3 & 0 \\ 0 & 0 & \cdots & -\mu_1 & \mu_3 \end{pmatrix}.$$

Similarly, the flow-balance equations for states $(0, N_2), \ldots, (N_1 + 1, N_2)$ can be written as

$$(\mathbf{A} + \mu_2 \mathbf{I})\tilde{\underline{\mathbf{P}}}(N_2) = \mu_2 \tilde{\underline{\mathbf{P}}}(N_2 - 1) + \mu_3 \tilde{\underline{\mathbf{P}}}(N_2 + 1), \qquad (6.2)$$

where $\mathbf{I}$ is the identity matrix of an appropriate size.

Equation (6.2) remains unchanged, except that N_2 is replaced by n_2 for the sets of states $(0, n_2), \ldots, (N_1 + 1, n_2)$ for $n_2 = N_2 - 1, \ldots, 1$. This behavior is due to the symmetry in the transition diagram shown in Figure 6.3. The flow-balance equations for states $(0, 0), \ldots, (N_1 + 1, 0)$ are different:

$$[\mathbf{A} + (\mu_2 - \mu_3)\mathbf{I}]\tilde{\underline{\mathbf{P}}}(0) = \mu_3 \tilde{\underline{\mathbf{P}}}(1). \qquad (6.3)$$

The special structure of the balance equations gives rise to a recursive method similar to the one used in Chapter 4 to analyze the two-node production line.

6.2. An Asynchronous Assembly System

However, the forward recursion in the present system of equations involves singular matrices that need to be inverted. For this reason, the backward recursion is chosen here and $\tilde{\mathbf{P}}(j)$, $j = N_2, \ldots, 1, 0$, is obtained as a function of $\tilde{\mathbf{P}}(N_2 + 1)$. For instance, from (6.1), which is written for states $(0, N_2 + 1), \ldots, (N_1 + 1, N_2 + 1)$, we have

$$\tilde{\mathbf{P}}(N_2) = \mu_2^{-1}\mathbf{A}\tilde{\mathbf{P}}(N_2 + 1) = \mathbf{Z}(N_2)\tilde{\mathbf{P}}(N_2 + 1), \tag{6.4}$$

where $\mathbf{Z}(N_2) = \mu_2^{-1}\mathbf{A}$.

Similarly, from (6.2) we can write

$$\begin{aligned}\tilde{\mathbf{P}}(N_2 - 1) &= \mu_2^{-1}[(\mathbf{A} + \mu_2\mathbf{I})\tilde{\mathbf{P}}(N_2) - \mu_3\tilde{\mathbf{P}}(N_2 + 1)] \\ &= \mu_2^{-1}[(\mathbf{A} + \mu_2\mathbf{I})\mathbf{Z}(N_2) - \mu_3]\tilde{\mathbf{P}}(N_2 + 1) \\ &= \mathbf{Z}(N_2 - 1)\tilde{\mathbf{P}}(N_2 + 1). \end{aligned} \tag{6.5}$$

Consequently, the following system of recursive equations is obtained:

$$\tilde{\mathbf{P}}(n) = \mathbf{Z}(n)\tilde{\mathbf{P}}(N_2 + 1), \qquad n = N_2, N_2 - 1, \ldots, 0, \tag{6.6}$$

where

$$\mathbf{Z}(N_2 + 1) = \mathbf{I}$$

and

$$\mathbf{Z}(n) = \mu_2^{-1}[(\mathbf{A} + \mu_2\mathbf{I})\mathbf{Z}(n+1) - \mu_3\mathbf{Z}(n+2)], \qquad n = N_2 - 1, \ldots, 0. \tag{6.7}$$

Notice that we have not yet used the balance equations for states $(0, 0), (1, 0), \ldots, (N_1 + 1, 0)$ that can be used to obtain $\tilde{\mathbf{P}}(N_2 + 1)$. The flow-balance equations for these states can be written in matrix form as follows:

$$[\mathbf{A} + (\mu_2 - \mu_3)\mathbf{I}]\tilde{\mathbf{P}}(0) = \mu_3\tilde{\mathbf{P}}(1),$$

or

$$[\mathbf{A} + (\mu_2 - \mu_3)\mathbf{I}]\mathbf{Z}(0)\tilde{\mathbf{P}}(N_2 + 1) = \mu_3\mathbf{Z}(1)\tilde{\mathbf{P}}(N_2 + 1).$$

Hence, $\tilde{\mathbf{P}}(N_2 + 1)$ can be obtained using

$$\{[\mathbf{A} + (\mu_2 - \mu_3)\mathbf{I}]\mathbf{Z}(0) - \mu_3\mathbf{Z}(1)\}\tilde{\mathbf{P}}(N_2 + 1) = \mathbf{0}. \tag{6.8}$$

One of the equations in (6.8) can be replaced by $P(0, 0) = 1$ (an arbitrary choice), and the corresponding $\tilde{\mathbf{P}}(N_2 + 1)$ is obtained. The coefficient matrix of $\tilde{\mathbf{P}}(N_2 + 1)$ in (6.8) is invertible.

The above recursive relationship suggests that we first obtain $\mathbf{Z}(0)$ and $\mathbf{Z}(1)$ using Eq. (6.7) and then $\tilde{\mathbf{P}}(N_2 + 1)$ using (6.8). The rest of the probability vectors are obtained from (6.6), recursively. A normalization has to be carried out once all the probabilities are obtained. Note that $\tilde{\mathbf{P}}(N_2 + 1)$ is obtained using $\mathbf{Z}(0)$ and $\mathbf{Z}(1)$, which in turn are obtained using $\mathbf{Z}(2), \ldots, \mathbf{Z}(N_2 + 1)$. After $\tilde{\mathbf{P}}(N_2 + 1)$ is obtained, $\mathbf{Z}(\cdot)$s are still needed to evaluate (6.6). The recursive method to obtain $\mathbf{Z}(n)$ is computationally insignificant. For efficient management of the computer memory, we suggest recomputing the $\mathbf{Z}(\cdot)$ matrices to evaluate (6.6) rather than storing them when they are first obtained using (6.7). Thus, only $\mathbf{Z}(n)$, $\mathbf{Z}(n + 1)$, and $\mathbf{Z}(n + 2)$ are stored for any n at any point in time.

The performance measures of the assembly station are as follows: The long-run average utilization is

$$U_a = 1 - \sum_{i=1}^{N_1+1} P(i, 0) - \sum_{j=1}^{N_2+1} P(0, j), \qquad (6.9)$$

the output rate is

$$\bar{o} = U_a \mu_3; \qquad (6.10)$$

the average number of jobs of type 1, for example; in the system is

$$\overline{N}_1 = \sum_{i=1}^{N_1} \sum_{j=0}^{N_2+1} i P(i, j) + N_1 \sum_{k=0}^{N_2+1} P(N_1 + 1, k), \qquad (6.11)$$

and the average number of pairs in the matching buffers is

$$\overline{N}_p = \sum_{i=1}^{N_1} \sum_{j=1}^{N_2} \min(i, j) P(i, j)$$

$$+ \sum_{k=1}^{N_1} \min(k, N_2) P(k, N_2 + 1) + \sum_{m=1}^{N_2} \min(N_1, m) P(N_1 + 1, m). \quad (6.12)$$

6.2.2 Characteristics of Assembly Stations

Let us study the properties of this system before we get into numerical examples. Intuitively, the flow-conservation law tells us that the throughput of each of the subassembly stations should be the same as that of the assembly station since no jobs are lost in the matching buffers. This may be shown as follows. By adding the balance equations in (6.1), we have

$$\Pr(n_2 = N_2)\mu_2 = \mu_3 \Pr(\text{station 3 is busy}, n_2 = N_2 + 1).$$

Similarly, from (6.2), we have

$$\Pr(n_2 = N_2 - 1)\mu_2 = \mu_3 \Pr(\text{station 3 is busy}, n_2 = N_2).$$

If we add these equations written for $n_2 = N_2, N_2 - 1, \ldots, 1, 0$, we obtain

$$\mu_2[1 - P_2(N_2 + 1)] = U_a\mu_3 = \bar{o}, \tag{6.13}$$

where $P_i(k)$ is the marginal distribution of the number of units in B_i.

This process can be repeated for station 1, resulting in

$$\mu_1[1 - P_1(N_1 + 1)] = \bar{o},$$

showing that the output rates of all three stations are identical. This is a direct consequence of the fact that there is one unit of each subassembly in a unit of the assembled product. The output rates would not be the same if this were not the case. Basically, the ratio of the subassembly units in the assembled unit characterizes the relationship between the throughputs of subassembly and assembly stations. For instance, in the system in Figure 6.2, if a unit of finished product were produced using two units from B_1 and one unit from B_2, then we would expect the throughputs of stations 2 and 3 to be the same and equal to half of the throughput of station 1. In this case, the transition diagram in Figure 6.3 and consequently the steady-state probabilities would be different.

Similar analysis can show that the steady-state average number of pairs in the system is less than or equal to the average number of jobs in each buffer, that is,

$$\overline{N}_p \leq \min(\overline{N}_1, \overline{N}_2).$$

Also, the average number of excess jobs of type i (waiting for their matching units), that is, $\overline{N}_{ie}$, is

$$\overline{N}_{ie} = \overline{N}_i - \overline{N}_p.$$

6.2.3 Analysis of the System Behavior

Let us consider the system shown in Figure 6.2 with $\mu_2 = 2$ and $\mu_3 = 3$. The values of the performance measures for varying values of μ_1, N_1 and N_2 are obtained using the above recursive method and are given in Table 6.1. Let us study the behavior of the assembly system by analyzing these numerical results. First of all, in the symmetric case ($\mu_1 = \mu_2$, $N_1 = N_2$), the steady-state distribution of the number of units in B_1 and that of B_2 are identical, and consequently the measures of the two buffers are the same. If $\mu_1 = \mu_2$, regardless of the values of N_1 and N_2, the blocking probabilities of the matching buffers are equal since the throughputs of the subassembly stations are identical.

As the capacity of a buffer increases, it is clear that the long-run average number of units in the buffer increases due to less blocking. Let us observe how the capacity of one matching buffer affects the average contents of the other. When $\mu_1 = \mu_2$, as (N_1, N_2) changes from (5,5) to (5,15), $\overline{N}_1$ decreases and $\overline{N}_2$ increases. The reason is that more units are available in B_2 for assembly at station 3. Consequently, the output rate of station 3 increases and $\overline{N}_1$ decreases since N_1 remains the same. The increase in $\overline{N}_2$ is solely in the number of units waiting for their matching units in B_2. The slight increase in $\overline{N}_p$ is again due to the increase in N_2. Similar behavior can be observed when (N_1, N_2) changes from (5,15) to (15,15).

The average number of units in the matching buffers are more sensitive to the changes in the buffer capacities in the asymmetric case, where $\mu_1 \neq \mu_2$. The increase in $\overline{N}_2$ is at extreme levels when (N_1, N_2) is changed from (5,5) to (5,15). Furthermore, $\overline{N}_1$ does not adjust when N_1 is also increased to 15. The reason is as follows: As N approaches infinity, the behavior of the system asymptotically converges to that of a station assembling units from buffers with infinite capacities.[1] When $\mu_1 \neq \mu_2$, regardless of the value of μ_3, the buffer with the larger arrival rate keeps accumulating the incoming units, which will wait for their matching units to arrive. Let us look at the case with $\mu_1 = 1$ and $N_1 = N_2 = 20$ in Table 6.1. $\overline{N}_1 = 0.5$ and $\overline{N}_2 = 19.50$, indicating that B_1 is almost always empty whereas B_2 is almost always full. Hence, as soon as a unit arrives at B_1, the assembly operation starts immediately since there are always units in B_2. Also notice that as the buffer capacities increase, the output rate of the assembly station approaches the minimum of the processing rates of the subassembly stations.

The instability built into the assembly systems may be better explained by the following new scenario. Let us assume that the units wait for their matching units in buffer C_i of capacity N_i. As soon as a pair is formed, it moves to a different buffer of infinite capacity. That is, buffers C_is operate as matching buffers but with zero assembly time. Consequently, C_1 and C_2 cannot have units at the same time.

TABLE 6.1. Numerical results of the recursive method.

μ_1	μ_2	μ_3	N_1	N_2	$\overline{N}_1$	$\overline{N}_2$	$\overline{N}_p$	$\overline{o}$	$P_1(B)$	$P_2(B)$
2	2	3	5	5	2.4525	2.4525	1.2228	1.7782	0.1109	0.1109
2	2	3	5	15	1.8968	7.8571	1.4130	1.8749	0.0626	0.0626
2	2	3	15	15	5.4117	5.4117	1.7263	1.9308	0.0346	0.0346
2	2	3	20	20	6.7176	6.7208	1.7971	1.9486	0.0256	0.0256
1	2	3	5	5	0.5161	4.5275	0.4916	0.9987	0.0013	0.5000
1	2	3	5	15	0.4959	14.5042	0.4959	0.9991	0.0009	0.5000
1	2	3	15	15	0.5000	14.5000	0.4999	0.9999	~ 0.0	0.5000
1	2	3	20	20	0.5004	19.4982	0.5001	1.0000	~ 0.0	0.5000

---

[1] Harrison (1973) showed that the delay of an arriving unit may have a very large value in an assembly system with infinite-capacity matching buffers, no matter what the arrival and the assembly rates are.

6.2. An Asynchronous Assembly System

Let $\overline{W}_b$ and $\overline{W}_c$ be the average delay of units in buffers B_i (our original buffers) and C_i, respectively. Clearly, $\overline{W}_b > \overline{W}_c$ due to the positive assembly time in the system with buffers B_i as opposed to the zero assembly time in the system with buffers C_i. This relationship also holds for the average number of units in the two systems. Let us denote the number of units in C_i at time t by $E_i(t)$ and show that its steady-state average approaches infinity as its capacity N_i does so. This in turn will show that our original system becomes unstable as N_i approaches infinity, due to the above relation between the average number of units in the two systems.

Let us study the excess process $E(t) = E_1(t) - E_2(t)$ whose domain is the set of integers $N_1, N_1 - 1, \ldots, 1, 0, -1, \ldots, -N_2$. Since the assembly time is zero, $E(t)$ is only affected by the arrivals at C_1 and C_2 (that is, job departures from stations 1 and 2). $E(t)$ is a continuous-time Markov chain whose transition diagram is given in Figure 6.4. The steady-state probabilities of E(t) can easily be found from the balance equations. Let us focus on the steady-state probability that buffer C_2 has N_2 units at any point in time, that is

$$\Pr(E(t) = -N_2) = \frac{\rho^{N_1+N_2}}{1 + \rho + \cdots + \rho^{N_1+N_2}}, \quad (6.14)$$

where $\rho = \mu_2/\mu_1$. As N_1 or N_2 (or both) approaches infinity, $\Pr(E(t) = -N_2)$ approaches a limit (by the L'Hospital rule). If $\rho \geq 1$, including in the symmetric case, $\Pr(E(t) = -N_2)$ converges to 1.0; it converges to 0.0 if $\rho < 1$. That is, infinitely many units accumulate in C_2, awaiting their matching units to arrive at C_1 when $\mu_2 > \mu_1$. The accumulation takes place in C_1 when $\mu_2 < \mu_1$. Both of the matching buffers experience the buildup in the symmetric case. The average number of excess units in any buffer C_i in the symmetric case is given by

$$\overline{N}_c = \frac{N(N+1)}{2(2N+1)} \quad (6.15)$$

which approaches $N/4$ for large N. Consequently, no matter what the values of μ_1, μ_2 and μ_3 are, our assembly system shown in Figure 6.2 is unstable for large N due to $\overline{N}_b > \overline{N}_c$. Thus, these systems should be designed with small buffers to eliminate large accumulations.

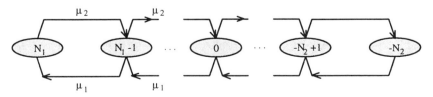

Figure 6.4. Transition diagram of the excess process $E(t)$.

6.2.4 Equivalence Properties

So far we have looked at the two-buffer assembly system with the after-processing-type blocking mechanism. In case the stations in the system cannot store the jobs during blocking, they get blocked immediately after the downstream buffer becomes full, experiencing the blocking-before-processing phenomenon. This new system has some interesting properties. It can again be studied through the analysis of the stochastic process $\{N_1(t), N_2(t)\}$, where $N_i(t) = 0, 1, \ldots, N_i$ with $N_i(t) = N_i$ representing the ith node being blocked. It is clear from the transition diagram of this stochastic process, shown in Figure 6.5, that its steady-state probabilities can be obtained using the recursive method developed earlier with the capacities of the matching buffers reduced by one. The equivalence of these two blocking mechanisms in two-node systems was already mentioned in Chapter 4 in the context of production lines.

Another property is the equivalence of the assembly system in Figure 6.2 to a production line in case both systems operate with the blocking-before-processing mechanism. Let us consider the two systems shown in Figure 6.6. In both systems, station 1 processes jobs as long as there is space in the downstream buffers B_1 and B'_1. Station 1 is blocked as soon as the downstream buffer is full in both systems. Station 3 in the production line operates as long as there are jobs in B'_1, and B'_2 is not full. Station 3 in the assembly system operates as long as there are jobs in both B_1 and B_2. Finally station 2 in the production line operates as long as B'_2 is not empty and station 2 in the assembly system operates as long as B_2 is not full.

Let n_i and n'_i represent the number of units in buffers B_i and B'_i at any point in time. Clearly, the number of units in B_1 and in B'_1 correspond to each other, and n_1 jobs in B_1 indicates $n'_1 = n_1$ jobs in B'_1. However, n_2 jobs in B_2 indicates $n'_2 = N_2 - n_2$ jobs in B'_2, implying that the contents of B'_2 are the empty cells in B_2. The transition diagrams of the underlying stochastic processes in both of

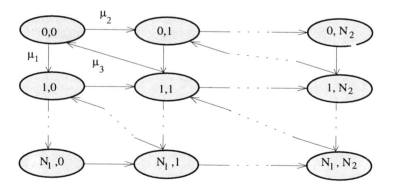

Figure 6.5. Transition diagram of $\{N_1(t), N_2(t)\}$ with blocking before processing.

6.2. An Asynchronous Assembly System   253

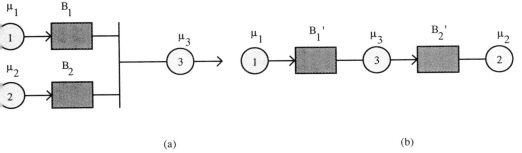

(a)   (b)

Figure 6.6. (a) and (b) are stochastically equivalent.

these systems have the same number of states, transitions and transition rates, but they have different states, as shown in Figure 6.7. Through this equivalence, with a reinterpretation of the state description, the steady-state probabilities of one system can be obtained from the probabilities of the other. For instance, $P_1(n_1) = \Pr(n_1' = n_1)$ and $P_2(n_2) = \Pr(n_2' = N_2 - n_2)$. Keep in mind that B_1 and B_2' both include the position on the processor at station 3 in the assembly system. Similarly, B_1' includes the position at station 3, and B_2' includes that at station 2 in the flow line. We already know how to analyze, exactly or approximately, the production lines with different types of blocking mechanisms. Thus, the steady-state performance measures of the assembly system can be obtained from the analysis of the equivalent production line. Unfortunately, this equivalence does not exist for the blocking-after-processing mechanism since the stochastic processes representing the two systems are slightly different, resulting in a different number of states and transitions.

EXAMPLE 6.1. Consider the two systems in Figure 6.6. Let us assume that the processing rates are $\mu_1 = 1.5$, $\mu_2 = 2.0$, and $\mu_3 = 0.75$, and the buffer capacities are $N_1 = 4$ and $N_2 = 6$. The blocking mechanism is the BBP type in both systems. The values of the measures of performance and the steady-state probabilities of the buffer contents in the two systems shown in Figure 6.6 are given in Table 6.2. The results clearly show the equivalence and the relationship between the contents of the

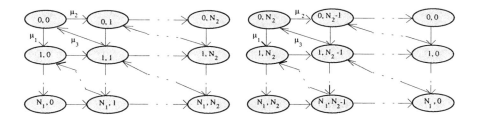

Figure 6.7. Transition diagrams of the systems shown in Figure 6.6.

TABLE 6.2. Comparison of the two systems in Figure 6.6

Assembly system

n_1	$P_1(n_1)$	n_2	$P_2(n_2)$
0	0.03209	0	0.00088
1	0.06439	1	0.00294
2	0.12898	2	0.00969
3	0.25814	3	0.02957
4	0.51641	4	0.08486
$\bar{N}_1$	3.16241	5	0.23475
		6	0.63731
		$\bar{N}_2$	5.44808

Output rate = 0.72539

Utilizations: $\underline{U} = (0.48359, 0.36269, 0.96719)$

Tandem System

n'_1	$P_1(n'_1)$	n'_2	$P_2(n'_2)$
0	0.03209	0	0.63731
1	0.06439	1	0.23475
2	0.12898	2	0.08486
3	0.25814	3	0.02957
4	0.51641	4	0.00969
$\bar{N}'_1$	3.16241	5	0.00294
		6	0.00088
		$\bar{N}'_2$	0.55192

Output rate = 0.72539

Utilizations: $\underline{U} = (0.48359, 0.96719, 0.36269)$

two buffers and between the processors. Notice that buffers B_1 and B'_1 have exactly the same probability distribution, whereas the probability distribution of buffer B'_2 is the reverse of that of buffer B_2. Consequently, $\overline{N}_1 = \overline{N}'_1$ and $\overline{N}_2 = N_2 - \overline{N}'_2$. The utilizations also reflect the topological relationship between the two systems. As expected, the two systems have the same output rate. The numerical results for the assembly system in Table 6.2 were obtained using the recursive method presented earlier, by reducing the capacities of the matching buffers by one. The results for the tandem system were obtained using the iterative **power method** presented in Section 4.3.3.

6.3 Approximate Analysis of the Two-Station Assembly System

Experiments with the preceding recursive method show that it may be unstable when N_i assumes large values (i.e., ≥ 30). This is due to the subtractions involved in the equations, which result in negative probabilities when the state space of the Markov chain is very large. Furthermore, the generalization of the recursive procedure for systems with more than two matching buffers is quite difficult, again due to the large state space of the Markov chain. Therefore, approximations and numerical methods are useful in the analysis of larger systems.

Similar to the approximations developed for the analysis of production lines in Chapter 4, the concept of decomposition can be used to approximate the behavior of the assembly systems. In fact, we have just mentioned the equivalence between assembly systems and production lines. Using the concept of two-node decomposition introduced in Chapter 4, one can isolate the matching buffers in the assembly system shown in Figure 6.2 by assigning each an imaginary downstream station. The processing times of the imaginary stations are very similar to those of the upstream nodes in the two-node decomposition of production lines. Upon departure of an assembled unit from station 3, the time until the next departure occurs is exponentially distributed with rate μ_3 if both buffers have units waiting to be assembled. It is likely that upon a departure, the next assembly cannot be started immediately due to a lack of units in either or both of the matching buffers. For instance, in case a departure leaves B_2 empty, the next assembly has to wait until a unit arrives at B_2. In case both buffers are empty, the next assembly can start only after units arrive at both buffers. Let Δ_i be the probability that an assembled unit leaves B_i empty. The processing time of the imaginary downstream station may be modeled using the MGE distribution shown in Figure 6.8. Consequently, the system is decomposed into two subsystems, $\Omega(1)$ and $\Omega(2)$, as shown in Figure 6.9. Furthermore, as mentioned earlier in Chapter 4, stations 1 and 2 can be replaced by the Poisson arrival processes with rates μ_1 and μ_2 provided that the capacities of B_1 and B_2 are increased by 1. This equivalence uses the memoryless property of the exponential distribution. Then, $\Omega(i)$ can be studied as an M/MGE-2/1/$N_i + 1$

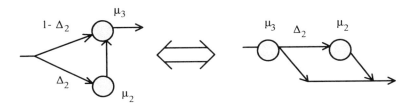

Figure 6.8. Processing time of the imaginary station in $\Omega(1)$.

256   6. Assembly Systems

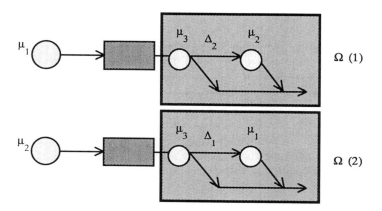

Figure 6.9. Decomposition of the assembly system.

work station from which Δ_i can be obtained using

$$\Delta_i = \frac{\mu_i}{\bar{o}_i} P_i(0) = \frac{P_i(0)}{1 - P_i(N_i + 1)}. \qquad (6.16)$$

The left-hand side of (6.16) is merely Little's law. Also, the right-hand side can be directly written using the following probabilistic argument. The idleness probability upon departure, that is, Δ_i, is the ratio of the number of times a departure leaves an empty system to the number of times a departure may leave any possible number of units in the system. Clearly, the possible numbers are $0, 1, 2, \ldots, N_i$, with the total probability of $1 - P_i(N_i + 1)$.

Then, a simple iterative method can be constructed to relate the subsystems to each other as Table 6.3 shows.

The M/MGE-2/1/N_i + 1 work station can easily be analyzed by studying the underlying Markov chain $\{N_t, J_t\}$ where N_t is the number of units in the system at time t and J_t is the indicator of the phase the machine is in at time t. We refer the reader to Chapter 3 for a recursive or a matrix-geometric analysis of the M/MGE-2/1/N_i+1 work station.

TABLE 6.3. Decomposition algorithm for the two-node assembly system.

INITIALIZATION:	$\Delta_2^{(o)} = 0.1$ (arbitrary), $j = 1$.
STEP 1:	Analyze $\Omega(1)$ for $P_1^{(j)}(0)$, $P_1^{(j)}(N_1 + 1)$, $\Delta_1^{(j)}$ and $\bar{o}_1^{(j)}$.
STEP 2:	Analyze $\Omega(2)$ for $P_2^{(j)}(0)$, $P_2^{(j)}(N_2 + 1)$, $\Delta_2^{(j)}$ and $\bar{o}_2^{(j)}$. If $|\bar{o}_1^{(j)} - \bar{o}_2^{(j)}| \le \varepsilon$, stop. Else, $j = j + 1$, go to STEP 1.

The above iterative procedure converges very fast. The approximate values of the performance measures for the problems presented in Table 6.1 are shown in Table 6.4. The accuracy of the results are acceptable possibly with the exception of the symmetric case with large N (e.g. > 15). As N gets larger, contrary to the instability argument mentioned in Section 6.2.2, the steady-state average number of units in the approximate subsystems converges to a finite value. That is, the approximation does not behave well in the symmetric case with large N. The reason is that the processing time of the imaginary stations cannot capture the dependence among the matching buffers in this case. The following simple approximation can be used to circumvent this situation.

6.3.1 Approximation for the Symmetric Case with Large N

We have mentioned earlier that as N increases, the number of units in the system becomes dominated by the units (the excess process) waiting for their matching units to arrive. The expected number of excess units in B_i approaches the number of units in buffer C_i in the scenario mentioned earlier to explain the instability issue. It is $N/4$ for large N, as can be obtained from (6.15). Furthermore, in the symmetric case with $\mu_1 = \mu_2 = \mu$, the steady-state expected number of pairs in the system is bounded above by the average number of units in an M/M/1 work station with $\rho = \mu/\mu_3 < 1$. That is, the average arrival rate of pairs is less than μ due to the fact that upon a unit's arrival, say to C_1, instantaneously a pair arrives at the assembly station only if there is a job in C_2 waiting for its matching unit. The complementary event, that an arrival does not see its matching unit, ensures the pair arrival rate to be less than μ. The average number of units in an M/M/1 system is $\overline{N}(M/M/1) = \rho/(1 - \rho)$. Notice that $\overline{N}(M/M/1) > \overline{N}_p$ in the symmetric cases in Table 6.4. Consequently, the average number of units in matching buffers B_1 and B_2 can be asymptotically approximated by

$$\overline{N}_i \cong \frac{N}{4} + \frac{\rho}{(1-\rho)}, \quad i = 1, 2, \tag{6.17}$$

TABLE 6.4. The approximate results of the problems in Table 6.1.

μ_1	μ_2	μ_3	N_1	N_2	$\overline{N}_1$	$\overline{N}_2$	$\overline{o}$
2	2	3	5	5	2.4362	2.4362	1.7992
2	2	3	5	15	1.8289	7.4994	1.8986
2	2	3	15	15	4.3610	4.3611	1.9772
2	2	3	20	20	4.7462	4.7464	1.9911
1	2	3	5	5	0.5157	4.5404	0.9989
1	2	3	5	15	0.4959	14.500	0.9991
1	2	3	15	15	0.5000	14.500	0.9999
1	2	3	20	20	0.5004	19.500	1.0000

258   6. Assembly Systems

which results in 5.75 (6 percent relative error) for $N = 15$, and 7.00 (4 percent relative error) for $N = 20$. Clearly, these results are acceptable and they become more accurate as N increases. The choice between the two approximations can be done as follows: For varying values of N, the difference between the $\overline{N}$ values produced by the two approximations is small in the middle, and it is large at either ends. We know that the decomposition results are accurate for smaller values of N, and the asymptotic results are more accurate for larger values of N. Thus, a few runs of both of the methods can indicate the break-even point according to which one may select the method for a given value of N.

If the assembly operation requires a different number of units from the matching buffers, then the individual subsystems have to be modeled as batch-service queues. For instance, if one unit from B_1 is assembled with two units from B_2, the characterization of $\Omega(1)$ needs modification while that of $\Omega(2)$ remains the same. The processing time at the second node of $\Omega(1)$ has an additional phase due to the arrival of another unit to B_2, as shown in Figure 6.10. The probabilities $\Delta_{2,0}$ and $\Delta_{2,1}$ are the probabilities that a departing unit leaves no units or one unit at B_2, respectively. Also, the second node in $\Omega(1)$ becomes a batch server that processes two units at a time. The analysis of batch-service (or bulk service) queues[1] can again be done by studying the underlying Markov chains.

6.3.2 The Behavior of the Approximate Method for Large N

We have shown the existence of instability in the assembly systems with infinite buffer capacities. For instance, we know by now that the first buffer $\Omega(1)$ is going to be unstable when $\mu_1 > \mu_2$. Let us study the behavior of $\Omega(1)$ under this condition. Remember that Δ_i is the departure-point probability of leaving behind an empty system. We have mentioned in Chapter 3 that $\Delta_i = P_i(0) = 1 - \rho_i$ in an M/G/1 system with an infinite queue capacity[2] where $P_i(0)$ and ρ_i are the

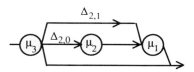

Figure 6.10. Processing time at $\Omega(1)$ if two units are required from B_1.

---

[1] In bulk-service queues, the service is given to a group of units at once. The server waits for the group to form, and the whole group departs after the service is completed. The group (or the bulk) size may be a random variable.

[2] In any queueing system where arrivals and departures occur one at a time, the steady-state distribution of the number of units seen by an arrival is identical to the distribution of the number of units left behind by a departure. In steady state, the number of arrivals seeing n in the system can differ only by one from the number of departures leaving n behind in any interval. Therefore, for a large number

steady-state probabilities that the server is idle and busy (utilization) at any point in time in $\Omega(1)$, respectively. The utilization at $\Omega(1)$ is

$$\rho_1 = \mu_1 \left(\frac{1}{\mu_3} + \frac{1 - \rho_2}{\mu_2} \right). \tag{6.18}$$

If μ_2 in the denominator is replaced by μ_1, (6.18) can be rewritten as

$$\rho_1 > \frac{\mu_1}{\mu_3} + 1 - \rho_2 = 1 + \frac{\mu_1}{\mu_3}, \tag{6.19}$$

showing that the utilization at $\Omega(1)$ is greater than 1.0, and as a result $\Omega(1)$ is unstable. (6.19) utilizes the expression for ρ_2 given below.

Let us observe how $\Omega(2)$ behaves when $\mu_1 > \mu_2$. The traffic intensity in $\Omega(2)$ is

$$\rho_2 = \mu_2 \left(\frac{1}{\mu_3} + \frac{1 - \rho_1}{\mu_1} \right).$$

By replacing μ_2 by μ_1, we obtain

$$\rho_2 < \mu_1 \left(\frac{1}{\mu_3} + \frac{1 - \rho_1}{\mu_1} \right) = \frac{\mu_1}{\mu_3} + 1 - \rho_1. \tag{6.20}$$

Substituting ρ_1 by its expression, we obtain $\rho_2 < 1$. This shows that for large N, the buffer having the smaller arrival rate experiences insignificant levels of accumulation.

To summarize, the buffers in the original system and those in the approximate systems behave quite analogously when $\mu_1 \neq \mu_2$. When $\mu_1 = \mu_2 = \mu$, clearly the original system as well as the approximate subsystems will be unstable when $\mu > \mu_3$ for large N. When $\mu_1 = \mu_2 = \mu < \mu_3$, we have already shown the unacceptable behavior of the approximate subsystems and have proposed to use a simpler approximation in this case.

6.4 Analysis of Larger Systems

Assembly systems with a larger number of matching buffers or networks of assembly stations can also be studied through decomposition approximations. These systems can be decomposed into several subsystems to be analyzed in isolation.

---

of units, the proportion of the number of arrivals seeing n becomes identical to the proportion of the number of departures leaving n behind. In case the arrivals follow a Poisson distribution, the above distribution is identical to the distribution of the number of units in the system at any point in time. See Cooper (1990) for details.

260   6. Assembly Systems

Each subsystem is equipped with imaginary stations whose behavior approximately replicates the behavior of some part of the original system. This is in fact a standard decomposition of any large-scale system. Next, we introduce decomposition approximations for different assembly systems, starting with what we have analyzed earlier but with more than two matching buffers.

6.4.1. More Than Two Matching Buffers

Consider an assembly station with three matching buffers, as shown in Figure 6.11. The assembly operation starts only if there are units in all three buffers. The finished unit is assembled by using one unit from each of the buffers. The characterization of the subsystems for buffers 1, 2, and 3 are identical. Therefore, we will focus on one of them, i.e., $\Omega(1)$. Just after an assembled unit departs from the system at the "marked time," leaving B_1 nonempty, the time until the next departure occurs at $\Omega(1)$ is exponentially distributed with rate μ_3 if both B_2 and B_3 are not empty at the marked time. On the other hand, the next assembly operation may be delayed by an exponentially distributed amount of time with mean μ_2 if B_2 is empty and B_3 is not at the marked time. It may be delayed by an exponential period with rate μ_3 if B_3 is empty and B_2 is not at the marked time. Finally, the next assembly may start only after units arrive at both B_2 and B_3 if they are both left empty at the marked time. Clearly, in this last case, the time until the next assembly starts is the maximum of the processing times at stations 2 and 3. Let X_i represent the processing time at the ith subassembly station and $Z_1 = \max(X_2, X_3)$. Then, $\Omega(1)$ can be characterized as shown in Figure 6.12.

To analyze $\Omega(1)$, we first need to deal with Z_1, the delay experienced in $\Omega(1)$. Notice that in the case of four matching buffers, we have to deal with Z_2, Z_3, Z_4 and $\max(X_2, X_3, X_4)$, as well as $\max(X_2, X_3)$, $\max(X_2, X_4)$, and $\max(X_3, X_4)$ in

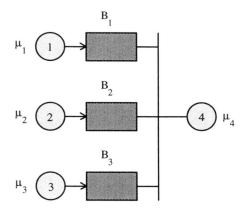

Figure 6.11. An assembly system with three matching buffers.

6.4. Analysis of Larger Systems

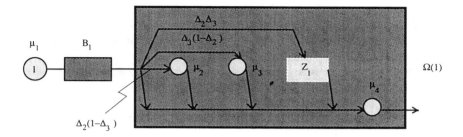

Figure 6.12. $\Omega(1)$ in isolation.

the analysis of $\Omega(1)$. As shown in Chapter 2, the CDF of Z_1 is

$$P(Z_1 \leq z) = 1 - e^{-\mu_2 z} - e^{-\mu_3 z} - e^{(-\mu_2 + \mu_3)z}, \qquad z \geq 0. \qquad (6.21)$$

Its first and second moments are

$$m_1 = \frac{1}{\mu_2} + \frac{1}{\mu_3} - \frac{1}{\mu_2 + \mu_3}, \qquad m_2 = \frac{2}{\mu_2^2} + \frac{2}{\mu_3^2} - \frac{2}{(\mu_2 + \mu_3)^2} \qquad (6.22)$$

It can be shown that $0.5 \leq \text{Cv}_{Z_1}^2 \leq 1.0$. Moreover, it can be shown that the squared coefficient of variation of the maximum of m independent exponential random variables is in $(1/m, 1.0)$. Thus, Z_1 can be modeled by using the two-moment approximations depicted in Figure 6.13. Notice that the analysis of $\Omega(2)$ requires $Z_2 = \max(X_1, X_3)$, and $Z_3 = \max(X_1, X_2)$ has to be obtained for the analysis of $\Omega(3)$.

The analysis of $\Omega(i)$ can be done by studying the underlying M/MGE-k/1/N_i + 1 system. The steady-state balance equations of the Markov chain $\{N_t, J_t, t \geq 0\}$ incorporates all the transitions due to the phase structure of the processing time at the imaginary station. Similar to the case where $k = 2$, it can again be analyzed through a matrix-recursive or matrix-geometric approach. The iterative procedure that relates all the subsystems to each other has exactly the same structure as the one in the two-node case.

Figure 6.13. Two-moment representation of Z_1.

6.4.2 A Network of Assembly Stations

In networks of assembly stations, the approximations are in the form of decompositions into subnetworks that in turn may be analyzed using any of the above-mentioned approaches. For instance, let us concentrate on the network given in Figure 6.1. Stations 3 and 5 are assembly stations. Station i has an exponentially distributed processing time with rate μ_i. Stations 1, 2, and 4 are always busy except when blocked, and the blocking mechanism is the BAP type. Each assembly station uses one unit from each of its matching buffers to assemble a semifinished or finished unit. One way of decomposing this system is to isolate stations 1, 2, and 3 to form $\Omega(1)$, and stations 3, 4, and 5 to form $\Omega(2)$ by characterizing an effective processing time for station 3 in both $\Omega(1)$ and $\Omega(2)$, as shown in Figure 6.14. Further decompositions may be employed in the analysis of Ωs in isolation.

The effective processing time at station 3 of $\Omega(1)$ is constructed by incorporating the effect of blocking coming from buffers B_3 and B_4 into the processing time at this station. A departing assembly may get blocked at station 3 due to lack of space in B_3. At this instance, station 5 may be operating, or it may be idle waiting for a unit to arrive at B_4. The time until blocking is over is exponentially distributed with rate μ_5 in the former case. It has an MGE-2 distribution in the latter case, as shown in Figure 6.14. The idleness experienced at station 3 in the original system is modeled as part of its effective processing time in $\Omega(2)$. This is similar to the blocking that the stations experience in the three-buffer system. Let ℓ_i, $i = 1, 2$, be the number of units left in B_i by a departing assembly at station 3. Upon a departure, station 3 may become idle due to $\ell_1 = 0$, $\ell_2 > 0$ for an exponentially distributed time with mean μ_1^{-1}, or due to $\ell_1 > 0$, $\ell_2 = 0$ for an exponentially distributed period with mean μ_2^{-1}. Finally, upon a departure, it is possible to have $\ell_1 = 0$, $\ell_2 = 0$ such that the idle period is the maximum of the processing times at stations 1 and 2. Notice that the throughputs of all the stations are expected to be identical for the same reason mentioned earlier for the two-buffer simpler systems.

An iterative procedure similar to the one employed for the two-buffer systems can be developed to analyze the above network of assembly systems. It starts with the analysis of $\Omega(1)$ assuming values for Π_3 and Δ_4. $\Omega(1)$ can be analyzed either exactly or approximately to obtain the steady-state probabilities of the number of units in its buffers. From this analysis, Δ_1 and Δ_2 are obtained to be used in the analysis of $\Omega(2)$ which in turn provides the values of Π_3 and Δ_4. The procedure continues until the throughputs of $\Omega(1)$ and $\Omega(2)$ are sufficiently close to each other. The subsystems in $\Omega(1)$, in case $\Omega(1)$ is approximated, are M/MGE-k/1/$N + 1$ work stations that can be analyzed as mentioned earlier. On the other hand, the subsystems in $\Omega(2)$ are more involved because of the phase structure of station 3. The first subsystem can be analyzed exactly or approximately as a two-node tandem system as we have done in Chapter 4. The second subsystem is again an M/MGE-k/1/$N + 1$ queueing system. In the analysis of large-scale systems, it is usually computationally beneficial to reduce the number of phases by moment-matching arguments discussed in Chapter 2, at the expense of insignificant deterioration in the accuracy.

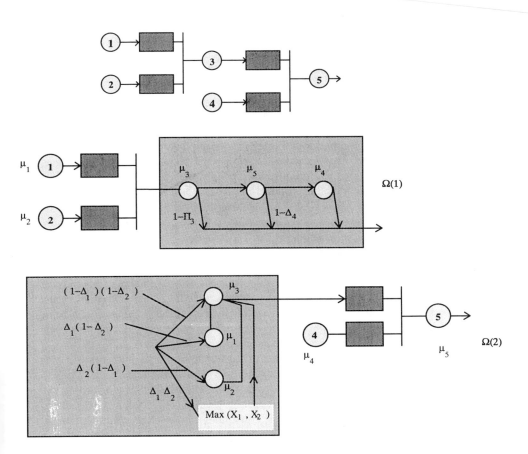

Figure 6.14. Decomposition of a network of assembly stations.

6.5 Synchronous Assembly Systems

In synchronous systems, similar to the production lines, every station processes a unit and transfers it to the downstream buffer within the same period of time. For instance, in the two-buffer system shown in Figure 6.2, at the end of the processing time, stations 1 and 2 each place a unit into B_1 and B_2, respectively. At the end of the same processing time, station 3 ejects an assembled unit. The concepts of conveyors and possibly the secondary buffers are still valid for the synchronous assembly systems. Each station in the system is subject to failures and the subsequent repairs. These systems can be analyzed by using approximation techniques similar to the ones used by Gershwin (1988) and the ones we have implemented for the synchronous production lines in Chapter 5. For instance, in the two-buffer system, there is a two-station synchronous subsystem for every buffer. The downstream stations in both subsystems are assigned fictitious failure and repair rates that relate the subsystems to each other. These rates are obtained in exactly the same way their counterparts are obtained in synchronous production lines. Also, an iterative method relates the subsystems to each other.

6.6 Relaxing the Assembly Assumption

In case the assembly assumption is relaxed, the system becomes a network of work stations linked to each other according to the process plans of the products. There are no assembly stations assembling units from different buffers. Each station has its own buffer, and all the products in a given buffer go through the same process that the station is designed for. Hence, the system looks like a group of arbitrarily connected work stations. Blocking still exists due to the limited space in the buffers. A simple example of such a system is the merge configuration shown in Figure 6.15. The processing time at station i is exponentially distributed with rate μ_i. Units departing from stations 3 and 4 join buffer B_3 to be individually processed at station 5. When station 3 is blocked, the time it remains so depends on whether or not station 4 is already blocked at this moment. When both stations are blocked, the one blocked earlier is released first when space becomes available in B_3. Clearly, it is possible to employ different blocking and release mechanisms such as last-blocked-first-released, or priority schemes, among others in this system.

The merge system is a special case of queueing networks with finite-queue capacities. These systems can be analyzed by studying the underlying Markov chains that keep track of the number of units in each buffer and of the server status. For instance, the stochastic process $\{N_1(t), N_2(t), N_3(t), t \geq 0\}$ where $N_i(t)$ is the number of units in buffer B_i constitutes a finite-state, continuous-time Markov chain. If (n_1, n_2, n_3) is a particular state of this process, then $n_i = 0, 1, \ldots, N_i, N_{i+1}$ for $i = 1, 2$, where N_{i+1} represents that station i is blocked. On the other hand, $n_3 = 0, 1, \ldots, N_3, N_3 + 1, N_3 + 2, N_3 + 3, N_3 + 4$, where $N_3 + 1$ represents that station 3 is blocked, $N_3 + 2$ represents that station 4 is blocked, and $N_3 + 3, N_3 + 4$ represent the cases in which both stations 3 and 4 are blocked in the order of 3-4 and 4-3, respectively. Any equation-solving technique including direct, numerical or recursive techniques briefly mentioned in Chapter 4 can be used to solve for the steady-state probabilities. Clearly, as the number of stations or the buffer capacities increase, these approaches become infeasible and approximations or simulation becomes preferable.

Let us briefly focus on the decomposition of the merge system into three subsystems, one for each buffer, as shown in Figure 6.16. The subsystems for buffers

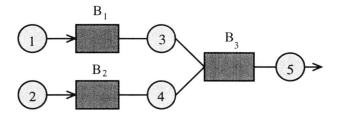

Figure 6.15. A network of work stations with no assembly operation.

B_1 and B_2, namely $\Omega(1)$ and $\Omega(2)$, are identical to each other. The time either of stations 3 and 4 remains blocked is an Erlang random variable with two phases with a phase rate μ_5 (if it is blocked second), or with one phase with the same rate (if it is blocked first). Let Π_k be the probability that station k ($k = 3, 4$) is blocked upon a departure. Also, let $\pi_{k,j}$ be the probability that station k is blocked given that station j is already blocked. Then, $\Omega(i)$, $i = 1, 2$, can be characterized as M/MGE-3/1/N_i + 1 work stations as shown in Figure 6.16. Let $P_i(n_i)$ be the steady-state probability of having n_i units in $\Omega(i)$, $i = 1, 2$, for future reference. Clearly, Π_i, $\pi_{3,4}$, and $\pi_{4,3}$ from the analysis of $\Omega(3)$ are necessary for the analysis of $\Omega(i)$. On the other hand, once $\Omega(i)$ is analyzed, Δ_i is obtained using (6.16) for the analysis of $\Omega(3)$.

The characterization of $\Omega(3)$ is more involved since it has two arrival streams. For the units departing from station 3, with probability $1 - \Delta_3$ (that a departure leaves station 3 busy), the time between two consecutive arrivals to B_3 is an exponential random variable with rate μ_3. It is an Erlang random variable with rates

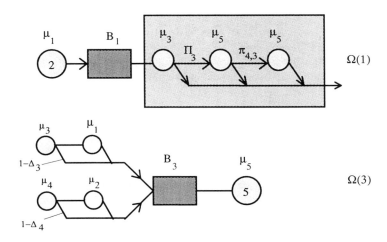

Figure 6.16. Decomposition of the merge system.

μ_1 and μ_3 with probability Δ_3 (that a departure leaves station 3 idle). The interarrival time of the units coming from station 4 has a similar construction with rates μ_4 and μ_2 and the departure-point probability Δ_4. Consequently, $\Omega(3)$ has two streams of MGE-2 arrivals with respective rates at buffer B_3, as shown in Figure 6.16. Then, $\Omega(3)$ is analyzed as a three-node system where nodes 3 and 4 are always busy except when blocked. They experience the same type of blocking mechanism as stations 3 and 4 do in the original system. The stochastic process $\{N_3(t), t \geq 0\}$ representing the number of units in B_3 gives rise to a Markov chain. Let n_3 be a particular state of $\{N_3(t)\}$ with $P_3(n_3)$ being its steady-state probability. The states in which nodes 3 and 4 are blocked are $n_3 = N_3 + 1$ (node 3 is blocked), $n_3 = N_3 + 2$ (node 4 is blocked), $n_3 = N_3 + 3$ (both nodes are blocked, node 3 is blocked first), and $n_3 = N_3 + 4$ (both blocked and node 4 is blocked first). Consequently, the steady-state probability that node 3 is blocked is $P_3(B) = P_3(N_3 + 1) + P_3(N_3 + 3) + P_3(N_3 + 4)$. This is also the approximate probability that station 3 is blocked in the original system. Similarly, the blocking probability of station 4 becomes $P_4(B) = P_3(N_3 + 2) + P_3(N_3 + 3) + P_3(N_3 + 4)$. Then the throughput of any of the nodes 3 and 4 with the mean effective processing times of $\bar{x}_3$ and $\bar{x}_4$, respectively, can be written as

$$\bar{o}_i = \frac{[1 - P_i(B)]}{\bar{x}_i}, \quad i = 3, 4. \tag{6.23}$$

The probability that node 4 is blocked given that node 3 is blocked, $\pi_{4,3}$, is obtained using the argument that the unit causing blocking at station 3 sees N_3 units in B_3. Now the question is whether or not it sees node 4 blocked. Hence, $\pi_{4,3}$ may be obtained by

$$\pi_{4,3} = \Pr(4 \text{ is blocked} \mid 3 \text{ is blocked})$$
$$= \frac{P_3(N_3 + 3) + P_3(N_3 + 4)}{P_3(N_3 + 1) + P_3(N_3 + 3) + P_3(N_3 + 4)}, \tag{6.24}$$

and Π_3 is obtained using

$$\Pi_3 \bar{o}_3 \left(\frac{1 + \pi_{4,3}}{\mu_4} \right) = P_3(B), \tag{6.25}$$

which is to be used in the analysis of $\Omega(1)$. Similarly, $\pi_{3,4}$ and Π_4 are obtained in the same manner for use in the analysis of $\Omega(2)$.

Next, it is not difficult to incorporate the above subsystems into an iterative algorithm. Each subsystem is analyzed by studying the associated queue in the order of $\Omega(1)$, $\Omega(2)$, $\Omega(3)$, and back to $\Omega(1)$ again. This procedure is similar to the ones employed earlier in the assembly systems and therefore will not be repeated here.

The analysis of assembly systems becomes more difficult in the case of nonexponential processing times. Phase-type processing times necessitate either the exact

Markovian approach or an approximate decomposition approach. The Markovian approach may be feasible through the use of numerical methods, which will have large computer memory requirements. Extensions of the approximations presented in this chapter are still feasible provided that a well-thought validation process is implemented. It will require extensive testing against simulation results. Cases with deterministic processing times will necessitate a treatment similar to the one developed for the production lines with deterministic processing times (see Chapter 4). Considering the complexity of the problem, simulation modeling may be the only choice for the analysis of large-scale assembly systems with general processing times.

Related Bibliography

The literature on assembly-related systems may be grouped under a few categories. These may be the assembly systems, assembly/disassembly networks, and the general queueing networks with finite queue capacities. In the assembly systems, Harrison (1973), Crane (1974), Latouche (1981), and Neuts (1981) considered infinite-capacity, single- or multiple-server assembly queues consisting of independent arrival streams. These studies deal with the inherent instability problem in the assembly queues and obtain the steady-state distribution of the excess process or the number of groups ready for assembly in the system. Kashyap (1965), Bhat (1986), and Lipper and Sengupta (1986) studied the assembly system where each arrival stream has a finite-capacity buffer. The server takes one unit from each buffer and assembles the final product. Lipper and Sengupta showed that the average number of pairs in the system in Figure 6.2 is bounded above by the average number of jobs in an M/M/1 queueing system with a traffic intensity of $\rho = \mu/\mu_3$, where $\mu_1 = \mu_2 = \mu$. The asymptotic approximation (6.17) is a consequence of this bound. The excess process $E(t)$, (6.14), and (6.15) are due to Lipper and Sengupta (1986). Sengupta (1989) also looked at an assembly queue with an arrival process where units arrive with some deviation from the scheduled arrival points in time.

The equivalence between assembly systems and production lines with BBP-type blocking mechanism were shown by De Koster (1987), Ammar and Gershwin (1989), and Dallery et al. (1991), among others.

Assembly/disassembly networks (ADN) or fork/join systems consist of assembly and disassembly nodes where upon departure a unit is placed into every connecting downstream buffer. Mandelbaum and Avi-Itzhak (1968), Flatto and Hahn (1984), Nelson and Tantawi (1988), Rao and Posner (1985), and Baccelli et al. (1989) studied the ADN systems where the arriving units split into a number of units that go through several processes and are merged into a single departing unit when all the tasks are completed. Avi-Itzhak and Halfin (1991a and b) extended the ADN systems to include different classes with priorities. Gershwin (1991), Di Mascolo et al. (1990), and Lipper and Sengupta (1986) developed decomposition approximations for the steady-state distribution of the number of jobs in a network of synchronous, unreliable assembly/disassembly systems. De Koster (1987) developed approximating arguments based on the continuous flow assumptions. Other related studies include Adan and Van Der Wal (1988), Baccelli and Makowski (1985), Baccelli, Massey, and Towsley (1989), and Konstantopoulos and Walrand (1989).

Finally, as we have mentioned earlier, once the assembly and disassembly assumptions are relaxed, the assembly/disassembly systems become equivalent to queueing networks with an arbitrary configuration and with finite queue capacities. Altiok and Perros (1987), Kerbache and Smith (1987), Jun and Perros (1988), and Lee and Pollock (1990) proposed decomposition approximations similar to the ones employed in Chapters 4 and 6 for these queueing networks.

References

Adan, I., and J. Van Der Wal, 1988, Monotonicity of the Throughput in Single Server Production and Assembly Networks with Respect to the Buffer Sizes, *Proc. 1st Int. Workshop, Queueing Networks with Blocking* (Perros and Altiok, eds.), North Carolina State Univ., Raleigh, NC.

Altiok, T., and H. G. Perros, 1987, Approximate Analysis of Arbitrary Configurations of Open Queueing Networks with Blocking, *Annals of O.R.*, Vol. 9, pp. 481–509.

Ammar, M. H., and S. B. Gershwin, 1989, Equivalence Relations in Queueing Models of Assembly/Disassembly Networks, *Performance Evaluation*, Vol. 10, pp. 233–245.

Avi-Itzhak, B., and S. Halfin, 1991a, *Non-Preemptive Priorities in Simple Fork-Join Queues, Queueing, Performance and Control in ATM, ITC-13* (Cohen and Pack, eds.). Elsevier Science Publishers, New York.

Avi-Itzhak, B., and S. Halfin, 1991b, Priorities in Simple Fork-Join Queues, RUTCOR Research Report No. 32–91, Rutgers University, Piscataway, New Jersey.

Baccelli, F., and A. M. Makowski, 1985, Simple Computable Bounds for the Fork-Join Queue, *Proc. 19th Ann. Conf. Inform. Sci. and Systems*, The Johns Hopkins University Press, Baltimore, MD, pp. 436–441.

Baccelli, F., A. M. Makowski, and A. Schwartz, 1989, The Fork-Join Queue and Related Systems with Synchronization Constraints: Stochastic Ordering, Approximations and Computable Bounds, *Adv. Appl. Prob.*, Vol. 21, pp. 629–660.

Baccelli, F., W. A. Massey, and D. Towsley, 1989, Acyclic Fork-Join Queueing Networks, *J. Assoc. Comp. Mach.*, Vol. 36.

Bhat, U. N., 1986, Finite-Capacity Assembly-Like Queues, *Queueing Systems*, Vol. 1, pp. 85–101.

Cooper, R. B., 1990, *Introduction to Queueing Theory*, 3rd edition, Ceep Press, Washington, DC.

Crane, M. A., 1974, Multi-Server Assembly Queues, *J. Appl. Prob.*, Vol. 11, pp. 629–632.

Dallery, Y., Z. Liu, and D. Towsley, 1991, Equivalence, Reversibility, Symmetry, and Concavity in Fork/Join Queueing Networks with Blocking, *To Appear in JACM*.

De Koster, M. G. M., 1987, Approximation of Assembly/Disassembly Systems, Rep. BDK-ORS-87-02, Dept. of IE, Eindhoven University of Tech., The Netherlands.

Di Mascolo, M., R. David, and Y. Dallery, 1991, Modeling and Analysis of Assembly Systems with Unreliable Machines and Finite Buffers, *IIE Transactions*, Vol. 23, pp. 315–331.

Flatto, L., and S. Hahn, 1984, Two Parallel Queues Created by Arrivals with Two Demands I, *Siam J. Appl. Math.*, Vol. 44, pp. 1041–1053.
Gershwin, S. B., 1991, Assembly/Disassembly Systems: An Efficient Decomposition Algorithm for Tree-Structured Networks, *IIE Transactions*, Vol. 23, pp. 302–314.
Harrison, J. M., 1973, Assembly-Like Queues, *J. Appl. Prob.*, Vol. 10, pp. 354–367.
Jun, K. P. and H. G. Perros, 1988, Approximate Analysis of Arbitrary Configurations of Queueing Networks with Blocking and Deadlock, *Proc. 1st Int. Workshop, Queueing Networks With Blocking* (Perros and Altiok, eds.), North Carolina State Univ., NC.
Kashyap, B. R. K., 1965, A Double Ended Queueing System with Limited Waiting Space, *Proc. of Natl. Inst. Sci. India*, A31, pp. 559–570.
Kerbache, L., and J. M. Smith, 1987, The Generalized Expansion Method for Open Finite Queueing Networks, *E. J. Oper. Res*, Vol. 32, pp. 448–461.
Konstantopoulos, P., and J. Walrand, 1989, Stationary and Stability of Fork-Join Networks, *J. Appl. Prob.*, Vol. 26, pp. 604–614.
Latouche, G. 1981, Queues with Paired Customers, *J. Appl. Prob.*, Vol. 18, pp. 684–696.
Lee, H., and S. M. Pollock, 1990, Approximation Analysis of Open Acyclic Exponential Queueing Networks with Blocking, *Operations Research*, Vol. 38, pp. 1123–1134.
Lipper, E. H., and B. Sengupta, 1986, Assembly-Like Queues with Finite Capacity: Bounds, Asymptotics and Approximations, *Queueing Systems*, Vol. 1, pp. 67–83.
Liu, X. G., 1990, Toward Modeling Assembly Systems: Applications of Queueing Networks with Blocking, Ph.D. Thesis, University of Waterloo, Waterloo, Ontario.
Mendelbaum, M., and B. Avi-Itzhak, 1968, Introduction to Queueing with Splitting and Matching, *Isr. J. Tech.*, Vol. 6, pp. 376–382.
Nelson, R., and A. N. Tantawi, 1988, Approximate Analysis Fork/Join Synchronization in Parallel Queues, *IEEE Transactions*, Vol. 37, pp. 739–743.
Neuts, M. F., 1981, *Matrix-Geometric Solutions in Stochastic Models–An Algorithmic Approach*, The Johns Hopkins University Press, Baltimore, MD.
Rao, B. M., and M. J. M. Posner, 1985, Algorithmic and Approximation Analysis of the Split and Match Queue, *Stochastic Models*, Vol. 1, pp. 433–456.
Sengupta, B., 1989, On Modeling the Store of an Assembly Shop by Due Date Processes, *Operations Research*, Vol. 37, pp. 437–448.

Problems

6.1. Consider the two-node assembly system shown in Figure 6.2. The assembly times are all exponentially distributed with rates $\mu_1 = 2$, $\mu_2 = 3$, and $\mu_3 = 4$. The buffer capacities are $N_1 = N_2 = 3$.

(a) Compute the values of the steady-state average number of units of type 1 and 2 in the system, the average number of pairs, and the average number of excess units in the system using the recursive method presented in Section 6.2.

(b) Compute the steady-state output rate and the average number of units in the system using the approximation method introduced in Section 6.3.

(c) How would the exact and the approximate methods be revised for an Erlang assembly time at station 3, with two phases and a phase rate of 2?

270   6. Assembly Systems

(d) For buffer capacities of 10 each, carry out both the decomposition and the Lipper and Sengupta approximations to obtain the steady-state average number of units in the system. Which one is more accurate?

6.2. Again consider the system in Figure 6.2 with $\mu_1 = 1, \mu_2 = 2, \mu_3 = 1, N_1 = N_2 = 3$ but with the following exception: One unit from B_1 is assembled with three units from B_2. The assembly operation does not start if there are less than two units in B_2. How should the recursive and the approximate methods be revised? Compute the performance measures.

6.3. Assume that when an assembly leaves station 3 (of Figure 6.2), the next assembly is made up of all the units in B_1 and B_2. The assembly operation does not start until each of the matching buffers has at least one unit in them. The processing times are exponentially distributed. How does this new scenario affect the recursive and the approximate methods?

6.4. Consider the three-station assembly system shown in Figure 6.10 with exponentially distributed processing times and with BBP-type blocking mechanism. The buffer capacities are $N_1 = N_2 = 2$.

(a) Is there another system that is equivalent to this system with fewer subassembly stations for station 4? Explain why they are equivalent.
(b) Develop a decomposition approximation to obtain the steady-state measures of our interest. The blocking mechanism is still BBP. The processing rates are all unity. For simplicity, you may replace the phase-type distributions in the isolated systems by exponential distributions with the same mean. Compute the approximate values of the performance measures.

6.5. Develop a decomposition approach for the approximate evaluation of the steady-state measures of performance of the following network of assembly systems. Assume that the processing times are exponentially distributed with rates $\underline{\mu} = (1, 1, 1, 1.5, 0.5)$.

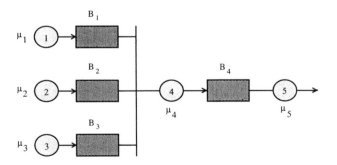

The buffer capacities are $\underline{N} = (2, 2, 2, 5)$. The blocking mechanism is of BAP type. What happens to the output rate of station 5 and the average accumulation in B_4 if the capacity of buffers B_1, B_2, and B_3 each approaches infinity? What happens if the capacity of B_4 also approaches infinity?

6.6. Consider the following assembly/disassembly system. The processing times are exponentially distributed with rates μ_1 and μ_2. Every time a departure occurs at station 1, a unit is added to either B_1 with probability q or to B_2 with probability $1 - q$. At the moment of a departure, if the destination buffer is full, station 1 gets blocked and remains so until there is space in the destination buffer. Station 2 is an assembly station that assembles the finished product by taking one unit from each buffer.

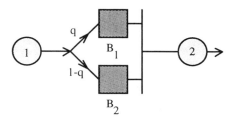

Let $N_1 = N_2 = 3$, $\mu_1 = 1$, and $\mu_2 = 2$. Develop a Markovian approach to obtain the steady-state output rate and the average number of pairs in the system. Is the system stable if the buffer capacities approach infinity?

6.7. The matching buffers of an assembly station are fed by two production lines as shown in the following diagram. The first line is producing type-A products, the second line is producing type-B products, and the products are stored in B_1 and B_2 to be assembled at assembly station 6. There are no intermediate buffers in the line, and the blocking mechanism is of BAP type. The assembly station assembles finished products by taking one unit from each of the matching buffers.

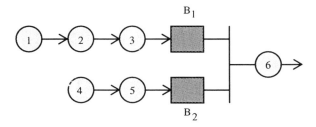

The processing times are exponentially distributed with rates $\mu = (2, 2, 2, 3, 3, 4)$. The buffer capacities are $N_1 = N_2 = 3$. Combining the concepts of production lines and assembly systems, develop an approximation method to obtain the steady-state output rate and the average number of units in each buffer.

6. Assembly Systems

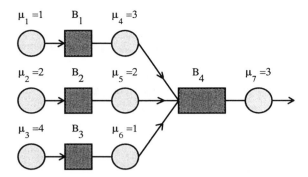

6.8. Consider the following network of work stations where the assembly assumption is relaxed. The processing times are exponentially distributed. The buffer capacities are $N = (2, 4, 3, 5)$. The blocking mechanism is of BAP type. Define the underlying stochastic process for the exact analysis of the network. Develop a decomposition approximation to obtain the steady-state average number of jobs in the system and the output rate of every station.

7

Pull-Type Manufacturing Systems

7.1 Introduction

In Chapters 4 and 5, we studied multistage, push-type manufacturing systems in the context of production lines. In push-type systems, the production schedule is decided based on the demand forecasts, in an upstream part of the system. That is, attention is not given to the immediate market response which may have a significant impact on the finished-product inventory levels. Therefore, push-type systems have difficulties in promptly adapting to the changes in the market, such as demand fluctuations. The forecasting effort itself is a source of considerable error in the master schedule. In practice, to circumvent this difficulty, the inventories are kept at high levels. Notice, for instance, that the production lines we have studied earlier assume a finished-product warehouse of unlimited capacity.

In pull-type systems, the downstream in the hierarchy of production control, being closer to the market, has the final authority on how many items to produce. That is, the production schedule is usually not set beforehand, rather, the process is triggered as the finished-product inventory level reaches a certain threshold. In a multistage pull-type production system, each stage produces as much as the downstream stage(s) requests. Thus, the general philosophy of pull systems is to produce as much as needed, while that of push systems is to produce as much as possible. It is conceivable that pull systems may adapt to external changes faster than push systems and may have a smaller inventory accumulation.

The simplest possible pull-type manufacturing system may be thought of as a single-stage production facility and a finished-product warehouse, as shown in Figure 7.1.

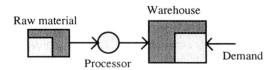

Figure 7.1. A single-stage, pull-type manufacturing system

We may assume an infinite supply of raw material for the facility to process. The finished products are stored in the warehouse until they are used to satisfy customer demand. Production may continue as long as the inventory level in the warehouse is below a target level. That is, it stops as soon as the inventory level reaches its target, and it restarts when the level drops below the target. This principle of operation is known as the **base-stock policy** in the classical inventory theory (see, for example, Johnson and Montgomery (1974)).

The difference between the above production/inventory system and a pure inventory system is that there is a production activity producing products one by one in a production/inventory system. The products are assumed to become part of the finished inventory as soon as they leave the manufacturing processor. Pure-inventory systems however, have a vendor-driven replenishment process, which results in mostly random arrivals of the ordered quantities from the vendors. The pure inventory problems in their broad context are usually distribution problems. Under certain circumstances, the two types of systems may coincide. For instance, batch manufacturing with a time lag from the shop-floor to the warehouse may also be considered as a pure inventory system.

In this chapter, we will study single-stage as well as multistage pull-type manufacturing systems, their properties, and their performance measures. We will generally assume that products are produced one by one with random unit processing times. For simplicity, we will assume that the customer demand received in a certain period of time follows a Poisson distribution. Demand is immediately satisfied from the available stock on hand, if possible. Otherwise, either a back order or a lost sale occurs. Our goal is to obtain the steady-state system performance measures such as the average inventory levels, the average back order levels, the shortage probability, and the processor utilization for a given set of system parameters. We will also discuss some design (optimization) issues.

7.2 Production Control Policies

In general, manufacturing systems operate according to a set of production control policies. These policies usually depend on the manufacturing environment and may vary from one system to another. Clearly, it is desirable to operate a manufacturing system with the most suitable production control policy. At the shop-floor level, these policies identify when to start and stop producing a product and when to switch from one product to another for every machining center or processor. In pull-type systems, the former issue may be handled by assigning inventory-level-dependent production schedules (or production rates) to processors such as $\mu_i(n_i)$ for product i. That is, the production rate for product i changes according to its inventory level. It can be zero at times, implying no production, and it can have varying values at other times. The latter issue is simply a matter of priority among the products. Let us be more specific about $\mu(n)$ and consider some well-known special cases.

7.2.1 Single-Stage Manufacturing Systems

Naturally, a single-stage facility producing a product may cease its operation upon reaching a sufficient level of inventory. Let us refer to this level as the **target inventory level** and denote it by R. If there is only one type of product, the processor will most likely be dedicated to it and therefore will always be setup for it. In this case, the processor restarts producing as soon as the level of inventory drops below the target level R and continues until the inventory level reaches back R again. A simple representation for $\mu(n)$ may be

$$\mu(n) = \begin{cases} \mu & \text{if } n < R, \\ 0 & \text{if } n = R, \end{cases}$$

where a fixed production rate μ is chosen for all possible levels of inventory, except the target level R. During those periods where the inventory level is below R, an outstanding order always exists with a quantity equal to the difference between R and the current inventory level, as depicted in Figure 7.2. This policy is referred to as **the continuous-review order up-to-R policy** or **the base-stock policy** in inventory theory.

In a production environment, a manufacturing cell or a station is usually paying attention to many types of products, and a delay is inevitable in diverting production from one product to another. Furthermore, a set-up may be necessary before the production of a particular product starts. A set-up may be necessary every time production restarts even in some single-product systems. In this case, the production may be deferred to a future point in time so that a sufficient amount of work accumulates, since it involves an unproductive period every time it starts. The mechanism that restarts production is usually the depletion of the finished-product inventory and its reach to a threshold level called **the reorder point**, denoted by r. Thus, our inventory-level-dependent production rate becomes

$$\mu(n) = \begin{cases} \mu & \text{as soon as } n \text{ downcrosses } r, \\ 0 & \text{when } n \text{ upcrosses } R. \end{cases}$$

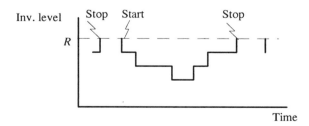

Figure 7.2. Production and inventory behavior under the base-stock policy.

276   7. Pull-Type Manufacturing Systems

This policy is a two-level control policy known as **the continuous-review (R, r) policy**. Clearly, the base-stock policy is a special case of the (R, r) policy where $r = R - 1$. In a periodic-review version of this policy, the inventory level is checked only at certain review points. A production order is placed if $n \leq r$ at the review point. The order quantity may be $R - r$ or $R - n$. Since the changes in inventory levels are detected rapidly in continuous-review policies, they are more responsive in placing production orders than the periodic-review policies.

The implementation of the (R, r)-type policies in various environments can be achieved by using an information system involving posting cards on a bulletin board or sending electronic messages between stages of manufacturing. This type of information systems are usually referred to as **Kanban systems**. "Kanban" means "card" in Japanese. Kanban in manufacturing originates from the Toyota production facility where it has been used to implement the "pull concept." As Figure 7.3 shows, customer demand arrives at queue D and waits there if it is not satisfied immediately. That is, any accumulation in queue D is considered as back order. Each finished unit (shaded box) in the warehouse is attached a production card (black box). As soon as a unit becomes available in the warehouse, W, it is assembled with a unit of backordered demand from queue D and it leaves the system. Immediately after its departure, the production card attached to it is sent back to the order pool, PO. Finished units accumulate in W if D is empty. Depending on the policy being implemented, the number of outstanding production orders is the key factor in deciding when to start and stop producing. For instance, as soon as the number of production orders reaches a threshold, say $R - r$, a set-up may take place, and production starts and continues until all the orders are cleared. Let N_t be the number of finished units in W, and let L_t be the number of production orders in PO, including the one being processed at time t. Keep in mind that L_t is the same as the number of empty cells (or holes) in the warehouse. Since R is the target inventory level or the warehouse capacity, we have $R = N_t + L_t$.

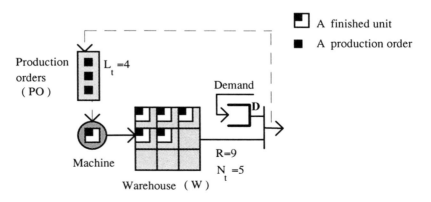

Figure 7.3. Implementation of an (R, r)-type control policy.

7.2. Production Control Policies

An alternative to defining critical inventory levels, and controlling production accordingly, may be to specify periods at the end of which the production starts or stops and a possible switch-over to another product may take place. However, this approach causes problems in scheduling due to the uncertainties inherent in the system (failures, stochastic demand, etc.).

In a **multiproduct** environment, each product has its own pool of production orders, as shown in Figure 7.4. The machine acts as a single-server attending the requests for production of different types of products. A production request is placed for product i when its production orders reach a sufficient level such as $Q_i = R_i - r_i$. Its set-up starts immediately if the machine is available. However, it is likely that there may be two or more products waiting for the machine's attention at any point in time. The two important issues in a multiproduct environment are "when to switch to a new product" and "what product to switch to". The first issue can be handled by an (R, r)-type policy. That is, the processor will be ready to switch to another product after all the orders of the currently produced product are processed. On the other hand, the issue of switching from one product to another can be resolved by using priorities. Naturally, some products are more important than others for various reasons. A reasonable switching policy should pay attention to products in such a way that higher-priority products get more attention and lower-priority products are not totally ignored.

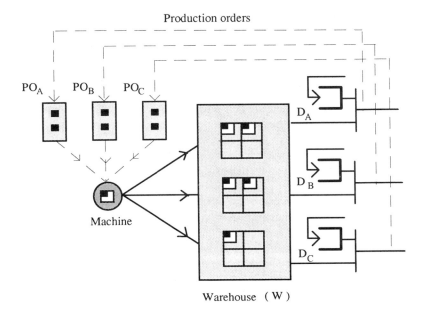

Figure 7.4. A single-stage, multiproduct system (C is being produced).

There are many switching policies that may be employed in a **single-stage** system. The machine responds to production requests according to an order-priority structure that may or may not be the same as the product-priority structure. Once the machine is ready to start processing a new product, the highest-priority order is selected and the respective set-up process takes place. Production starts only after the set-up is completed. The amount of time during which the machine is dedicated to a product depends on the switching policy. One possible scenario is that production may continue until the inventory level of the selected product reaches its target level R. During this period, the processor is assumed to be totally dedicated to the current product, ignoring the status of the other products. When processing of the current product is completed, in case none of the other products has placed a request for production, the processor may remain inactive until a product requests its attention. Notice however that there may be production orders, but they may not be sufficient to place a production request. We will refer to this policy as the **exhaustive production policy**. In a variant of this policy, in case none of the products needs attention, the machine may remain setup for the product it has just completed and continue producing it as demand arrives.

The order-priority structure is said to be **static** if it remains the same as the product-priority structure. A drawback of using static priorities is that depending upon the values of the system parameters, the lower-priority orders may experience longer delays in getting the machine's attention. That is, the machine may switch among the higher-priority orders many times, such that the others may be ignored for long periods. This problem can be avoided by altering the order priorities based on the system status, and thereby creating a **dynamic order-priority structure**. One way to create dynamic order priorities is to limit the number of switchings between any two higher-priority orders to a small number (say 1) while a lower-priority order is waiting for attention. For instance, while a production request exists for product C, the machine may switch between products A and B only once. Then it has to switch to product C even if products A or B may again generate production requests. Thus, the order priorities of products A and B are reduced to less than that of product C. This does not make products A and B less important than product C, but it delays their production temporarily. The order priorities are reinitialized as soon as the production of C is completed. As special cases of the dynamic priority structure, the production orders may be processed on a first-come-first-served or cyclic basis. These priorities may also be altered in various other ways.

In **nonexhaustive production policies,** the machine may stop producing the current product before its target inventory level is attained, and it may switch to another product. For instance, it is possible to switch to a higher-priority product immediately after its production request is placed. In some cases, it is also possible to keep producing the current product until its inventory level reaches a threshold (not necessarily R) and then switch to a higher-priority product. Clearly, nonexhaustive policies result in a larger number of switchings than exhaustive policies. Consequently, exhaustive policies may be appropriate for systems with high set-up costs, whereas nonexhaustive policies may be appropriate for systems with high back ordering costs. For a given set of system parameters, the justification of one policy over another should be achieved by comparing the associated expected total-cost figures, which consist of the expected holding, backordering, and setup costs.

7.2.2 MultiStage Manufacturing Systems

Multistage systems are usually extensions of single-stage systems shown in Figures 7.3 and 7.4 to tandem, and possibly to arbitrary configurations, as shown in Figures 7.5 through 7.7. In these systems, production orders at each stage generate the demand for the upstream stage(s). Again, production cards or Kanbans are used in transferring information from downstream stages to the upstream ones to implement the "pull concept." Let us focus on the system given in Figure 7.5 consisting of K stages where demand for the finished products at a stage is generated by the production in the immediate downstream stage. We denote the machine at stage k by M_k as we have done earlier. Each stage has a distinct set of production cards attached to its finished-products awaiting orders from the next stage. Let B_k be the finished product buffer at stage k with $N_t(k)$ being the number of finished units at time t. Let PO_k be the pool of production orders at stage k with $L_t(k)$ being the number of production orders waiting for attention at time t. Also, let Q_k be the threshold level to place a production request at stage k and R_k be the target finished-product level at stage k. Consequently, $L_t(k)$ and $N_t(k+1)$ are bounded above by R_{k+1}. Then, production starts at stage k at time t, if the machine is available and if $L_t(k) \geq Q_k$ and $N_t(k-1) \geq 1$. A unit from the inventory of stage $k-1$ is transferred to stage k, its production card is sent back to the pool of production orders at stage $k-1$, and a production card from stage k is attached to it. After its processing is completed, the finished unit with the attached production card is placed into the buffer at stage k. Notice that production orders and production cards are the same and used interchangeably for convenience. Also, the production cards remain in the stages that they are assigned to. Again, the (R, r) production control policy or its variations can be used in responding to production requests at each stage. That is, production at each stage is **locally controlled** by the inventory level in its own buffer. Locally controlled multistage pull systems are nothing but flow lines (Chapter 4) driven by the demand at the last stage. A difference in flow lines is that $Q_k = 1$ for all k indicating that $(R_k, R_k - 1)$ is implemented at all stages.

Let us briefly focus on the information transfer or the backward propagation of the production orders in locally controlled systems. Let k^* be a production stage with $L_t(k^*)$ and $N_t(k^*) > 0$ such that M_{k^*} is busy producing, and $M_{k^*+1}, \ldots, M_K$ are idle, with $N_t(k) = R_k$ at time t. Also, assume that $Q_k = 1$ for $k > k^*$. A customer demand received at time $t + \Delta t$ initiates a propagation of production orders from stage K through stages $K - 1, \ldots, k^* + 1$, and finally to stage k^*. Immediately after the demand is satisfied, we will have $L_{t+\Delta t}(k) = L_t(k) + 1$ and $N_{t+\Delta t}(k) = N_t(k) - 1$ for all $k \geq k^*$. Since $Q_k = 1$ for $k > k^* + 1$, $M_{k^*+1}, \ldots,$

M_K will start processing immediately. Thus, a unit of demand that is received at $t + \Delta t$ initiates production in stages $k > k^* + 1$. This backward propagation of orders stops at k^* since k^* has already been producing at time t. It may also stop due to starvation at a stage.

An alternative to local control in the above system may be to use a **global control** policy that considers the total number of production orders in the lower echelon, that is $\Theta_t(k) = \sum_{i=k}^{K} L_t(i)$. For instance, at time t, production starts at stage k if M_k is available, $\Theta_t(k) \geq Q_k$ and $N_t(k-1) \geq 1$, where Q_k is a preset threshold to trigger production at stage k. Similarly, the amount of the total lower-echelon work-in-process inventory can also be used in triggering production at any stage. In this approach, production at stage k may start at time t if $\Theta_t(k) = \sum_{i=k}^{K} N_t(i) \leq Q_k$ and $N_t(k-1) \geq 1$. Note that both local and global control policies are different implementations of the base-stock policy.

On the other hand, we know that the production at each stage is indirectly affected by the demand for the final product, which drives the whole system. Therefore, another approach to control a multistage system may be to directly reflect the impact of the demand for the final product on the production decision at stage 1. As shown in Figure 7.6, as soon as a unit of demand is satisfied, its production card is placed at the order pool PO_o at stage 1. The first stage continues processing units as long as both PO_o and PO_1 have units in them. Notice that PO_K does not exist in this policy. Production cards of stage K wait in PO_o as production orders. A card from PO_o is assembled with a production card from PO_1 and is attached to each unit at the first stage. In this system, production cards of stages 1 through $K - 1$ are only an indication of available space in their respective buffers. Units move from stage 1 to K as if it is a push system. Each stage produces as long as it is not blocked (meaning $L_t(k) > 0$ for stage k) and there are jobs at the upstream stage. A unit that arrives at the warehouse, after being processed on M_k, immediately departs from the system since the demand unit that has initiated its production order is still waiting for it. That is, demand is always backordered

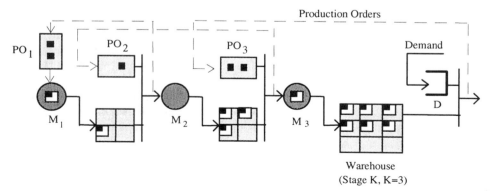

Figure 7.5. A multistage system with local control.

7.2. Production Control Policies 281

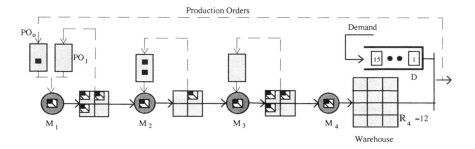

Figure 7.6. A backorder-driven system.

under this policy, indicating an always-empty warehouse as shown in Figure 7.6. Still, the (R, r) policy may be implemented at each stage due to possible set-ups.

Since the system shown in Figure 7.6 is driven by backorders, it starts with zero finished and work-in-process inventories at time $t = 0$, and initiates production at stage 1 when demand arrives at the warehouse. Thus, stage 1 starts producing only when a demand arrives at the system, and *not earlier*. This is where the key role of PO_o is. It forces the inventory buildup (if any) to occur at the raw material level rather than at the work-in-process or the finished-product level. Consequently, the system has a low average inventory level at the expense of having zero customer-service level (probability of satisfying customer demand upon arrival). Notice that the total number of units either being processed or waiting in a buffer in the entire system is limited by R_K. It is not equal to R_K at all times since some of the production cards from stage K may be waiting in PO_o. In the particular scenario presented in Figure 7.6, we have $R_K = 12$, where $K = 4$. We also have 15 units of demand of which 12 already placed its production order and 3 units are waiting to do so. We have assumed that $Q_k = 1$ for all k. Out of 12 orders already placed at stage 1, 11 of them are being processed and 1 that is in PO_o is yet to be started.

Backorder driven control policies are common in industries where production takes place only after orders are received. Many examples can be seen in process industries such as food and pharmaceutical.

We mentioned earlier the blockage of information transfer in locally controlled systems when a production card is passed to a stage already producing or starving. A more involved card-passing policy may circumvent the situation and allow each stage to anticipate the future demand from the immediate downstream stage. Knowing the number of outstanding production orders at stage $k + 1$, it may be possible to place an equal number of orders at stage k, even if stage k is already producing. Passing the production cards to the upstream stage faster will definitely get the finished units to the warehouse quicker and eventually reduce the average backorders and the backorder probability. The mechanism that we refer to as **the generalized**

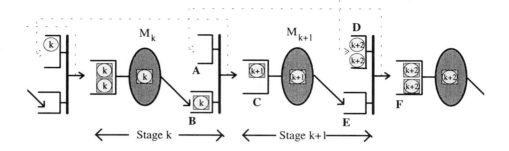

Figure 7.7. The generalized Kanban scheme.

Kanban scheme is demonstrated in Figure 7.7[1]. We assume a fixed number of production cards (or Kanbans), say C_k, for stage k, which may be independent of R_k. In Figure 7.7, production cards or orders are represented by circles with the station index in the middle, and products are represented by boxes. Queues A and D are the pools of production cards for stages $k+1$ and $k+2$, respectively. The finished products of stage k are stored in queues B and C, and those of stage $k+1$ are stored in queues E and F. C and E may be considered as the input and the output buffers of stage $k+1$. Queues A, B, D and E are assembly queues. That is, as soon as the units in queues A and B are synchronized, a production card of stage $k+1$ from queue A is attached to a finished unit from queue B and is placed into queue C. The production card that was on the finished unit in queue B is sent back to the pool of cards of stage k. Similarly, as soon as there are units in queues D and E, the card of stage $k+2$ in queue D is attached to a unit in queue E and is placed into queue F. The card of stage $k+1$ from queue E is sent back to queue A instantaneously. Notice that units do not reside in queues A and B (or in queues D and E) simultaneously. Both queues may be empty at times, but both cannot have units at any time. As soon as a unit arrives at the empty queue (assuming the other have units), a unit from each queue forms a pair, which proceeds to queue C (or F). Thus, queues C and F hold units for the foreseen demand from the immediate downstream stage. As soon as a unit is marked with the card of the next stage, the card of the upstream stage is released to initiate an order. We should mention that if queues C and F are eliminated, the systems in Figures 7.5 and 7.7 become identical. It is difficult to compare the two systems since the choice of C_k affects the system behavior.[2] If $C_k = R_k$ for any k, then it is clear that queue B may hold at most R_k units and queue C may hold at most $R_{k+1} - 1$ units (one unit is

---

[1] The reason for changing the representation of the cards in Figure 7.7 is to associate them with stations by placing station indices in them.

[2] An appropriate way of comparison would be first to allocate the same amount of buffer storage between stages k and $k+1$ and then expedite the differences in their behavior for different values of C_{k+1}, for all k.

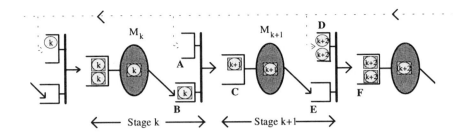

Figure 7.8. Passing production orders from the last stage to every other stage.

being processed at stage $k + 1$). That is, there may be more than R_k units between processors k and $k + 1$. Hence, the system with a faster card-passing mechanism will have fewer backorders at the expense of higher average inventory.

The Kanban system shown in Figure 7.7 may still fail in passing a production order at stage k when queue B is empty upon order arrival at queue A. In this case, the order gets blocked at station k and waits until a unit arrives at B. Perhaps a solution to the problem of blocking the backward propagation of information transfer is to send a copy of the production order from stage K to every upstream stage simultaneously, as shown in Figure 7.8. This is a generalization of the backorder-driven system shown in Figure 7.6. Under this card-passing policy, each stage, say stage $k + 1$, operates as long as there is work at queue C and space in queue E. Thus, job movements from one stage to another are synchronized with the demand arrivals, and upstream card passing is never blocked.

The objective of the above Kanban-passing policies is to coordinate the response of the upstream stages to the needs of the warehouse or the last stage. It is not difficult to design variations of the above policies and others for this purpose. For instance, for stages experiencing long set-ups or significant set-up costs, it may be crucial to look back at the upstream buffer (or echelon) to see if there are sufficient units to process. The set-up process may start only after a threshold level of work-to-process is achieved. This reduces the number of set-ups per unit time but in turn increases the backorder probability. As mentioned earlier, each policy has advantages and drawbacks with revenues and costs attached to it. Therefore, the choice of a production control policy for a given system has to be a part of the design process, and it has to be handled in connection with the cost analysis.

In **multi-product, multi-stage systems**, each stage deals with several types of products and involves production-card-passing policies, switching, and possible look-back policies simultaneously. The (R, r) policy and its variations and the exhaustive and nonexhaustive switching policies can be implemented at each stage. These systems are quite complex and very difficult to analyze.

7.3 Analysis of Single-Stage, Single-Product Systems

The single-stage, single-product manufacturing system given in Figure 7.3 under the continuous-review (R, r) policy operates as shown in Figure 7.9. Production stops as soon as the inventory level reaches R. After some time, due to customer demand, the inventory level drops to the reorder point r and a production request is initiated. Let T_I represent the length of this period and I be the idle state of the machine. When the inventory level drops to the reorder level, the machine goes through a delay period D of length T_D at the end of which the set-up process starts.[1] Let us assume that T_D is independent of the inventory level as well as the rest of the system parameters. Let S and T_S represent the status of the processor during the set-up process and the set-up time, respectively. Production starts as soon as the set-up process is over, and it continues until the inventory level reaches R again. Let P represent the state of the processor while producing and T_P the length of the production period. Notice that production is exhaustive. Thus, the status of the processor belongs to the set of $\{I, D, S, P\}$ at any point in time.

Let $\{N_t, t \geq 0\}$ represent the inventory level of the finished products in the warehouse at time t. After achieving the target value, the next change in the inventory level occurs only after the first demand arrival occurs. Since the demand interarrival times are independent and identically distributed (not necessarily exponential, as in our case), every time N_t reaches R, it probabilistically regenerates itself. That is, the points in time N_t reaches R are the regeneration points, and the time between any two successive points of this type becomes the regeneration

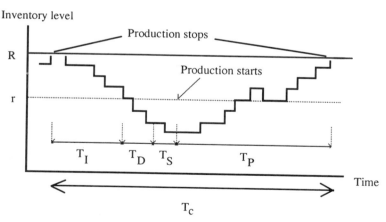

Figure 7.9. Behavior of a single-stage manufacturing system under (R, r) control policy.

---

[1] The delay period T_D will be useful in generalizing the problem to the one with multiple products. In single-product systems, one may assume $T_D = 0$.

7.3. Analysis of Single-Stage, Single-Product Systems

cycle denoted by T_c. Clearly, in any given regeneration cycle, we have

$$T_c = T_I + T_D + T_S + T_P. \tag{7.1}$$

Using the regeneration argument, one can compute the steady-state probabilities of the machine's status. Since the inventory level goes back to R at the end of a cycle, the customer demand that arrives during a cycle must be cleared during the production period in that same cycle. That is,

$$\lambda E[T_c] = \mu E[T_P], \tag{7.2}$$

from which the processor utilization, denoted by ρ_p, is obtained by

$$\rho_p = \frac{E[T_P]}{E[T_c]} = \frac{\lambda}{\mu}. \tag{7.3}$$

Since $T_p < T_c$, the condition for $E[T_c]$ to be finite is $\rho_p < 1.0$. That is, when $\rho_p < 1.0$, the machine is able to cope with the demand and clear all the production orders in the system from time to time, thereby creating a sequence of regeneration points. In case $\rho_p = 1$, we have $E[T_P] = E[T_C]$, indicating that the processor is always busy, which results in no regeneration points. Thus, the following analysis is not valid if the condition $\rho_p < 1.0$ is not satisfied. Let us denote the steady-state probabilities of the processor's status by $\mathbf{P}(I), \mathbf{P}(D), \mathbf{P}(S)$ and $\mathbf{P}(P)$, respectively. Now, we can rewrite (7.3) as

$$\rho_p = 1 - \mathbf{P}(I) - \mathbf{P}(D) - \mathbf{P}(S) = \mathbf{P}(P), \tag{7.4}$$

where $\mathbf{P}(I), \mathbf{P}(D), \mathbf{P}(S)$ are obtained using the regeneration argument shown below. Let us denote the steady-state distribution of the finished-product inventory level by $\lim_{t \to \infty} P(N_t = n) = P_n$. First of all, N_t takes the values of $R, R-1, \ldots, r+1$, during the idle period T_I, and the average time it spends at each level is simply the mean demand interarrival time. That is,

$$P_R = P_{R-1} = \cdots = P_{r+1} = \frac{1}{\lambda E[T_c]}, \tag{7.5}$$

and consequently, $\mathbf{P}(I)$ is given by

$$\mathbf{P}(I) = (R - r) P_R. \tag{7.6}$$

On the other hand, the expected amount of time the machine spends in state D is simply the mean delay, $E[T_D]$, and it is the mean set-up time $E[T_S]$ in state S. Then, using (7.5), we can write

$$\mathbf{P}(D) = E[T_D]/E[T_c] = \lambda E[T_D] P_R = \rho_d P_R. \tag{7.7}$$

Similarly, the set-up state probability $\mathbf{P}(S)$ is given by

$$\mathbf{P}(S) = \lambda \mathrm{E}[T_S] P_R = \rho_s P_R. \tag{7.8}$$

Now, we can rewrite (7.4) as

$$\rho_p = 1 - (R-r)P_R - \rho_d P_R - \rho_s P_R,$$

from which we obtain

$$P_R = \frac{1 - \rho_p}{R - r + \rho_d + \rho_s}. \tag{7.9}$$

Hence, $\mathbf{P}(I)$, $\mathbf{P}(D)$, and $\mathbf{P}(S)$ are obtained using (7.6)–(7.9), and $\mathbf{P}(P) = \rho_p$. Note that ρ_d and ρ_s may assume any nonnegative value, while $\rho_p < 1$. Also, P_R is insensitive to the higher moments of the delay, set-up and the processing times. That is, (7.9) is valid no matter what the distribution of these random variables are.

In summary, with the help of a regeneration argument, we have obtained the steady-state probabilities of the processor's status. Moreover, we have found the steady-state probabilities of the inventory level in period T_I as given by (7.5) and (7.9). That is, $P_R, \ldots, P_{r+1}$ are obtained independently from the rest of the state probabilities of the inventory process. This leads to a stochastic decomposition argument in manufacturing systems, which we will introduce later in this chapter and use in obtaining performance measures.

7.3.1 A Single-Stage Manufacturing System with Exponential Processing Times

Consider the single-product system shown in Figure 7.3 with the delay, set-up, and unit processing times assumed to be exponentially distributed with rates β, γ, and μ, respectively. Customer demand arrives according to a Poisson distribution with rate λ. The inventory level in the warehouse and the operation of the processor are controlled by the continuous-review (R, r) policy. Also, let us consider the backordering case in which the unsatisfied demand is not lost but is expected to be satisfied as soon as possible. The exponential case is rather easier to follow, and the approach remains almost the same for cases with more general assumptions that may include phase-type set-ups, delays, and the unit processing times.

Let us observe the inventory level in the warehouse as well as the machine's status over time as shown in Figure 7.9. Consider the stochastic process $\{N_t, J_t, t \geq 0\}$ where $N_t = R, R-1, \ldots$ is the finished-product inventory level, and $J_t = i, d, s, p$ is the status of the machine at time t. $\{N_t, J_t\}$ is a continuous-time, discrete-state Markov chain having a unique steady-state distribution for $\rho_p < 1$. The transition diagram of $\{N_t, J_t\}$ is given in Figure 7.10. Notice from the figure that $\{N_t, J_t\}$ goes from the idle state $(r+1, i)$ to the delay state (r, d) and from the delay states to the set-up states, and eventually the process visits the production

7.3. Analysis of Single-Stage, Single-Product Systems   287

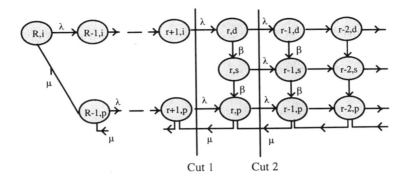

Figure 7.10. The transition diagram of the exponential case.

states. The inventory level during the delay, set-up, and production periods may in principle approach $-\infty$, indicating significant backorders.

Let (n, j) be a particular state of the process and $P(n, j)$ be its steady-state probability. We know from (7.5) that $P(R, i) = \cdots = P(r + 1, i)$, which are equal to P_R given by (7.9). The steady-state flow-balance equations of the production states $(R - 1, p), (R - 2, p), \ldots$ are

$$(\lambda + \mu)P(R - 1, p) = \mu P(R - 2, p),$$
$$(\lambda + \mu)P(n, p) = \lambda P(n + 1, p) + \mu P(n - 1, p),$$
$$r + 1 < n < R - 1, \qquad (7.10)$$
$$(\lambda + \mu)P(n, p) = \lambda P(n + 1, p) + \mu P(n - 1, p) + \gamma P(n, s), \quad n \leq r,$$

from which one can obtain $P(n, p)$ for $n = R - 1, \ldots, r$, explicitly.

$$P(n, p) = \frac{\rho_p(1 - \rho_p^{R-n})}{1 - \rho_p} P(R, i)$$
$$= \frac{\rho_p(1 - \rho_p^{R-n})}{R - r + \rho_d + \rho_s}, \quad n = R - 1, \ldots, r. \qquad (7.11)$$

Notice that (7.11) is explicit because P_R is known explicitly in the backorder case.

Next, we focus on the probabilities of delay, set-up, and production states simultaneously for $n < r$. For convenience, we define the following vector of joint probabilities:

$$\underline{\tilde{P}}(n) = \begin{pmatrix} P(n, d) \\ P(n, s) \\ P(n, p) \end{pmatrix}, \quad n = r, r - 1, \ldots,$$

such that the marginal probability of having n finished products in the system is $P_n = \underline{e}^T \underline{\tilde{P}}(n)$, where $\underline{e}^T$ is a vector of ones. To be able to obtain an expression for $\underline{\tilde{P}}(n)$, we suggest using cuts 1 and 2 in Figure 7.3, where the flow rates from both ends into each cut must be identical in steady-state. This argument can be easily proven by using the flow-balance equations of states (n, i) and (n, p) for $n > r$. For instance, the flow equivalence in cut 1 is

$$\rho_p[P(r+1, i) + P(r+1, p)] = P(r, p).$$

The same argument is valid for cut 2 as well as for any similar cut in steady-state. Using the flow-balance equations for states (r, d), (r, s) and cut 1, we obtain

$$\underline{\tilde{P}}(r) = \begin{pmatrix} \frac{\lambda}{\lambda+\beta} \\ \left(\frac{\lambda}{\lambda+\beta}\right)\left(\frac{\beta}{\lambda+\gamma}\right) \\ \frac{\rho_p(1-\rho_p^{R-r})}{1-\rho_p} \end{pmatrix} P_R = \underline{\sigma} P_R. \qquad (7.12)$$

On the other hand, the flow equivalence in cut 2 is

$$\rho_p[P(n, d) + P(n, s) + P(n, p)] = P(n-1, p), \qquad (7.13)$$

which is valid for any $n \leq r$. Then, the flow-balance equations for states (n, d), (n, s), and (n, p) for $n < r$ coupled with (7.13) can be written in the following matrix form:

$$\underline{\tilde{P}}(n-1) = \lambda \mathbf{A}^{-1}\underline{\tilde{P}}(n), \qquad n = r-1, r-2, \ldots,$$

and can be generalized to

$$\underline{\tilde{P}}(n) = (\lambda \mathbf{A}^{-1})^{r-n}\underline{\tilde{P}}(r), \qquad n = r-1, r-2, \ldots, \qquad (7.14)$$

where

$$\mathbf{A} = \begin{pmatrix} \lambda+\beta & 0 & 0 \\ -\beta & \lambda+\gamma & 0 \\ -\lambda & -\lambda-\gamma & \mu \end{pmatrix}.$$

Thus, we have developed explicit expressions for the steady-state joint distribution of the stochastic process $\{N_t, J_t\}$. Furthermore, from this joint distribution, we can obtain the following marginal distribution of the inventory and backorder

7.3. Analysis of Single-Stage, Single-Product Systems

levels in the warehouse.

$$\left. \begin{array}{l} P_n = \dfrac{1 - \rho_p^{R-n+1}}{R - r + \rho_d + \rho_s}, \qquad n = R, \ldots, r+1 \\[2mm] P_r = \underline{\mathbf{e}}^T \underline{\sigma} P_R \\[1mm] P_n = \underline{\mathbf{e}}^T (\lambda \mathbf{A}^{-1})^{r-n} P_r, \qquad n = r-1, \ldots, \end{array} \right\} \quad (7.15)$$

where $\underline{\sigma}$ is defined by (7.12). The above procedure can be systematically generalized to accommodate phase-type processing, set-up, and delay times. From the above marginal distribution of the finished products, we can now obtain the measures of our interest in the usual way. The average finished-product inventory and the backorder levels in the warehouse are given by

$$\overline{N} = \sum_{n=1}^{R} n P_n, \qquad \overline{B} = \sum_{n=-\infty}^{0} -n P_n,$$

respectively, and the probability of encountering a backorder upon a demand arrival is given by

$$\Pr(\text{backorder}) = 1 - \sum_{n=1}^{R} P_n,$$

which is also the arbitrary-time probability of having backorders in the system, due to the Poisson demand arrival process.

One may also be interested in the distribution and the average number of production orders in the system at any time. We will leave the discussion of this measure to the next section for a better understanding of the analytical form of the number of production orders.

The summation of the infinite terms in $\overline{B}$ is troublesome from a computational point of view. Although truncation may be a practical choice to evaluate infinite summations, it is usually difficult to know where to truncate. In the next section, we will develop explicit expressions that do not involve infinite summations for the above measures, utilizing queueing analogy of manufacturing systems.

In some cases, the unsatisfied demand is lost and is referred to as the **lost sale**. In this case, the stochastic process $\{N_t, J_t\}$ becomes a finite-state, continuous-time Markov chain due to the fact that the inventory level in the warehouse may vary only in $[0, R]$. The steady-state flow-balance equations of the lost sale case are the same as the ones in the backordering case, except that $\tilde{\mathbf{P}}(0)$ is obtained using the following boundary conditions:

$$\begin{aligned} \beta P(0, d) &= \lambda P(1, d), \\ \gamma P(0, s) &= \lambda P(1, s) + \beta P(0, d), \\ \mu P(0, p) &= \lambda P(1, p) + \gamma P(0, s). \end{aligned} \quad (7.16)$$

7. Pull-Type Manufacturing Systems

Notice that the argument of all the demand arriving in a cycle being cleared in the same cycle is not valid in the lost sale case due to the lost demand. Therefore, neither the expression (7.9) for $P(R, i)$ nor (7.6)–(7.8) for the probabilities of the machine's status are valid in the lost sale case. Nevertheless, one can proceed by assuming that $P(R, i) = 1.0$ and use the same approach coupled with (7.16) to obtain $\tilde{\mathbf{P}}(0)$. The probabilities of the machine's status have to be obtained from the joint probabilities. At the end of the procedure, we have to go through a normalization process since the true probabilities must add up to 1.0. Until the normalization is carried out, the probabilities may take any positive value depending on the initial value assigned to $P(R, i)$. This value may become crucial in the Markov chains with a large number of states. Before being normalized, the so-called probabilities may take very large values such that they may cause numerical problems. In these situations, truncation of the state-space may be unavoidable.

In the special case of negligible (or zero) delay and set-up times, the expressions for the steady-state marginal probabilities of the number of units in the warehouse simplify to the following:

$$P_n = \begin{cases} \dfrac{1 - \rho^{R-n+1}}{1 - \rho} P_R, & r < n \le R, \\ \dfrac{\rho^r - \rho^R}{1 - \rho} \rho^{1-n} P_R, & n \le r, \end{cases}$$

where $\rho = \lambda/\mu$ and $P_R = (1 - \rho)/(R - r)$ for the backordering case. P_R is found through normalization in the lost sale case.

EXAMPLE 7.1. Consider a single-stage, single-product manufacturing system where the demand arrival process is Poisson with rate $\lambda = 1.0$, and the processing and set-up times are exponentially distributed with rates $\mu = 2.0$ and $\gamma = 0.5$, respectively. There exists a start-up delay aside of the set-up time, which is expected to be exponentially distributed with a mean of 5.0 time units. The continuous-review (R, r) policy is used to control the production process and the inventory levels. The target inventory level is $R = 10$ units, and the reorder point is $r = 5$ units. The unsatisfied demand is backordered. Let us obtain the steady-state probabilities of the machine's status, the finished inventory level, and the average values of the associated measures.

First of all, $\rho_p = 0.5$, $\rho_d = 5.0$, and $\rho_s = 2.0$. Using (7.6)–(7.9), we obtain P_R and the steady-state probabilities of the machine's status:

$$P_R = 0.0417, \quad \mathbf{P}(I) = 0.2084, \quad \mathbf{P}(D) = 0.2084,$$
$$\mathbf{P}(S) = 0.0833, \quad \mathbf{P}(P) = 0.5.$$

The steady-state joint and the marginal probabilities of the stochastic process $\{N_t, J_t\}$ are obtained using (7.15) and are given in Table 7.1, for $n = 10, \ldots, 0$. It is clear from (7.11) that $P(n, p)$ for $n \ge r$ is increasing in n and therefore P_n for $n \ge r + 1$ is increasing in n. This argument explains the behavior of P_n that it increases to a point and decreases afterward as observed in Table 7.1. By the way, it has to decrease due to the fact that the probabilities add up to 1.0. We leave the issue of where the peak of P_n can be to the reader as an exercise (see Problem 7.4).

The resulting average inventory level is given by $\overline{N} = \sum_{n=1}^{10} n P_n = 3.4820$, and the probability of encountering a backorder is Pr(Backorder)$= 1 - \sum_{n=1}^{10} P_n = 0.3201$. For the average backorder level, we have no choice but truncation. We leave its discussion to the next section.

7.3.2 A Queueing Analogy

With a proper interpretation, it is possible to characterize a queueing system that is implicitly built into the above single-stage manufacturing system. Consequently, the analysis of the manufacturing system can be carried out through the analysis of this queueing system which can be described as follows:

As soon as a unit of demand is satisfied, a production order is placed into the pool of orders. This implies that the arrival of a unit of demand results in the arrival of a production order, and therefore the order arrival process is equivalent to the demand arrival process. On the other hand, the processing time of a physical unit can be visualized as the service time of a production order. Since a production order is transformed into a unit of inventory, it is assumed that an order departs from the system at the end of its service. The way the orders are served is solely dictated by the inventory control policy used in the physical system.

The waiting line in the queueing system is in fact our pool of production orders. The one-to-one correspondence between the two systems is shown in Figure 7.11. When the warehouse is full, the queueing system is empty; and when the warehouse is empty, the queueing system has R units, including the one in service. In case the unsatisfied demand is backordered, the queue has no capacity limitation and the orders in excess of R units in the system will represent the backorders. In the lost sale case however, the unsatisfied demand is lost, and consequently there can be at

TABLE 7.1. The steady-state probabilities in Example 7.1

n	$P(n, i)$	$P(n, d)$	$P(n, s)$	$P(n, p)$	P_n
10	0.0417	—	—	—	0.0417
9	0.0417	—	—	0.0208	0.0625
8	0.0417	—	—	0.0313	0.0729
7	0.0417	—	—	0.0365	0.0781
6	0.0417	—	—	0.0391	0.0807
5	—	0.0350	0.0046	0.0404	0.0797
4	—	0.0289	0.0070	0.0399	0.0758
3	—	0.0241	0.0079	0.0379	0.0698
2	—	0.0201	0.0079	0.0349	0.0629
1	—	0.0168	0.0075	0.0315	0.0557
0	—	0.0140	0.0069	0.0279	0.0487

292   7. Pull-Type Manufacturing Systems

most R orders in the queueing system. Practically, the queueing system becomes a finite-capacity queue in the lost sale case. The graphical representation of the sample path behavior of each of the systems further emphasizes their relationship, as shown in Figure 7.12.

During period T_I, the server in the queueing system waits for $Q = R - r$ orders to accumulate. T_D and T_S are the delay and set-up periods for the machine in both systems. At the end of the set-up process, the server starts serving the customers. Queueing systems of this sort have been studied under the title of queues with vacations or with removable servers (or with servers of walking type). The main principle is that the server takes a vacation upon the occurrence of a particular event. The length of the vacation is either a random variable or is defined by the number of units accumulating in the system. Our manufacturing system relates to the latter case. That is, the time it takes for the inventory level to go from R to r in the physical system is the vacation time in the queueing system.

Queues with vacations possess a useful property that is referred to as the stochastic decomposition property. We observed it earlier in a simple form. That is, we obtained the idle-period probabilities $P(R, i), \ldots, P(r+1, i)$ independently of the rest of the probabilities in Section 7.3. In the next section, we will generalize this property and use it to obtain the measures of performance related to production orders as well as backorders. The queueing approach helps us in understanding the behavior of the single-stage system. The immediate benefit of this approach is to be able to develop expressions for the measures of our interest. We will also

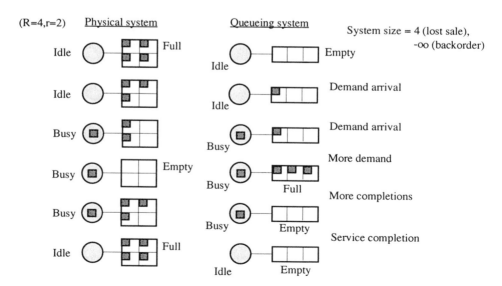

Figure 7.11. The correspondence between the physical system and the queueing system.

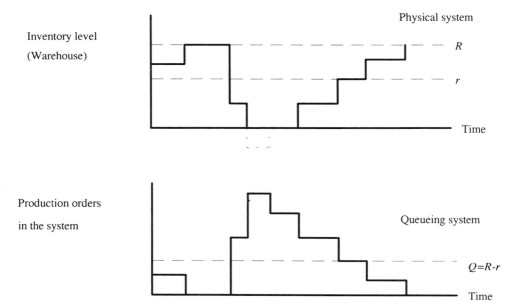

Figure 7.12. Sample path behaviors of the accumulation in the two systems.

employ the queueing approach in the analysis of multiproduct as well as multistage systems.

7.3.3 The Stochastic Decomposition Property in Queues with Vacations

The property says that the number of units present in an M/G/1-type work station with vacations at any time is the sum of the number of units at any time during the vacation and the number of units in the underlying regular M/G/1 system at any time.[1] Let us first analyze the property in an M/G/1 system with no delay or set-up periods before the service starts. That is, as soon as Q units accumulate in the system, the server starts serving units one by one and continues doing so until the system is cleared. Let us refer to this system as the M/G/1 station with the Q-policy and let

$Q(z)$ = generating function of the distribution of the number of units in the M/G/1 station with the Q-policy at any time,

---

[1] In fact, the property applies to various queueing systems with different vacation mechanisms (see Related Bibliography).

$Q_v(z)$ = generating function of the distribution of the number of units in the M/G/1 station with the Q-policy at any time during a vacation,

$\Pi(z)$ = generating function of the distribution of the number of units in the system in the underlying regular M/G/1 station at any time.

These generating functions are also valid for the distributions of the number of units seen by arrivals (PASTA property). Let us refer to the units that arrive during a vacation as the primary units and to the ones that arrive during the busy period as the secondary units. A primary unit automatically becomes a secondary unit as soon as its service starts. If $\chi(z)$ is defined as the generating function of the distribution of the number of units seen by an arrival at a busy system in a regular M/G/1 system, we can write

$$\Pi(z) = \rho\chi(z) + 1 - \rho,$$

yielding

$$\chi(z) = \frac{\Pi(z) - (1 - \rho)}{\rho}, \qquad (7.17)$$

where $\rho = \lambda/\mu < 1$ is the traffic intensity, λ is the arrival rate, and μ is the mean service rate. On the other hand,

$$\text{Pr(an arriving unit is a primary unit)} = 1 - \rho,$$
$$\text{Pr(an arriving unit is a secondary unit)} = \rho.$$

As mentioned earlier, $Q_v(z)$ is also the z-transform of the distribution of the number in the system as seen by a primary customer. If the LIFO service rule is adopted, which will not affect the stationary distribution[1] of the number in the system, the number of units seen by a secondary unit will be the sum of the primary units and the secondary units arrived earlier in the present busy period. Hence, we can write

$$Q(z) = (1 - \rho)Q_v(z) + \rho Q_v(z)\chi(z). \qquad (7.18)$$

Substituting (7.17) into (7.18), we obtain

$$Q(z) = Q_v(z)\Pi(z), \qquad (7.19)$$

which is an important relationship since it is expressed in terms of the generating functions of the distributions of the number of units in the system. Moments of the number in the system can be obtained using the derivatives of $Q(z)$.

---

[1] As mentioned in Chapter 3, the stationary distribution of the number of units in the system is independent of the service discipline (FIFO, LIFO, or SIRO, etc.) as long as the units are served in an order that is independent of their service times, and the service is nonpreemptive.

7.3. Analysis of Single-Stage, Single-Product Systems

Let us emphasize that, in terms of our manufacturing system, (7.19) says that the number of outstanding production orders (including the one being processed) at any time is the sum of the orders during the idle period under an (R, r) policy and the orders in the system if the system were to operate under an $(R, R-1)$ policy. Using this property, we will develop simple expressions for the performance measures for single as well as multiproduct systems.

The long-run average number of units (or production orders) in the above queueing system, that is, $\bar{L}$, is obtained by taking the derivative of (7.19) with respect to z and is given by

$$\bar{L} = Q'(1) + \Pi'(1). \tag{7.20}$$

An immediate consequence of (7.20) is the mean number of units in the M/G/1 station with the Q-policy, that is

$$\bar{L} = \frac{Q-1}{2} + \bar{L}_{M/G/1}. \tag{7.21}$$

The first term in (7.21) is the long-run average number of units in the system given that the system is in vacation. $\bar{L}_{M/G/1}$ is the average number of units in the system in the underlying regular M/G/1 work station, given by the Pollacek-Khinchine formula

$$\bar{L}_{M/G/1} = \rho + \frac{\rho^2(\text{Cv}_X^2 + 1)}{2(1-\rho)},$$

with Cv_X^2 being the squared coefficient of variation of the service time.

The vacation period becomes $T_I + T_D + T_S$ in case delay and set-up times are involved. It is still valid that $Q(z) = Q_v(z)\Pi(z)$ where $Q_v(z)$ is the weighted sum of the generating functions of the distributions of the number of units in the individual periods during a vacation. That is,

$$Q_v(z) = Q_I(z)\frac{P(I)}{1-\rho} + \tilde{Q}_I(z)Q_D(z)\frac{P(D)}{1-\rho} + \tilde{Q}_I(z)\tilde{Q}_D(z)Q_S(z)\frac{P(S)}{1-\rho}, \tag{7.22}$$

where

$Q_i(z)$ = generating function of the distribution of the fresh accumulation (secondary units) at any time in period T_i,
$\tilde{Q}_i(z)$ = generating function of the distribution of the total number of arrivals at the end of T_i,
$P(i)$ = steady-state probabilities of the processor's status given by (7.6)-(7.8).

Taking the derivative of (7.22) and substituting it into (7.21), we obtain

$$\bar{L} = \bar{L}_v + \bar{L}_{M/G/1}, \tag{7.23}$$

296   7. Pull-Type Manufacturing Systems

where $\overline{L}_v$ is the average number of units during a vacation, given by

$$\overline{L}_v = \frac{1}{1-\rho}\left[\overline{L}(T_I)P(I) + \overline{L}(T_D)P(D) + \overline{L}(T_S)P(S)\right].$$

$\overline{L}(T_i)$ is the average number of units in the M/G/1 system with the Q-policy given that the system is in period T_i:

$$\begin{aligned}\overline{L}(T_I) &= \overline{M}(T_I) = \frac{Q-1}{2}, \\ \overline{L}(T_D) &= Q + \overline{M}(T_D), \\ \overline{L}(T_S) &= Q + \lambda E[T_D] + \overline{M}(T_S).\end{aligned} \qquad (7.24)$$

Finally, $\overline{M}(T_j)$ is the time-average accumulation of the secondary units during the random period T_j and is a function of the moments of T_j. The residual time of T_j at the time of an arrival can be considered as the holding time of the arrival until the end of T_j. Then, via Little's law, we can write

$$\overline{M}(T_j) = \frac{\lambda E[T_j^2]}{2E[T_j]}, \qquad j = D, S.$$

Notice that $\overline{M}(T_j) = \lambda E[T_j]$ if T_j is exponentially distributed. Thus, $\overline{L}$, the average number of units in the system with the Q-policy, is explicitly obtained by using (7.23) and (7.24).

The number of units in the system during a production period can be obtained using

$$\overline{L} = \overline{L}_v(1-\rho) + \overline{L}(T_P)\rho$$

and (7.23), resulting in

$$\overline{L}(T_P) = \frac{\overline{L}_{M/G/1}}{\rho} + \overline{L}_v. \qquad (7.25)$$

Now let us turn our attention to the average backorders. Let q_n be the stationary probability that there are n production orders in the system at any time. Then, the average backorder level in the warehouse can be written as follows:

$$\overline{B} = \sum_{n=R+1}^{\infty}(n-R)q_n = \overline{L} - \sum_{n=0}^{R}nq_n - R[1 - \sum_{n=0}^{R}q_n],$$

or simply

$$\overline{B} = \overline{L} + \overline{N} - R. \qquad (7.26)$$

7.3. Analysis of Single-Stage, Single-Product Systems

$\overline{N}$ is the average inventory level in the warehouse and is obtained using a finite number of probabilities, as shown in Section 7.3.1. Since there is a one-to-one correspondence between the manufacturing system and the queueing system, $\{q_n\}$ can be obtained using the matrix-recursive method introduced in Section 7.3.1. That is, $q_n = P_{R-n}$ for all n. Clearly, this method needs to be revised for more general processing, set-up, or delay times such as the ones of phase-type. Thus, utilizing the queueing system representing the behavior of the production orders, we have developed an explicit expression for the average backorder level in our manufacturing system. This approach does not involve infinite summations and avoids truncation. Now let us apply what we have developed in this section to compute the average production orders and backorders in Example 7.1.

EXAMPLE 7.2. Let us consider the queueing system of production orders in Example 7.1. The steady-state probabilities of the processor's status remain unchanged. That is,

$$P(I) = 0.2084, \quad P(D) = 0.2084, \quad P(S) = 0.0833, \quad P(P) = 0.5.$$

Using (7.24), we obtain the expected number of production orders in the system given that the processor's status is I, D, and S:

$$\overline{L}(T_I) = 2, \quad \overline{L}(T_D) = 10, \quad \overline{L}(T_S) = 12$$

resulting in $\overline{L}_v = 7.0008$. Note that the expected fresh accumulation of production orders (the secondary units) during the delay and set-up periods (exponentially distributed) are

$$\overline{M}(T_D) = 5 \quad \text{and} \quad \overline{M}(T_S) = 2,$$

respectively. The underlying queueing system is an M/M/1 system with $\overline{L}_{M/G/1} = 1$; and from (7.25), the average number of production orders in the system while the processor is producing is $\overline{L}(T_P) = 9.0008$. Finally, from (7.23), the overall average number of production orders in the system amounts to $\overline{L} = 8.0008$. It is expected that $\overline{L}(T_P) > \overline{L}$, as implied by (7.25).

The average finished=product inventory level from Example 7.1 is $\overline{N} = 3.4820$. Finally, using (7.26) we obtain the average backorder level, $\overline{B} = 1.4820$.

7.3.4 Optimizing the Parameters of the Control Policy

As part of the overall design of a manufacturing system, we want to obtain the best possible values of R and r with respect to a given criterion which is normally the highest possible output (or sales) with the minimum possible investment. However, for many reasons, less investment (i.e., insufficient inventory or low quality) results in less sales due to the unsatisfied demand. That is, the smaller the amount of inventory held in the warehouse is, the higher the backorder level will be. Clearly, this presumption does not take into account any scheduled demand arrivals or any type of planned arrangement between the system and the customer. An oversimplified approach to solve the problem would be to employ a cost-minimizing objective function that assigns penalties for holding both inventory and shortages. Also, a penalty per set-up should be charged to avoid excessive set-ups. Let us introduce the following notation:

$TC(R, r)$ = the steady-state expected total cost per unit time,
K_s = set-up cost per set-up,
h = unit holding cost per unit time,
g = unit backordering cost per unit time,
π = shortage cost per unit short.

Notice that g is the counterpart of h for shortages, which is the cost of carrying a backorder to the next period as opposed to satisfying it in the current period. On the other hand, π is the instantaneous cost incurred at the time a demand is not satisfied.

In the case with backorders, there may be any number of outstanding backorders in the system at any point in time. It is clear that a positive inventory level indicates no backorders, and backorders indicate no finished products in the system. The expected total cost per unit time includes the set-up cost per unit time, the holding and the backordering costs per unit time, and the shortage penalty per period. We know from the earlier analysis that there exists only one set-up per regenerative cycle, that is, the set-up cost per unit time can be expressed as $K_s/E[T_c]$. Thus, the expected total cost per unit time can be written as

$$TC(R, r) = K_s/E[T_c] + h\overline{N} + g\overline{B} + \pi\lambda \Pr(backorder), \qquad (7.27)$$

where $\overline{N}$, $\overline{B}$, and the Pr(backorder) are discussed in the previous section and $E[T_c]$ is obtained using (7.5).

In the lost sale case, there are at most R finished units and no negative inventory in the system. Backordering cost disappears from the objective function, and the shortage cost per unit time will be due only to the lost demand in a unit period. Hence, the cost-minimizing objective function for the lost sale case becomes

$$TC(R, r) = K_s/E[T_c] + h\overline{N} + \pi\lambda P_o, \qquad (7.28)$$

where λP_o is the expected number of demands lost per unit time.

Hence, we are able to evaluate the objective function for the backordering case in a finite number of operations without any truncation in the infinite summations. Also, the objective function in the lost sale case has finite summations by its nature. The optimization problem is not difficult since it involves only two discrete decision variables. Once $TC(R, r)$ is assigned a large value for $R < 0$ and also for $r \geq R$, any unconstrained discrete optimization procedure can be used to obtain (R^*, r^*). However, the process of searching for the optimal solution can be made more

7.4. A Single-Stage System with Two Products

efficient if some special properties that the problem under study may have are exploited. In our case, the problem possesses the following important property: The probabilities $\{q_n, n = 0, \ldots, \infty\}$, that is, the probability distribution of the production orders, remain the same for fixed $Q = R - r$, no matter what the values of R and r are. This should be obvious now due to the queueing analogy in Section 7.3.2. Clearly, it is more involved to compute $\{q_n\}$ for a given Q than to evaluate (7.27) for a given R, once $\{q_n\}$ is computed. Moreover, the computational effort to find $R^*(Q)$ for a given Q, using a one-dimensional search procedure, is quite insignificant. Therefore, in light of this observation, one may use a one-dimensional search procedure for each Q and R and proceed in the direction of finding $R^*(Q)$ for a given Q. This approach to obtain (R^*, r^*) (or Q^*) will be computationally more efficient than a direct application of a multivariable search method. As Q approaches Q^*, a fewer number of steps will be required to find $R^*(Q)$, and it will be quite fast to obtain $R^*(Q^*)$. On the other hand, a multivariable search method may be implemented more efficiently by storing $\{q_n\}$ for all n for every Q. However, this will increase the computer memory requirements considerably and eventually will be undesirable. Unfortunately, the above-mentioned property holds only for the backordering case and not for the lost sale case in which a multivariable search procedure has to be used to obtain (R^*, r^*). Clearly, it is also possible to implement any type of heuristic search or even exhaustive search procedures in the above problem since there are only two decision variables.

When we look at the relationship between the optimal values of the model parameters R^* and $Q^* = R^* - r^*$, and the cost coefficients in the above objective functions, namely K_s, h, g, and π, the following observations can be made. For a fixed Q, the average holding cost per unit time increases and the average shortage cost per unit time decreases as R increases. On the other hand, the complete opposite is observed as Q increases. Also, the average set-up cost per unit time decreases as Q increases.

From the opposite end, we observe the following. As the set-up cost K_s increases, $Q^*(K_s)$ increases to achieve less-frequent set-ups. However, $R^*(K_s)$ increases to compensate the lost demand due to fewer set-ups. This indicates that as the set-up cost decreases, the (R, r) policy converges to the $(R, R - 1)$ policy.

As the holding cost h increases, $R^*(h)$ decreases to reduce the average holding cost per unit time. However, $Q^*(h)$ decreases to generate more-frequent set-ups to compensate for the lost demand due to lower $R^*(h)$. As h continues to increase, the (R, r) policy converges to the $(R, R - 1)$ policy. Thus, the set-up and the holding costs have opposite impacts on the type of policy to use.

As the backordering cost g increases, $R^*(g)$ increases to reduce the chance of a backorder, while the behavior of $Q^*(g)$ depends on the rest of the cost parameters. For instance, as g approaches zero while π remains the same, $Q^*(g)$ may take a value less than $R^*(g)$ or it may take a value greater than $R^*(g)$, forcing the reorder point to have a negative value depending on the values of K_s, h, and π. The impact of π on R^* and Q^* also exhibits a similar behavior.

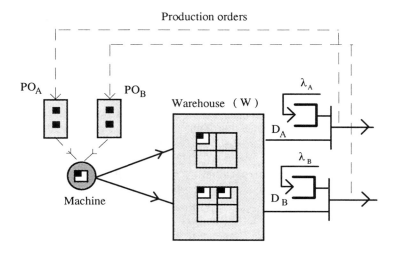

Figure 7.13. A single-stage system producing two types of products (A is being produced).

7.4 A Single-Stage System with Two Products

Consider a system producing two products and storing them in a finished-product warehouse as shown in Figure 7.13. For each type of product, the demand arrival process is assumed to be Poisson and independent of the demand arrival process of the other product. The unsatisfied demand for any product is assumed to be backordered. Production of a particular product requires the machine to go through a set-up process that is unique to that product. The unit processing times are independent and arbitrarily distributed. Let us use the (R, r) continuous-review policy to control the production and the inventory levels of the products. The set-up process of a product starts as soon as its inventory level reaches its reorder point and the machine is idle. Notice that it is not possible for both of the products to reach their reorder levels or to be below their reorder levels and waiting for the machine's attention simultaneously. For this reason, assigning priorities is meaningless in the two-product case. One of the products will reach its reorder level before the other one does and will start its set-up. Production starts after the set-up is completed and continues until the inventory level of the selected product reaches its target level R. While processing, the machine is assumed to be totally dedicated to the current product, and it ignores the status of the other. When processing of a product is completed, in case the other product does not need to be produced, the machine will remain inactive until a product requests its attention again.

The machine utilization in the multiproduct environment is the percentage of time that the machine spends in producing each product. In the case of backorders, the utilization can be expressed as $\rho = \sum_{i=1}^{K} \rho_i$ if there are K products, with $\rho_i = \lambda_i/\mu_i$ and μ_i being the processing rate for product i. This can again be verified using the regeneration argument. The points in time where the inventory level of a product attains its target value when the rest of the products are already at their target levels constitute the regeneration points. Clearly, there may be many production periods of a product in a regeneration cycle. However, since the products are produced only during their production periods, ρ_i will be the ratio of the expected total production time of product i and the expected cycle time. Then, the overall utilization of the machine becomes $\rho = \sum_{i=1}^{K} \rho_i$. In the case of lost sales, ρ should be expressed in terms of the steady-state probabilities of the machine's status.

In the above two-product problem, the underlying stochastic process is quite complicated due to the interactions among the products, and it possesses an infinite number of states. Therefore, in the next section, we will introduce an approximation method to obtain the performance measures of our interest.

7.4.1 Motivation for the Approximation

In a multiproduct, single–machine environment, there is clearly a high level of interaction among the inventory levels of the products. This is simply due to the fact that having placed a request for production, a product will wait for the machine to complete the production of the current product, in case it is busy at the time of the request. Hence, this is a situation where a random delay besides the set-up time may occur before starting the production of a product. The following observation can be made when the system is looked at from the single-product point of view. The delay and the production periods (T_D) and (T_P) of a product are affected by each other. For instance, a longer production period for product A will most likely result in an accumulation of a higher number of production orders for product B. This in turn will most likely increase the delay of product A. That is, with a high possibility, a longer production period will be followed by a longer delay period in each so-called single-product system. Furthermore, due to the policies employed in controlling the system's behavior, the products can influence each other's behavior only through their delay periods. That is, the interdependencies among the products are confined into their delay periods. Hence, if the delay period of each product is characterized, at least through its mean, the analysis of the original problem may approximately be reduced to the analysis of a set of single-product manufacturing systems. In this setting, each single-product system will act the same as the single-product problem we have studied in Section 7.3.

It is possible to develop an iterative procedure that relates these single-product problems to each other. The iterative procedure can be carried out in such a way that the results of the analysis of one system provide approximate input values for the analysis of the others. Each single-product problem can be studied through the analysis of the underlying stochastic process as we have done earlier. Below, we will first concentrate on characterizing the mean delay of each product in the two-product problem, considering the possible interactions among them. Then, we will introduce an approximate iterative procedure that assumes independence among the products to compute the mean of the delay periods.

7.4.2 Analysis of the Delay Period

Let us consider the above single-stage system with products A and B. During its delay period, product i waits until the inventory level of the currently produced product, say product j, reaches its target level. We denote this period by Y_i, which consists of the remaining processing time of product j and the processing time of all the work-to-process of type j until its inventory level reaches its target level. Let X_i be the unit processing time and $F_X^*(s)$ be the LST of its density function. Also, let U_i be the set-up time with $F_U^*(s)$ being the LST of its density function. In the same manner, let X_i^r and U_i^r be the remaining processing and set-up times with $F_{X^r}^*(s)$ and $F_{U^r}^*(s)$ being their LSTs, respectively.

When product A (or B) needs attention, the cases of positive delay include that the machine may be producing B (or A) or may be in a set-up for product B (or A) at the time of the request. Let us start with the case where product A needs attention and the machine is producing product B. Let the inventory level of product B be $N_B = j$ at the moment product A requests production. Clearly, $j \leq R_B - 1$, since the machine is producing product B. It has to complete processing B and process at least another $R_B - j - 1$ units before it switches to product A. This number will be $R_B - j - 1$ if no further demand arrives during the production of B. Otherwise, every future demand for B also has to be satisfied before R_B is attained. Let T_o^j represent the time consisting of the remaining processing time of the current job plus $R_B - j - 1$ processing times. Also, let T_1^j be the time the machine needs to produce for the demand that arrived during T_o^j. In general, T_i^j is the time to produce for the demand that arrives during T_{i-1}^j. For the sake of simplicity, we suppress the indices A and B in the definition of T_i^j. Now, the delay of A while B is produced can be expressed as

$$Y_A^B = \sum_{i=0}^{\infty} T_i^j, \qquad (7.29)$$

given that $N_B = j$ at the time of the request.

Let us denote the distribution function and the LST of the density function of T_i^j by $F_{T_i^j}(t)$ and $F_{T_i^j}^*(s)$, respectively, and establish a relationship between $F_{T_i^j}^*(s)$ and $F_{T_{i+1}^j}^*(s)$. The LST of the density function of T_i^j is given by

$$F_{T_{i+1}^j}^*(s) = E(e^{-sT_{i+1}^j}) = \int_{t=0}^{\infty} e^{-\lambda_B t} \sum_{n=0}^{\infty} \frac{\left[\lambda_B t F_{X_B}^*(s)\right]^n}{n!} dF_{T_i^j}(t)$$

$$= F_{T_i^j}^*(\lambda_B - \lambda_B F_{X_B}^*(s)), \qquad (7.30)$$

7.4. A Single-Stage System with Two Products

where

$$F^*_{T_0^j}(s) = F^*_{X_B^r}(s)[F^*_{X_B}(s)]^{R_B - j - 1}.$$

It can be verified that

$$E[T^j_{i+1}] = \rho_B E[T^j_i] = \rho_B^i E[T^j_0], \qquad (7.31)$$

where $\rho_B = \lambda E[X_B]$.

The delay of product A conditioned on N_B is given by

$$E[Y^B_A \mid N_B = j] = E[\sum_{i=0}^{\infty} T^j_i] = \sum_{i=0}^{\infty} \rho_B^i E[T^j_0] = \frac{E[T^j_0]}{1 - \rho_B}. \qquad (7.32)$$

On the other hand, for $i = 0$, we have

$$E[T^j_0] = E[X^r_B] + (R_B - j - 1)E[X_B], \; j \leq R_B - 1. \qquad (7.33)$$

Hence, the expected delay of product A while B is produced, $E[Y^B_A]$, can be rewritten as

$$E[Y^B_A] = \sum_{j=-\infty}^{R_B - 1} E[Y^B_A \mid N_B = j] P(I_B = j \mid N_A = r_A + 1, M = B)$$

$$= \frac{E[X^r_B]}{1 - \rho_B} + \frac{E[X_B]}{1 - \rho_B} [R_B - 1 - \sum_{j=-\infty}^{R_B - 1} j P(N_B =$$

$$j \mid N_A = r_A + 1, M = B)]. \qquad (7.34)$$

The variables N and M in the above conditional probabilities represent the inventory level of either product A or B and the product type being produced, respectively. Notice that the above conditional probabilities are in fact the arrival-point probabilities. For instance, the event that triggers the request of product A is the marked arrival of a unit of demand. The marked arrival sees $r_A + 1$ units of inventory in the warehouse and reduces it to r_A which in turn initiates a production request for product A. A delay occurs if the marked arrival also sees that the machine is producing product B, that is $M = B$.

In case the machine is in set-up for product B at the moment product A needs attention, the delay experienced by product A, $Y^{S_B}_A$, consists of another element, namely the remaining set-up time of product B. In this case, we have $N_B \leq r_B$, since the set-up process can only start after the reorder point is downcrossed. Again, let $N_B = j$ when product A requests attention. We can write

$$E[T^j_0] = E[U^r_B] + (R_B - j)E[X_B]. \qquad (7.35)$$

Through similar arguments used for $E[Y_A^B]$, we obtain the expected delay of product A while the machine is being setup for B, that is, $E[Y_A^{S_B}]$:

$$E[Y_A^{S_B}] = \frac{E[U_B^r]}{1-\rho_B} + \frac{E[X_B]}{1-\rho_B}[R_B - \sum_{j=-\infty}^{r_B} jP(N_B = j \mid N_A = r_A + 1, M = S_B)], \quad (7.36)$$

Consequently, the expected delay of product A, upon a marked demand arrival initiating its request for production, is given by

$$E[Y_A] = E[Y_A^B]P(M=B \mid N_A = r_A+1) + E[Y_A^{S_B}]P(M = S_B \mid N_A = r_A+1), \quad (7.37)$$

where $P(M = B \mid N_A = r_A + 1)$ is the conditional probability that the machine is producing product B given that a marked demand arrival for product A sees $r_A + 1$ units in the warehouse. Notice that (7.37) has another term, which is the delay of product A when a marked arrival sees the machine idle, in which case the delay is zero.

In the two-product case, the arguments used to develop the expression for the expected delay are the same for both products, which in turn makes the delay expressions the same except that the product indices are interchanged. For instance, if the indices A and B are interchanged in (7.34)–(7.37), we obtain the delay expression for product B. However, these expressions will be different for each product in case there are more than two products in the system. This is due to the priorities and the switchovers among the products. For instance, let us assume that there are three products, namely A, B, and C. A is the most important and C is the least. At the time the production of B is completed, if both products A and C have been requesting attention, the production of A will start first. If A did not request attention, the production of C would be initiated. In fact, there may be many switchovers between A and B before the production of C may start. Thus, it is clear that the delay expressions for A, B, and C will be different due to priorities and will be more complicated.

7.4.3 Approximations

So far we have presented exact arguments and expressions for the expected delay experienced by products A and B, once they have placed a request for production. Unfortunately, these expressions, such as (7.36) and (7.37), involve conditional probabilities of the joint behavior of the products. These probabilities are not known. One possible way to evaluate them is to resort to approximations. A myopic approach for the approximation would be to assume independence among the products. The expected impact of the independence assumption would be a deterioration in the results as the congestion in the system climbs up. The accuracy of the approximation can be checked via simulation.

7.4. A Single-Stage System with Two Products

The independence assumption results in the following approximations:
In (7.34),

$$\sum_{j=-\infty}^{R_B-1} jP(N_B = j \mid N_A = r_A + 1, M = B) \cong \sum_{j=-\infty}^{R_B-1} jP(N_B = j \mid M = B). \tag{7.38}$$

In (7.36),

$$\sum_{j=-\infty}^{r_B} jP(N_B = j \mid N_A = r_A + 1, M = S_B) \cong \sum_{j=-\infty}^{r_B} jP(N_B = j \mid M = S_B). \tag{7.39}$$

In (7.37)

$$P(M = B \mid N_A = r_A + 1) \cong P(M = B) \tag{7.40}$$

and

$$P(M = S_B \mid N_A = r_A + 1) \cong P(M = S_B).$$

As a result, (7.37) is approximated by

$$E[Y_A] = \frac{E[Y_A^B]P(M = B) + E[Y_A^{S_B}]P(M = S_B)}{1 - P(M = A) - P(M = S_A)}. \tag{7.41}$$

The reason for the division by $1 - P(M = A) - P(M = S_A)$ is that we are concerned with the instances where a demand arrival initiates a request of production for product A. That is, the arrival will not see the machine producing or setting up for product A. The above division simply conditions on those possible views of the system at which a demand arrival may initiate a request for product A. $E[Y_B]$ can also be written in a similar manner.

Thus, one may visualize the original system with two products as decomposed into two systems each, having an imaginary machine, producing only one product. In each production cycle, before their set-up starts, these two systems, A and B, experience delays that are defined as functions of each other's inventory levels. Using the concept of production orders and the queueing analogy that we have developed in Section 7.3.2, we can write the following for system A:

$$\sum_{j=-\infty}^{R_A-1} jP(N_A = j \mid M = A) = R_A - \bar{L}_A(T_P), \tag{7.42}$$

where $\overline{L}_A(T_P)$ is the mean number of production orders during the busy period in system A. Similarly,

$$\sum_{j=-\infty}^{r_A} j P(N_A = j \mid M = S_A) = R_A - \overline{L}_A(T_S), \qquad (7.43)$$

where $\overline{L}_A(T_S)$ is the mean number of production orders during the set-up period in system A, and it is given by (7.24).

Finally, $P(M = A) = \rho_A$ and $P(M = S_A) = P_A(S)$ are the utilization and the set-up probability in system A, as given by (7.3) and (7.8). The evaluation of $E[Y_A]$ through (7.41) requires $\overline{L}_B(T_P)$ which in turn requires $\overline{L}_B$ and consequently $\overline{M}_B(T_D)$. However, in (7.25) $\overline{M}_B(T_D)$ requires $E[Y_B^2]$ (or $E[T_D^2]$ in system B) that is not known. The second moment of the delay period is highly complicated, therefore here we make the simplifying assumption that the delay period is exponentially distributed, which results in $\overline{M}_B(T_P) = \lambda_B E[Y_B]$.

Summarizing the above discussion, we have the following two expressions for the expected delay each product experiences:

$$E[Y_A] = \frac{\rho_B}{1 - \rho_A - P_A(S)} \left\{ \frac{E[X_B^r]}{1 - \rho_B} + \frac{E[X_B]}{1 - \rho_B} [\overline{L}_B(T_P) - 1] \right\}$$

$$+ \frac{P_B(S)}{1 - \rho_A - P_A(S)} \left\{ \frac{E[U_B^r]}{1 - \rho_B} + \frac{E[X_B]}{1 - \rho_B} \overline{L}_B(T_S) \right\} \qquad (7.44)$$

$$E[Y_B] = \frac{\rho_A}{1 - \rho_B - P_B(S)} \left\{ \frac{E[X_A^r]}{1 - \rho_A} + \frac{E[X_A]}{1 - \rho_A} [\overline{L}_A(T_P) - 1] \right\}$$

$$+ \frac{P_A(S)}{1 - \rho_B - P_B(S)} \left\{ \frac{E[U_A^r]}{1 - \rho_A} + \frac{E[X_A]}{1 - \rho_A} \overline{L}_A(T_S) \right\} \qquad (7.45)$$

where $E[Y_A]$ is written as a function of $E[Y_B]$ through $\overline{L}_B(T_S)$ and $\overline{L}_B(T_P)$, and similarly $E[Y_B]$ is written as a function of $E[Y_A]$. Actually, we have a set of two equations and two unknowns, but the structure of the expressions involved are not convenient for a simple solution. One can resort to an iterative procedure to solve for the expected delays as shown in Table 7.2.

The algorithm can also be stopped using $E[Y_B]$. Once it converges, two single-product systems emerge from the analysis, each having a random delay with an expected value of $E[Y_i]$, $i = A, B$. It does not seem possible to prove the convergence of the algorithm; however, numerical experience shows that it has a fast and monotonic convergence. Note that a drawback of the independence approximations is the loss of information related to the joint behavior of the two systems. We may have to resort to simulation modeling for the analysis of their joint behavior.

Once the mean delay is obtained, one can use the matrix-recursive procedure presented in Section 7.3.1 to compute P_n and consequently $\overline{L}, \overline{N}, \overline{B}$, and Pr(Backordering) for the exponential or phase-type set-up and processing times

TABLE 7.2. Approximation algorithm for the two-product problem

INITIALIZE $E[Y_B] = 0, k = 1$

STEP 1: Obtain $P_B^{(k)}(S)$, $\overline{L}_B^{(k)}(T_S)$, and $\overline{L}_B^{(k)}(T_P)$ using (7.24) and (7.25).
Evaluate $E[Y_A]^{(k)}$ using (7.44).
If $|E[Y_A]^{(k)} - E[Y_A]^{(k-1)}| \leq \varepsilon$ (i.e., $\varepsilon = 10^{-5}$), stop; otherwise go to STEP 2.

STEP 2: Obtain $P_A^{(k)}(S)$, $\overline{L}_A^{(k)}(T_S)$ and $\overline{L}_A^{(k)}(T_P)$ using (7.24) and (7.25).
Evaluate $E[Y_B]^{(k)}$ using (7.45).
Set $k \leftarrow k + 1$; go to STEP 1.

and the exponential delay assumptions. The procedure can be generalized to accommodate phase-type distributions without much difficulty. In the following two examples, we first focus on the behavior of the iterative procedure and the results of the approximation.

EXAMPLE 7.3. Consider the above two-product problem with the input data given in Table 7.3.

For the values of $E[Y_A]^{(0)} = E[Y_B]^{(0)} = 1.0$ and $\varepsilon = 10^{-8}$, the convergence is obtained at iteration 9. The corresponding values of the expected delays in each iteration are shown in Table 7.4. The convergence is monotonic and only weakly dependent on the starting values, such that it takes the same number of steps to converge from $E[Y_A]^{(0)} = 100$ and $E[Y_B]^{(0)} = 50$. Also, it takes two more steps to converge from the very odd initial values of $E[Y_A]^{(0)} = 1000$ and $E[Y_B]^{(0)} = 500$.

EXAMPLE 7.4. This set of examples is intended to show the impact of the overall machine utilization and the magnitudes of R and r on the accuracy of the approximation. Table 7.5 shows the impact of the machine utilization ρ on the accuracy of the approximation. For this purpose, we let ρ vary from 0.4 to 0.9 with ρ_1 and ρ_2 alternating between low and high values. Hence, in some examples, product A has a higher intensity than product B, and vice versa in the other examples.

TABLE 7.3. Input data for Example 7.3

	Products	
	A	B
(R, r)	(20,10)	(30,10)
$\bar{x}$	2.0	5.0
$\bar{u}$	10.0	30.0
ρ	0.4	0.4

TABLE 7.4. The path of convergence of the mean delays

Iter. no. k	$E[Y_A]^{(k)}$	$E[Y_B]^{(k)}$
1	99.3836	60.9623
2	113.1466	66.5643
3	114.6823	67.1952
4	114.8576	67.2673
5	114.8777	67.2756
6	114.8800	67.2765
7	114.8802	67.2767
8	114.8803	67.2767
9	114.8803	67.2767

In general, a low value of ρ_i indicates less demand for product i and results in high average inventories. As ρ_i increases, the average inventory levels decrease. The impact of a change in ρ_i on the average inventory levels of both products depends on their parameters such as the arrival rates and the squared coefficient of variation of the processing and set-up times.

The simulation results were obtained using a GPSS model based on a run length of 50,000 process completions. The results of the approximation procedure are highly accurate for the average inventory levels. For the average backorder levels, smaller values tend to have a larger relative error. The large values still have an acceptable relative error considering the complexity of the problem. Notice that since the products do not have priorities over each other, if the input data are swapped, the results will be swapped also.

Table 7.6 shows the impact of the magnitudes of R and r on the accuracy of the approximation. Clearly, high values of R and r result in high average inventory levels and low average backorders. Again, the average inventory levels are accurate. The highest relative error in the average inventory levels is in the case with $(R, r) = (5, 2)$, which is 0.26. The average inventory levels tend to have larger error for small values for R and r. Fortunately, in application, the target and reorder levels are usually not chosen to be very small, especially for products that are of higher priority.

In practice, there are various situations where it is possible to cluster the products under a few main groups. Similar approximation procedures taking the product interrelationships into account can be developed for systems with a higher number of products. Although similar arguments apply, the algorithm will be a little more involved for a larger number of products due to priorities and the resulting switching patterns.

TABLE 7.5. The impact of ρ on the accuracy of the approximation

Input	Product A		Product B	
(R,r)	(20,10)		(20,10)	
$(\bar{u}, \bar{u}^2)$	(2,8)		(5,50)	
$(\bar{x}, \bar{x}^2)$	(1,1.5)		(3,13.5)	
Output	Appx.	Sim.	Appx.	Sim.
ρ_i	0.2		0.2	
$\bar{N}$	14.598	14.550	15.088	15.110
$\bar{B}$	1.39×10^{-4}	0.006	$\cong 0.0$	$\cong 0.0$
ρ_i	0.4		0.2	
$\bar{N}$	12.949	12.980	14.859	14.980
$\bar{B}$	0.041	0.182	2.5×10^{-6}	$\cong 0.0$
ρ_i	0.7		0.2	
$\bar{N}$	6.634	7.661	12.238	13.540
$\bar{B}$	4.977	3.809	0.216	0.062
ρ_i	0.2		0.4	
$\bar{N}$	13.326	13.130	14.384	14.580
$\bar{B}$	0.040	0.100	3.04×10^{-5}	6.6×10^{-4}
ρ_i	0.2		0.7	
$\bar{N}$	7.195	8.074	10.052	11.530
$\bar{B}$	6.161	4.793	0.495	0.325
ρ_i	0.4		0.4	
$\bar{N}$	9.400	9.806	13.154	14.020
$\bar{B}$	1.736	1.540	0.024	0.008

7.5 Multistage Pull-Type Systems

In this section, we turn our attention to multistage pull-type production/inventory systems consisting of stages in series with intermediate storage locations as shown in Figures 7.5 to 7.8. A production stage can be visualized not only as a machine or a work station but also as a department such as a press shop, a heat treatment facility, an assembly, or a painting department. For various reasons, such as cost consideration or as a matter of operating principle, the interstage inventories are kept below certain levels which we call target inventory levels. (R, r)-type policies can again be implemented in such a way that after reaching its target level, a stage may not process production orders awaiting attention until its finished inventory level drops to its reorder level r. It is clear from our earlier discussions in this chapter that the magnitude of $R - r$ at a stage depends on the existence and the magnitude of its set-up time or the set-up cost.

TABLE 7.6. The impact of R and r on the accuracy of the approximation

Input	Product A		Product B	
ρ	0.40		0.35	
$(\bar{u}, \overline{u^2})$	(2,7)		(4,48)	
$(\bar{x}, \overline{x^2})$	(1,1.5)		(2,6)	
Output	Appx.	Sim.	Appx.	Sim.
(R, r)	(5,2)		(20,10)	
$\overline{N}$	1.295	1.749	13.875	14.450
$\overline{B}$	3.296	3.617	0.003	0.004
(R, r)	(20,10)		(20,10)	
$\overline{N}$	10.931	11.330	13.559	14.180
$\overline{B}$	0.590	0.808	9.036×10^{-3}	0.004
(R, r)	(100,75)		(20,10)	
$\overline{N}$	83.039	85.260	12.679	13.380
$\overline{B}$	5.52×10^{-5}	4.280×10^{-4}	0.075	0.020
(R, r)	(20,10)		(5,2)	
$\overline{N}$	12.418	13.040	2.390	2.527
$\overline{B}$	0.111	0.258	0.402	0.412
(R, r)	(20,10)		(100,75)	
$\overline{N}$	8.430	8.882	85.660	86.850
$\overline{B}$	3.038	3.828	2.830×10^{-5}	$\cong 0.0$

Let us consider the system shown in Figure 7.5 consisting of K stages with intermediate storage locations and a warehouse. The warehouse is perceived as part of stage K. The first stage has raw material to process at all times. Production orders at each stage are processed on a FIFO manner. Upon achieving the target level in its storage, stage i enters a period during which it is idle or it may be viewed as working on auxiliary and interruptable tasks. During this period, the inventory in its buffer is depleted due to production at the next stage. As soon as the inventory level at stage i drops to r_i, a set-up is initiated and production resumes after the set-up is completed. For simplicity, assume that the unit processing and set-up times are exponentially distributed with rates μ_i and γ_i for stage i, respectively.

As mentioned in Section 7.2, the (R, r) policy is implemented by maintaining a pool of production orders awaiting attention. Each stage has a distinct set of production cards attached to its finished products waiting for orders from the next stage. Let $L_t(k)$ be the number of production orders, and let $N_t(k)$ be the number of finished units at stage k at time t. Then, at time t, production starts at stage k if the machine is available and if $L_t(k)$ and $N_t(k-1) \geq 1$. When a unit from the inventory of stage $k-1$ is transferred to stage k, its production card is sent back to the pool of production orders at stage $k-1$, and a production card from stage k is attached to it. After processing is completed, the finished unit with the attached production card is placed into the buffer at stage k. Notice that the production cards always remain in the stages to which they are assigned.

Multistage systems are similar in nature to the production lines studied in Chapter 4. As the system size (in terms of the buffer capacities or the number of stages) increases, the use of the exact methods becomes computationally infeasible. Therefore, other than computer simulation modeling, the approximation methods again remain the only viable approach to obtain the long-run average measures of system performance. Our interest is still in the inventory- and the production-order related measures, as in the case of the single-stage problems studied earlier. For this purpose, we are going to extend the decomposition approach introduced in Chapter 4 to analyze pull-type systems with (R, r) control policies.

It is possible to view a manufacturing system as a group of subsystems (stages) having a local operational control policy with its own demand arrival process. As shown in Figure 7.14, for every stage j, $j = 1, \ldots, K$, in the original system, there exists a subsystem $\Omega(j)$ where the first node represents the behavior of the upper echelon (production process), and the second node represents the behavior of the lower echelon (demand process). The buffer at the jth stage in the original system is represented by the buffer in subsystem $\Omega(j)$. Each subsystem uses the (R, r) policy associated with the corresponding original buffer. The production orders at the first node are generated by the demand process represented by the second node. The structure of the demand process generates demand with random interarrival times as long as there are units in the buffer. It is switched off when there are no units in the buffer. This is due to the stoppage of production at stage $j + 1$ when there are no units in the buffer at stage j.

The subsystem $\Omega(K)$ can be modeled as a single-stage system and analyzed as shown in Section 7.3 and in the Appendix. In Section 7.3, we have treated the single-stage system as an M/G/1 work station with units representing the production orders in the manufacturing system. The unit arrival process is equivalent to the demand arrival process, and the service time is the same as the effective processing time in $\Omega(K)$. The distinction here is that the server does not experience a delay before the set-up starts. The subsystems are linked to each other in an iterative manner. Next, we will focus on the characterization of the production and the demand processes in $\Omega(j)$, for $j = 1, \ldots, K - 1$.

7.5.1 Characterization of Subsystem $\Omega(j)$

Let U'_j and U''_j be the random variables representing the effective processing time and the demand interarrival time in $\Omega(j)$, respectively. U'_j incorporates into the processing time at stage j the effect of idleness due to starvation at the upstream stages. Similarly, U''_j incorporates into the processing time at stage $j + 1$ the effect of stoppages due to the target inventory levels in the storage locations at the downstream stages. The characterization of U'_j is quite similar to the one in

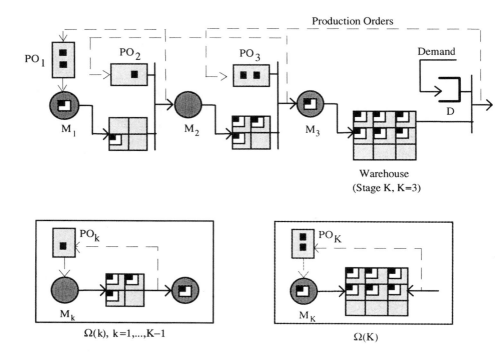

Figure 7.14. Decomposition of a pull-type manufacturing system.

the analysis of the two-node subsystems in Section 4.4.2. Here we will briefly summarize it. At a departure point at stage j, the time until the next departure occurs is the processing time at stage j, that is, X_j, if units are available in the buffer at stage $j - 1$. Otherwise, stage j starves until a unit arrives at the upstream buffer. Let $\Delta(j)$ be the conditional probability that there are no jobs in the upstream buffer given that a unit is just about to finish its processing time at stage j. Also, let V_j represent the period of time the jth stage remains idle upon a process completion. V_j is a phase-type random variable with a $(\underline{\theta}_j, I_j)$ representation with $j - 1$ phases. Then, the effective processing time in $\Omega(j)$ can be defined by

$$U'_j = \begin{cases} X_j & \text{wp } 1 - \Delta(j), \\ X_j + V_j & \text{wp } \Delta(j), \end{cases} \quad (7.46)$$

where X_j and V_j are assumed to be independent. This is an approximation, since V_j depends on X_j as well as on the processing times of all the other stages in the original system. Then, U'_j is a phase-type random variable represented by the pair $(\underline{\alpha}'_j, S'_j)$ with j phases.

As Figure 7.15 shows, with probability $1 - \Delta(j)$ a unit goes through the processing time at stage j. With probability $\Delta(j)$, it starts its processing in one of the phases of the idle period and moves according to the matrix of transition rates I_j. When the idle period is over, that is, when stage j no longer starves, the unit is transferred to the phases of the processing time. After the exponentially distributed processing time X_j, it is placed into the buffer of $\Omega(j)$. Thus, $\Omega(j)$ processes its production orders with processing time U'_j according to the (R_j, r_j) control policy. The above characterization of U'_j is valid for all the subsystems $\Omega(j)$, $j = 1, \ldots, K$. The particular forms of $\underline{\theta}_j$ and I_j are similar to the ones in Section 4.4.2. The difference here is the inclusion of the possibility of having stage $j - 1$ in its set-up at the start of V_j.

The demand at stage j is generated by the process completions at stage $k = j + 1$. Let $\Pi(k)$ be the conditional probability that there are no outstanding production orders given that the last order is just processed at stage k. Consequently, the inventory level in the buffer reaches its target value R_k, and stage k stops its production (orders are cleared). Let Z_k represent the duration stage k remains forced idle or blocked once R_k is attained. If we let $Q_k = R_k - r_k$, then Z_k represents the time for Q_k departures to occur from the buffer at stage k. Z_k is a phase-type random variable since all the downstream stages have exponential or phase-type processing times. Let the pair $(\underline{\nu}_k, B_k)$ represent the phase-type distribution of Z_k. The set-up process of length U_k follows Z_k.

With probability $1 - \Pi(k)$, the time a unit spends on the machine at stage k is its processing time X_j. With probability $\Pi(k)$, a unit fills up the buffer and initiates blocking for stage k. In this case, the time until the process restarts at stage k includes the period Z_k and the following set-up time once the inventory level in the buffer drops to the reorder level r_k. Then, the effective demand interarrival time U''_k is given by

$$U''_k = \begin{cases} X_k & \text{wp } 1 - \Pi(k), \\ X_k + Z_k + U_k & \text{wp } \Pi(k), \end{cases} \qquad (7.47)$$

where X_k, and Z_k are assumed to be independent. Again, this is an approximation since Z_k depends on the processing times of all the stages including stage k. U''_k

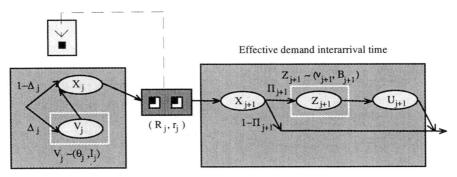

Figure 7.15. Characterization of $\Omega(j)$

is a phase-type random variable with an $(\underline{\alpha}_k'', S_k'')$ representation having a number of phases that depends on $Q = R - r$ of the downstream stages. The particular form of the vectors θ_j, α_j', ν_k, and α_k', and the matrices I_j, B_k, S_j', and S_k' will be clearer with the help of an example in the following section.

7.5.2 Iterative Analysis

An iterative method sequentially analyzes all the subsystems $\Omega(1), \Omega(2), \ldots$, $\Omega(M)$, which we call a **forward pass**. Then, it executes a **backward pass** from $\Omega(M-1)$ back to $\Omega(2)$. The analysis of $\Omega(k)$ utilizes information from $\Omega(k-1)$ in the forward pass and information from $\Omega(k+1)$ in the backward pass. The iterative method continues to execute these forward and backward passes until the throughputs of $\Omega(k)$ for all k are sufficiently close to each other.

The subsystem $\Omega(j)$ can be analyzed by studying the underlying Markov chain as we have done for the two-node subsystems in Chapter 4. It is a three-variable Markov chain $\{I_t, J_t, N_t, t \geq 0\}$ keeping track of the phases of the processing time, the demand interarrival time, and the number of units in the buffer at time t. Let us define $P_j(i, \ell, n), n < R_j$, as the stationary probability that the $\Omega(j)$ is in phase i of its processing time, the demand process is in phase ℓ of its interarrival time, and n units reside in the buffer in $\Omega(j)$. Also let $P_j(B, \ell, n), r_j < n \leq R_j$ be the stationary probability that M_j in $\Omega(j)$ is blocked, the demand process is in phase ℓ, and n units are in the buffer. Then, one can express $P_j(n)$ as a proper sum of the joint probabilities of $\Omega(j)$. We will not repeat the matrix-recursive procedure here. Rather, we leave it to the reader as an exercise (see Problem 7.11).

In a forward pass, after the analysis of $\Omega(j)$ is completed, new values of $\underline{\theta}_{j+1}$, I_{j+1} and $\Delta(j+1)$ are obtained and used in the characterization of $\Omega(j+1)$. For instance, if we let

$$\varpi_j'(i) = Pr(\Omega(j) \text{ is in phase } i \text{ of its processing time} \mid \text{a demand arrival is about to leave the buffer empty}),$$

then, we can write

$$\underline{\theta}_{j+1}^T = [\varpi_j'(1), \ldots, \varpi_j'(j-1)] \quad \text{and} \quad I_{j+1} = S_j'. \tag{7.48}$$

The conditional probability $\Delta(j+1)$ can again be obtained using Little's law, as we did in Chapter 4. The average number of departing units (demand arrivals) causing starvation at stage $j+1$ per unit time can be written as $\bar{\xi}\Delta(j+1)$, where $\bar{\xi}$ is the average throughput in the system. The expected length of the starvation period is $E[V_{j+1}]$, and the stationary probability that stage j is starving is $P_j(0)$. Then, $\Delta(j+1)$ is obtained using

$$\bar{\xi}\Delta(j+1)E[V_{j+1}] = P_j(0), \quad j = 1, \ldots, K-1, \tag{7.49}$$

where $P_j(0)$ can be obtained using the joint probabilities of $\Omega(j)$.

7.5. Multistage Pull-Type Systems

In a backward pass, after the analysis of $\Omega(k)$ is completed, the values of $\underline{\nu}_k$, B_k and $\Pi(k)$ are obtained to characterize $\Omega(k-1)$. For instance, if we define

$\varpi_k''(i) = \text{Pr}(\text{The demand process in } \Omega(k) \text{ is in phase } i \mid \text{a departing unit is about to block } M_k)$,

then, $\underline{\nu}_k^T$ is given by

$$\underline{\nu}_k^T = [\varpi_k''(1), \ldots, \varpi_k''(m_k), 0, \ldots, 0], \qquad (7.50)$$

where m_k represents the number of phases in the demand process in $\Omega(k)$.

A similar argument can be used to obtain $\Pi(k)$. The rate at which blocking occurs at stage j is $\overline{\xi}\Pi(k)$. The average time spent in this state is $\text{E}[Z_k]$ and the stationary probability that stage k is blocked is $P_k(B) \cong \sum_{l=1}^{m_k} \sum_{n=r_k+1}^{R_k} P_k(B, l, n)$. Then, again using Little's law, we can write

$$\overline{\xi}\Pi(k)\text{E}[Z_k] = P_k(B), \qquad k = 2, \ldots, K. \qquad (7.51)$$

The matrices **I** and **B** will be clarified through a graphical representation of their phase structures in an example given later in this section.

Thus, the iterative analysis of the two-node subsystems continues until their throughputs are sufficiently close to the true throughput of the system. Notice that the true throughput is known. It is $\overline{\xi} = \lambda$, since the system allows backorders. For $\Pi(K)$, the above expression simplifies to

$$\Pi(K) = \frac{1 - \lambda\text{E}[U_K']}{R_K - r_K + \lambda\text{E}[U_K]}, \qquad (7.52)$$

which we are already familiar with (see (7.9) and Problem 7.12). This is convenient because there is no need to compute the stationary probabilities of the inventory level in the warehouse until the iterative method converges. These probabilities are computed to obtain $\overline{N}_K$ and Pr(backordering) after convergence is achieved. $\overline{N}_k$ for $k = 1, \ldots, K-1$ are obtained using the probabilities of $\Omega(k)$. The iterative algorithm is given in Table 7.7.

The stability of multistage systems is an important issue. We want to know under what condition the system is able to cope with the demand such that the number of backorders does not keep increasing. Unfortunately, such a condition does not exist unless the system has very large target levels. In this case, $\lambda < \text{Min}(\mu_i)$ will be sufficient for stability. This relation is necessary but may not be sufficient for the case with finite target inventory levels.

EXAMPLE 7.5. Let us consider a three-stage production/inventory system with exponentially distributed processing and set-up times. The parameters of the (R, r) control policy implemented at each stage are $(R_1, r_1) = (5, 2)$, $(R_2, r_2) = (4, 1)$, and $(R_3, r_3) = (4, 2)$. For the purpose of obtaining the approximate values of the

TABLE 7.7. The iterative decomposition algorithm

INITIALIZATION: $k = 1, \Pi_i = 0, i = 2, \ldots, K$
STEP 1: FORWARD PASS: $j = 1$
    (i) Analyze $\Omega(j)$ to obtain the probabilities $P_j()$ and the throughput $\xi_j^{(k)}$.
    (ii) Obtain $\Delta(j+1), \varpi_j'()$ using (7.49).
    (iii) Let $j = j + 1$.
    (iv) If $j = K$, obtain Π_K using (7.52) and go to STEP 2.
        Else, go to (v).
    (v) Construct the arrays $\underline{\theta}_j$ and $\mathbf{I}_j$ using (7.48), and go to (i).
STEP 2: BACKWARD PASS: $j = K - 1$
    (i) Analyze $\Omega(j)$ to obtain the probabilities $P_j()$ and its throughput $\xi_j^{(k)}$.
    (ii) If $j = 1$, go to (v).
        Else, go to (iii).
    (iii) Obtain $\Pi(j), \varpi_j''()$ using (7.51) and go to (iv).
    (iv) Construct the arrays $\underline{\nu}_j$, and $\mathbf{B}_j$ using (7.50), let $j = j - 1$, and go to (i).
    (v) If $|\xi_j^{(k)} - \xi^{(k-1)}| < \varepsilon$, for all j, then stop, else $k \leftarrow k + 1$ and go to STEP 1.

---

performance measures, let us decompose the system as shown in Figure 7.16. There are three subsystems, one for each stage, $\Omega(j), j = 1, 2, 3$. The production process in $\Omega(1)$ is identical to that of stage 1 in the original system. The demand process in $\Omega(1)$ brings the effects of stoppages from stages 2 and 3. With probability $\Pi(2)$, upon reaching R_2, stage 2 is forced to stop for a period of Z_2 consisting of three effective demand arrivals in $\Omega(2)$ for the inventory level to drop to r_2. Production at $\Omega(2)$ restarts after the set-up process is completed with rate γ_2. A more rigorous analysis of Z_2 would include the phase the effective demand process in $\Omega(2)$ is in at the time Z_2 is initiated, with its associated probability.

$\Omega(2)$ models the effects of the stoppages from both echelons. The effective processing time U_2' models the impact of starvation, represented by V_2, and the demand process models the forced-down idleness, that is, Z_3. In case a departure at stage 2 leaves the buffer at stage 1 empty, with probability $\Delta(2)$, stage 2 will wait for a process completion at stage 1. This waiting time is represented by V_2. A more rigorous modeling of V_2 would include the possibility of having stage 1 in its set-up process at the start of V_2. In case a departure at stage 3 fills up the warehouse, with probability $\Pi(3)$, it initiates a forced idleness at stage 3, represented by Z_3. This period is the time for two units of demand to arrive at the warehouse.

$\Omega(3)$ models stage 3 with the warehouse. The effective processing time U_3' models the starvation time at stage 3, represented by V_3. Starvation starts at a process completion at stage 3 when no units are present in the buffer at stage 2,

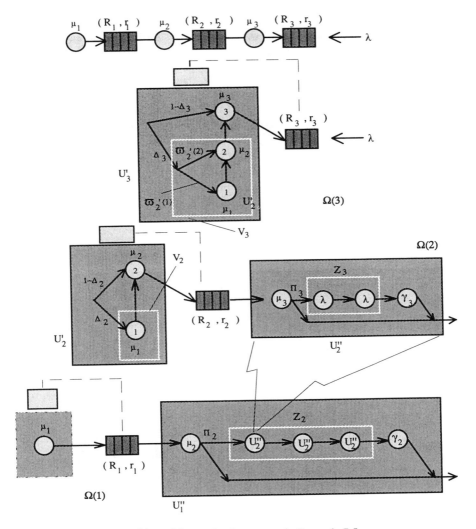

Figure 7.16. Decomposition of the production system in Example 7.5.

with probability $\Delta(3)$. This is the same event as a demand arrival leaving the buffer in $\Omega(2)$ empty. At this point, the effective production process in $\Omega(2)$ may be in its first or the second phase with probabilities $\varpi_2'(i)$, $i = 1, 2$. Hence, the phase structure of V_3 can be constructed accordingly. Notice that the phase structure of V_j is identical to that of U_{j-1}' but with its own starting probability vector $\underline{\theta}_j$ consisting of $\varpi_2'(i)$, for all i.

Hence, through this example, the phase structures of the blocking and starvation periods, V_k and Z_k, and their associated matrices $\mathbf{I}_k$ and $\mathbf{B}_k$, should be clear with the help of Figure 7.16. The above method can be applied to systems with phase-type processing or set-up times with minor modifications.

318   7. Pull-Type Manufacturing Systems

EXAMPLE 7.6. In this example, we analyze a more general scenario than the one outlined in the preceding approximation procedure. Consider a four-stage, pull-type manufacturing system with the following data. The target inventory and the reorder levels for the three interstage buffers and the warehouse are as follows:

	Stage 1	Stage 2	Stage 3	Warehouse
R	5	4	6	15
r	2	2	3	10

The demand size is a random variable with a discrete empirical distribution. The demand arrival rate is 0.03 with the following demand size probabilities:

d	1	2	3	4	5
$P(D = d)$	0.3	0.3	0.2	0.1	0.1

That is, when a customer arrives, his demand may be 1 unit with probability 0.3, 4 units with probability 0.1, and so forth.

The processing and the set-up times have the following mean and squared coefficient of variations:

		Stage 1	Stage 2	Stage 3	Stage 4
Proc. Times	$\bar{x}$	4.00	3.00	5.00	4.00
	Cv^2	0.25	0.35	0.25	0.30
Set-up Times	$\bar{u}$	1.00	2.50	1.50	2.00
	Cv^2	0.50	0.75	0.50	0.80

Notice that the demand arrival process is compound Poisson, which plays a role in the analysis of $\Omega(K)$ and $\Omega(K-1)$. Also, the processing and set-up times are nonexponential. In this example, the distributions of both processing and set-up times are represented by phase-type distributions using their first two moments. We refer the reader to Chapter 2 on phase-type representations of general distributions.

In this example, we have extended the above procedure to accommodate the phase-type assumptions. This can be done by carefully keeping track of the phases of each of the PH random variables. Compound Poisson demand arrivals and the phase-type assumptions are already incorporated into the procedure presented in

TABLE 7.8. The average inventory levels and the throughput in Example 7.6.

	Appx.	Sim.
$\overline{N}_1$	3.2000	3.2734 (± 0.0114)
$\overline{N}_2$	2.7255	2.7088 (± 0.0046)
$\overline{N}_3$	3.9096	3.9185 (± 0.0167)
$\overline{N}_4$	12.2680	12.4509 (± 0.0235)
Throughput	0.0695	0.0720

the Appendix. The numerical results and their comparison with the simulation results (with 95% confidence intervals) are given in Table 7.8.

In the case of lost sales rather then backorders, $\Omega(K)$ becomes a finite-capacity loss system that can also be analyzed by slightly modifying the procedure presented in the Appendix. The boundary equations must be added to the flow-balance equations.

7.5.3 Make-to-Order Systems

Consider next the multistage system shown in Figure 7.6 where a production order (or Kanban) is sent to stage 1 every time a unit of demand arrives at the warehouse. Let us assume a K-stage system with a total of M Kanban cards for the entire system. There are no finished products in the system. Stage 1 produces as orders arrive. Since the demand arrival process is independent of the system state, we can think of production orders, with rate λ, forming a queue at stage 1. This queue, again denoted by PO, has an infinite capacity due to the assumption of backorders. Figure 7.17 shows handling of the production orders. Orders wait in queue PO until a Kanban card becomes available in queue C. A Kanban arrives at queue C when an order departs from the system. Orders having Kanban cards attached to them proceed to stage 1 to start production. Notice that we have used Kanban cards as production orders themselves before. However, here we use them to limit the number of orders being processed in the system at any time. This is equivalent to having a token mechanism that lets orders enter the system at certain points in time. It is clear that orders may wait in queue PO even if stage 1 is available for production. This is completely dependent upon the number K, which has to be set according to some resource availability constraints in the system.

Let us assume deterministic processing times, x_j for stage j. The demand arrival process and equivalently the order arrival process can be arbitrary. Similar to flow lines with deterministic processing times, which we studied in Chapter 4, we have the following observations:

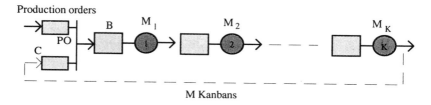

Figure 7.17. A make-to-order system.

Observation 1:

The flow time of a production order in the system is independent of the order of stages as well as of the capacities of the buffers. The buffer capacities are immaterial since each buffer space can be thought of as a stage with zero processing times. As in Chapter 4, the stage having the longest processing time, that is, x^*, can be placed as the first stage without causing any change in the flow time of a particular order. In this new arrangement, orders do not experience any delays in the rest of the system.

Observation 2:

The system in Figure 7.17 is equivalent to a two-stage system shown in Figure 7.18. The first stage has x^* as the processing time. The second stage has M parallel machines each having $x' = T - x^*$, where $T = \sum_{i=1}^{K} x_i$ and $x^* = \text{Max}\{x_i\}$.

Observation 3:

The delay of a production order (its waiting time) in the two-stage system is equivalent to the waiting time in a GI/D/1 work station if $Mx^* \geq \sum_{i=1}^{K} x_i$ holds.

The condition in Observation 3 is highly conceivable for manufacturing systems. Let us study this observation in detail. Let τ be the starting time of a busy cycle

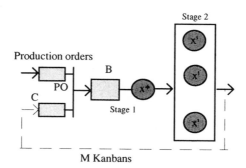

Figure 7.18. Two-stage system equivalent to the one in Figure 7.17.

7.5. Multistage Pull-Type Systems

for the entire two-stage system. Also, let a_i and w_i be the time between orders i and $i - 1$, and the delay of order i in the system, respectively. The waiting time w_i has two parts, w_i^{PO} and w_i^B. The orders arriving during this busy cycle are numbered starting from 1. Thus, for order 1, we have $w_1 = 0$, and $d_1 = \tau + T$ as its departure time from the network. Notice that there is no waiting in queue PO for the first M orders, that is $w_i^{PO} = 0$, for $i = 1, \ldots, M$. For the jth order, $2 \leq j \leq M$,

$$w_j^B = \text{Max}\{0, w_{j-1} + x^* - a_j\}; \tag{7.53}$$

the departure time from stage 1 and stage 2 are

$$d_j^1 = \tau + \sum_{i=2}^{j} a_i + w_j + x^* \quad \text{and} \quad d_j^2 = \tau + \sum_{i=2}^{j} a_i + w_j + T,$$

respectively. The first Kanban returns to queue C at d_1^2. The $(M+1)$st order arrives at $\tau + \sum_{i=1}^{M} a_i$. Stage 1 becomes available for the $(M+1)$st order at

$$d_M^1 = \tau + \sum_{i=2}^{M} a_i + w_M + x^* = \tau + \sum_{i=1}^{M} w_i + Mx^*.$$

Since we assume $Mx^* \geq T$, we have $d_M^1 \geq d_1^2$. This means that the machine at stage 0 finishes processing the Mth order only after the first order departs from the system. Then, w_{M+1} depends on if M_1 is available at a_{M+1}. This indicates that if an order waits in PO, it has to wait in queue B also. For the $(M+1)$st order, we can write

$$w_{M+1}^{PO} = \text{Max}\{0, T - \sum_{i=2}^{M+1} a_i\}, \qquad w_{M+1}^B = \text{Max}\{0, w_M + x^* - a_{M+1}\} - w_{M+1}^{PO}.$$

In general, for the kth order, we have

$$w_k^{PO} = \text{Max}\{0, T + w_{k-M} - \sum_{i=k-M+1}^{k} a_i\}, \quad w_k^B = \text{Max}\{0, w_{k-1} + x^* - a_k\} - w_k^{PO}. \tag{7.54}$$

On the other hand, in a GI/D/1 work station, the waiting time of the jth customer is

$$w_j = \text{Max}\{0, w_{j-1} + x^* - a_j\}, \tag{7.55}$$

which is identical to (7.53) and (7.54).

Thus, if $\overline{W}$ is the average waiting time in a GI/D/1 station, the average manufacturing lead time in the original system of K stages will be $\overline{W} + T$. $\overline{W}$ can be approximated by the Kramer and Langenbach-Belz formula given by (3.13) in Chapter 3. Consequently, the average number of production orders in the system can be obtained by $\overline{N} = \lambda \overline{W}$, where λ is the demand arrival rate.

Observation 4:

The delay of a production order in the two-stage system is independent of the number of Kanbans in the system when $M \geq M^* = \lceil T/x^* \rceil$. This can be shown by using Observation 3. When $Mx^* \geq T$, the delay of an order is the same as its waiting time in the GI/D/1 queue, which is independent of M. This argument is also valid for $M > K$.

Observation 5:

The stability condition for the K-stage system is $\lambda T/M < 1$.

7.5.4 Sending Production Orders to Multiple Stages

Sending production orders from the warehouse to a particular stage j will force it to work at the pace of the demand due to the fact that the stage produces only if there is a production order from the warehouse and $N_{j-1} > 0$. Consequently, there will be less inventory accumulation at the stage receiving the orders and in all succeeding stages, when compared to the case where orders arrive from the immediate downstream stage. Less inventory is desirable, but, as expected, it may result in a higher number of outstanding backorders in the warehouse. Hence, a comparison of behaviors of the systems in Figures 7.5 and 7.6 will show the tradeoff between the average inventory and backorder levels, making a cost or a profit objective the decisive factor. Reducing inventories will be highly desirable when its impact on the backorders is insignificant or when backordering is less costly. One may also think of sending orders from the warehouse to an intermediate stage or to every stage simultaneously, as indicated in Figure 7.8. Sending orders from the warehouse to stage j and to any other stage after j has simply no impact on the system behavior. These orders will wait for stage j, and perhaps the upstream stages, to finish their work. In the case of sending orders to every stage including stage 1, whether they receive orders from the warehouse or not, the downstream stages will wait for stage 1 and the intermediate stages to finish its work. Again, the argument stems from the assumption that the demand rate is less than the production rates, such that the pace of the orders coming from the warehouse slows the production at the stages.

In case a better performance is demanded in the backorder probability or in the average number of backorders, it may be necessary to resort to the card-passing mechanism the generalized Kanban scheme demonstrated in Figure 7.7.

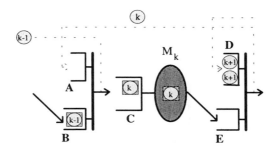

Figure 7.19. Stage k using the generalized Kanban scheme.

7.6 The Generalized Kanban Scheme

In this card-passing policy, stage k has a pool of production orders (queue A), a buffer for the unfinished inventory (queue C), and a buffer for the finished inventory (queue E), as shown in Figure 7.19. Production orders from stage $k + 1$ arrive at queue A. As soon as both queues A and B have units, a finished inventory is transferred from stage $k - 1$ to stage k with a production card from stage k on it. Note that queues D and E cannot have units simultaneously since they are the subassembly queues to the assembly operation of transferring a finished unit from stage k to stage $k + 1$. Let C_k be the maximum number of production cards in stage k. Thus, the total number of units in queues C and E can not exceed C_k.

This Kanban scheme is equivalent to the rule that allows a stage to keep the blocking unit in its input buffer and lets its machine process another item while there is no space in the downstream buffer. That is, the input buffer is "shared" by the arriving units as well as the finished, blocking units. Clearly, more than one unit may be processed and placed in the input buffer at any point in time. Then, at the first opportunity, the unit that has encountered blocking first is transferred to the next stage first. The shared-buffer rule is equivalent to the generalized Kanban scheme with C_k being the buffer capacity.

Also note that when queue C is eliminated in the above Kanban scheme, the governing message-passing policy is nothing more than the local control policy (see Figure 7.5).

An Isolated Kanban Cell

Due to the difficulties in analyzing the whole system at once, we look at an isolated approximate view of stage k, as shown in Figure 7.20. Let us assume a processing rate of μ_k at stage k. Queue B has a capacity of C_{k-1} and a job arrival process with rate λ_k. Also, queue D has a finite capacity C_{k+1} and a demand arrival process with rate d_k. Job and demand arrivals at both queues are lost when they are full, as the figure indicates. The impact of the isolated systems on each other are clearly

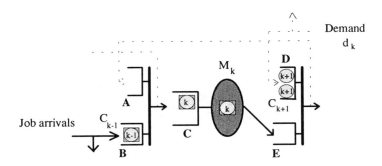

Figure 7.20. Isolated view of stage k.

confined into buffers B and D. The card-passing mechanism within each stage is the same as that of the original system.

With the assumption of exponential interarrival and processing times, each isolated Kanban stage (or cell) can be studied as a Markov process with a finite number of states. Let $I_{k,t}$ and $J_{k,t}$ be two integers such that $-C_{k-1} \le I_{k,t} \le C_k$ and $-C_{k+1} \le J_{k,t} \le C_k$ with $I_{k,t} + J_{k,t} \le C_k$. When $I_{k,t} < 0$, $-I_{k,t}$ is the number of jobs in queue B, and when $I_{k,t} > 0$, it is the number of production cards in queue A. Similarly, $J_{k,t} < 0$ indicates the demand quantity in queue D, and $J_{k,t} > 0$ indicates the finished units in queue E.

Let $\bar{o}_k$ be the throughput of finished units in isolated stage k. We can write

$$\bar{o}_k = \lambda_k [1 - \Pr(I_k = -C_{k-1})], \qquad k = 2, \ldots, K. \tag{7.56}$$

Note that stage 1 has sufficient raw material to avoid starvation. Also,

$$\bar{o}_k = d_k [1 - \Pr(J_k = -C_{k+1})], \qquad k = 1, \ldots, K-1. \tag{7.57}$$

Clearly, $\underline{\lambda} = (\lambda_2, \ldots, \lambda_K)$ and $\underline{d} = (d_1, \ldots, d_{K-1})$ are the vectors of unknowns in overall $K - 1$ stages. First of all, we expect that the throughputs of all the stages should be the same. That is,

$$\bar{o}_{k-1} = \lambda_k [1 - \Pr(I_k = -C_{k-1})], \tag{7.58}$$

or

$$\lambda_k = \bar{o}_{k-1} / [1 - \Pr(I_k = -C_{k-1})], \qquad k = 2, \ldots, K. \tag{7.59}$$

Similarly, we can write

$$d_k = \bar{o}_{k+1} / [1 - \Pr(J_k = -C_{k+1})], \qquad k = 1, \ldots, K-1. \tag{7.60}$$

7.6. The Generalized Kanban Scheme

We can also relate stages $k - 1$ and k using

$$\Pr(I_k = -C_{k-1}) = \Pr(J_{k-1} = C_{k-1}), \qquad k = 2, \ldots, K, \quad (7.61)$$

indicating that queue B in the isolated stage k is in fact queue E of stage $k - 1$. Thus, (7.56) and (7.61) give us

$$\lambda_k = \bar{o}_k / [1 - \Pr(J_{k-1} = C_{k-1})], \qquad k = 2, \ldots, K.$$

Similarly, $\Pr(I_{k+1} = C_{k+1}) = \Pr(J_k = -C_{k+1})$ gives us

$$d_k = \bar{o}_k / [1 - \Pr(I_{k+1} = C_{k+1})], \qquad k = 1, \ldots, K - 1. \quad (7.62)$$

Now, using the concept of throughput equivalence among the isolated stages, we can write the following sets of $2(K - 1)$ equations and $2(K - 1)$ unknowns:

$$\lambda_k = \bar{o}_{k-1} / [1 - \Pr(J_{k-1} = C_{k-1})], \qquad k = 2, \ldots, K, \quad (7.63)$$
$$d_k = \bar{o}_{k+1} / [1 - \Pr(I_{k+1} = C_{k+1})], \qquad k = 1, \ldots, K - 1. \quad (7.64)$$

The unknowns $\lambda_2, \ldots, \lambda_K$ and $d_1, \ldots, d_{K-1}$ are obtained using a fixed-point iteration algorithm that starts with some initial values of the unknowns. The probabilities $\Pr(J_{k-1} = C_{k-1})$ and $\Pr(I_{k+1} = C_{k+1})$ are obtained from the analysis of stages in isolation. Note that $\lambda_1 = \infty$, that is, the first stage never starves, and $d_K = D_K$, where D_K is the known demand rate at stage K. The algorithm is summarized in Table 7.9.

Once the convergence is achieved, the average inventory level at stage k is given by

$$\overline{N}_k = C_k - \sum_{n=1}^{C_k} n \Pr(I_k = n), \qquad k = 1, \ldots, K.$$

TABLE 7.9. The iterative algorithm that relates the isolated Kanban cells to each other

---

INITIALIZATION: $\ell = 1$, $\lambda_1 = \infty$, $d_k = \mu_{k+1}$, $k = 1, \ldots, K - 1$, $d_K = D_K$
STEP 1:   FORWARD PASS: $k = 2$
    (i) Analyze stage k to obtain $\Pr(J_{k-1} = C_{k-1})$ and $\bar{o}_k^{(\ell)}$.
    (ii) Obtain $\lambda_k^{(\ell)}$ using (7.63).
    (iii) Set $k \leftarrow k + 1$. If $k = K + 1$, go to STEP 2, else go to (i).
STEP 2:   BACKWARD PASS: $k = K - 1$
    (i) Analyze stage k to obtain $\Pr(I_{k+1} = C_{k+1})$ and $\bar{o}_{k+1}^{(\ell)}$.
    (ii) Obtain $d_k^{(\ell)}$ using (7.64)
    (iii) Set $k \leftarrow k - 1$. If $k = 0$, go to (iv), else go to (i).
    (iv) If $|d_1^{(\ell)} - d_1^{(\ell-1)}| \leq \varepsilon$ stop; else go to STEP 1.

Note that $\overline{N}_k$ is the average total inventory level, which includes the inventory in buffers C and E.

It is clear from the discussion in Section 7.2.2 that the generalized Kanban scheme improves the customer-service level when compared to the local control policy. Remember that the customer-service level is the probability that a customer's demand is satisfied upon arrival. This is achieved by sending the production cards (Kanbans) to the upstream stages faster than the one in the local control policy. The drawback of the Kanban scheme is the increase it causes in the average inventory levels. On the other hand, it is difficult to compare the performance of the two systems. In the local control policy, the space between stages k and $k+1$ is considered as the output buffer for stage k as well as the input buffer for stage $k+1$, with a capacity or a target level of R_{k+1}. In the generalized Kanban scheme, the space between stages k and $k+1$ is divided into two parts that are queues B and C, as shown in Figure 7.19. R_{k+1} of the local control policy refers to the sum of the capacities of buffers B and C, whereas C_{k+1} refers to the maximum level buffers C and E can hold together. Thus, R_{k+1} and C_{k+1} are not the same parameters. The comparison of the two systems also brings up the issue of how the space between the stages k and $k+1$ should be divided. This is a design issue. Inventory holding cost, backordering cost, as well as the revenue from the throughput are important factors in this design problem. One should find the best possible division of space between consecutive stages and then compare it to other systems.

7.7 Manufacturing Systems with Inventory Procurement

In the single-stage systems we studied earlier as well as in the first stages of the multistage systems, we have always assumed that there is an abundance of raw material to be processed. That is, we have not been concerned about the availability nor the delivery of the raw material in the system. However, it is likely that orders of some quantity are placed at certain points in time to procure raw material at stage 1, as shown in Figure 7.21 for a single-stage system. Thus, the level of raw material in the input buffer has to be controlled by using a pure inventory control policy. In pure inventory systems, there is no production, and products are procured from the suppliers through a distribution system. For instance, an order of size Q is placed at the supplier(s) when the on-hand raw material level drops to the reorder level r, where $Q = R - r$. It arrives back at the system as a batch of Q units at the end of the procurement lead time (the time for the supplier to deliver), and consequently the inventory level in the input buffer goes up by Q. This policy is known as the continuous-review (Q, r) policy in pure inventory systems. It is equivalent to the (R, r) policy in the pure inventory environment when demand (e.g., from stage 1 to stage 0) arrives in single units. In case demand arrives in random quantities, the order quantity will be a variable since the inventory level at the time it downcrosses the reorder point will no longer be r but will be any level below r.

7.7. Manufacturing Systems with Inventory Procurement

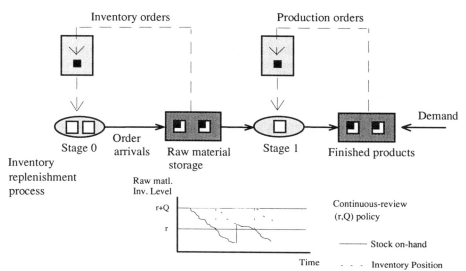

Figure 7.21. Raw material procurement.

The inventory replenishment process can be viewed as another stage, say stage 0, thus making it a two-stage system. Stage 1 is a regular production stage with its production orders initiated by the demand arrivals. One distinction is that units move in batches of size Q at stage 0 and as single units at stage 1. Notice that controlling the inventory in the input buffer by the "on-hand level" results in only one outstanding order at stage 0 at any time. This is because the decision for placing an order is made at the downcrossings of r or at the order arrivals to place the next order immediately if the inventory level is still below r.

An alternative way of controlling the raw material level would be to look at the "on-hand quantity plus the quantity on order," which is known as the "inventory position." In this case, it is possible that the inventory position downcrosses r before the outstanding order is delivered and initiates another order of size Q. Thus, there may be many outstanding orders of size Q waiting to be processed at stage 0. Depending on the supplier structure, this situation may be modeled viewing stage 0 as a single-processor batch system[1] (processing a batch at a time) with a queue of batches of size Q. Alternatively, it may be viewed as a multiprocessor batch system where a processor is turned on as soon as an order is placed. In the single-processor representation, the procurement lead time includes the waiting time of the order in the pool of stage 0. In the multiprocessor representation, it includes only the processing time.

The above inventory control and procurement issues can certainly be extended to the multistage systems studied earlier. Stages may have both input and output

---

[1] Single-processor work stations processing one batch at a time can be modeled as Markovian systems provided that job interarrival and processing times are phase-type random variables.

328   7. Pull-Type Manufacturing Systems

buffers, and it may take time to transfer units from one stage to another. However, modeling will not be much different since the transfer operation can be viewed as another stage and the inventory issues apply to it as discussed earlier for the single-stage system. Furthermore, the raw material orders can be transferred to the upstream stages faster by using the generalized Kanban mechanism.

7.8 A Two-Stage, MultiProduct System

Multistage, multiproduct systems with finite target inventory levels are very difficult to study analytically due to the complex interactions among the products and the stages. For this reason, in this section we will focus on the special case of a two-stage system with multiple products. Let us consider the manufacturing system shown in Figure 7.22 consisting of two stages with a finite capacity buffer in between and a finished product warehouse, producing different types of products. There exists distinct storage space for each product in the buffer as well as in the warehouse. Demand arrival processes are assumed to be Poisson and independent of each other. The processing times are random and we'll assume exponentially distributed for simplicity. The unsatisfied demand for any product is assumed to be backordered. Also, there exists a priority structure according to which the second processor's attention is allocated to the products in the warehouse. Production of a particular product requires the stages to go through a set-up process. Every product type has its own distinct set-up process. The set-up times are assumed to be independent of the processing times of the products, independent of the sequence in which the products are processed, and exponentially distributed.

We have again assumed finite target inventory levels in storage locations, for the stages have to switch back and forth among the products. Stage 2 responds to the requests for production according to the priority structure of the products. Once the processor is available, the highest-priority product that needs the processor's

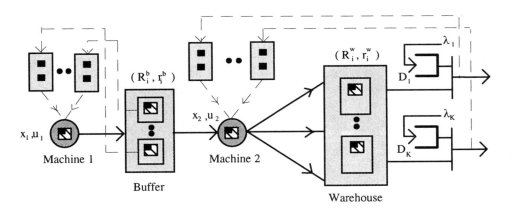

Figure 7.22. Two-stage, multiproduct production/inventory system.

7.8. A Two-Stage, MultiProduct System

attention is selected and the respective set-up process takes place. Production starts after the set-up is completed, and it continues until the inventory level of the selected product reaches its target level. While processing, the processor is assumed to be totally dedicated to the current product. When processing of a product is completed, in case none of the other products needs to be produced, the processor will remain inactive until a product requests its attention. The inventory level in the intermediate buffer is also controlled by using an (R, r) policy for each product. To simplify the problem, let us assume that $Q_i^w = R_i^w - r_i^w > R_i^b$ for each product i, where the superscript w stands for the warehouse and b for the buffer. The strongest implication of this assumption is that when stage 1 becomes available, there will be only one product (or none) requesting its attention. In the presence of this assumption, the concept of priority no longer exists in the buffer. We will elaborate more on the implications of this assumption in the following section.

The approximation algorithms that we presented earlier for multistage or multiproduct systems can be extended to analyze systems with both of these features. In this section, we will briefly introduce such an algorithm that decomposes the system into a set of subsystems each representing the behavior of the inventory of a product in the buffer or the warehouse. A scheme of nested fixed-point iterations relates these subsystems to each other.

The system produces products $1, 2, \ldots, N$, with product 1 being the highest-priority product. Let $\bar{x}_{1i}$ and $\bar{x}_{2i}$ be the mean processing times of product i at stages 1 and 2, respectively. Similarly, let $\bar{u}_{i1}$ and $\bar{u}_{i2}$ be the mean set-up times of product i at stages 1 and 2, respectively. The demand for product i arrives with rate λ_i and eventually depletes the inventory level to r_i^w, initiating a request to start production for product i at stage 2. Stage 2 switches among the products according to the priority structure in the warehouse with the following switching rule: The second stage cannot be set up for the same high-priority product while a low-priority product is waiting for the processor's attention. We have also incorporated into the system a simplifying assumption of $R_i^w - r_i^w > R_i^b$ for all i. This assumption has various implications. First, it states that the total number of units of product i in the buffer is always less than the number necessary to fill the warehouse at the time a production request is placed. It implies that at the moment a production request is made for product i at stage 2, the same request has already been placed at stage 1. Thus, stage 1 reacts to the needs of stage 2 in advance. It also implies that while stage 1 is producing a product, there may be only *one* other product asking for its attention. This is because stage 2 may finish producing i and switch to j while stage 1 may still be producing product i. In case the inventory level of product j in the buffer drops to zero, stage 2 will simply wait until more of product j appears in the buffer. During this period, stage 2 does not demand other products in the buffer until it finishes producing product j, and therefore product j remains to be the only product in the buffer requesting attention from stage 1. This in turn eliminates the priority concept in the buffer. In the absence of this simplifying assumption, the problem becomes highly complicated to study analytically.

330   7. Pull-Type Manufacturing Systems

Hence, the operating policy for stage 1 is as follows: Production of the current set-up (say product i) continues until the target level R_i^b is reached. Then, it sets up for another product requesting its attention, if there is any. Otherwise, it is likely that stage 2 still produces product i and may possibly deplete the inventory of product i in the buffer. In this case, stage 1 continues to supply product i.

As shown in Figure 7.23, the system is decomposed into two major sets of subsystems, one representing the buffer and the other representing the warehouse. Each subsystem is further decomposed into several nodes, each representing a single-product system denoted by $\Omega(j, i)$ with $j = 1, 2$ and $i = 1, \ldots, N$. Each subsystem is constructed in a manner that incorporates the interactions among the products and that of the stages into the analysis. Ahead, we briefly describe the construction and the analysis of the subsystems starting from the downstream end.

Characterization of $\Omega(2, i)$

$\Omega(2, i)$ is constructed in such a way that the impact of the other products in the warehouse as well as the impact of the idleness from the first stage are taken into account in the analysis. The impact of the other products in the warehouse appear in the form of a delay. On the other hand, the impact of stage 1 appears in the effective unit processing times of the products at stage 2. The analysis of $\Omega(2, i)$ is carried out utilizing independence assumptions and regeneration arguments. The imaginary processor at $\Omega(2, i)$ goes through idle, delay, set-up, and busy periods. This set of subsystems can be analyzed by using the method presented earlier in this chapter to study single-stage, multiproduct systems. The approximation method introduced earlier can be used to obtain the expected delay that product i experiences due to sharing the processor with the other products at stage 2. An iterative procedure again relates $\Omega(2, i)$s to each other.

Characterization of $\Omega(1, i)$

$\Omega(1, i)$ represents the inventory process of product i and takes into account the delays in the buffer due to sharing the processor at stage 1 and the demand process from stage 2. Let us select a particular product, say product 1, and explain how $\Omega(1, 1)$ is modeled and analyzed in detail. First, let us examine the demand arrival process of product 1 (from stage 2) in the buffer. The demand will come only if the

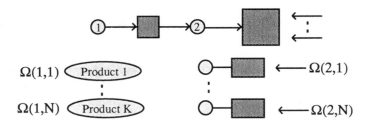

Figure 7.23. Decomposition of the two-stage, multiproduct system.

second stage is producing product 1 and the interarrival time will be the processing time of product 1 at stage 2. During other times while stage 2 is not producing product 1, there are no demand arrivals for product 1 in the buffer. This time period is composed of the idle time, the delays and the set-up time of product 1 at stage 2, which all may have phase-type representations. Consequently, the period during which no arrivals occur for product 1 may be modeled using a phase-type distribution with appropriate parameters.

The production of product 1 at stage 1 is controlled by the parameters R_1^b and r_1^b in the buffer. When the inventory level of product 1 in the buffer reaches its target level R_1^b, if the second processor is still producing product 1, we simply let the first processor continue producing product 1 to bring its inventory level up to its target value. Thus, the first stage stops producing product 1 only after the second stage does so.

Let us focus on the stochastic behavior of the inventory process of product 1. Suppressing the index for product 1, consider the stochastic process $\{N_t, M_t, t \geq 0\}$ where $N_t = 0, 1, \ldots, R_1^b$ represents the inventory level of product 1 in the buffer and M_t represents the status of the first stage at time t.

The particular realizations of M_t may be explained as follows: The states of this stochastic process can be divided into 6 groups. The first group of states represents that the arrival process is on and the inventory level is dropping ($M_t = 1$). As soon as its inventory level drops to r_1^b, product 1 initiates a request for the processor's attention. However, it may experience a delay before the processor starts setting up for product 1. The second group of states corresponds to this delay period with $M_t = 2$. Let us assume that this delay is exponentially distributed with an appropriate mean. This is an approximation, as we have done earlier. Once the delay period is over, the processor starts to set up for product 1, and the third group of states is introduced with $M_t = 3$. Production starts after the set-up is over. The demand arrival process is still on. These states are represented by the fourth group with $M_t = 4$. The fifth group of states captures the behavior of the inventory process while the demand arrival process is turned off, and production still continues until the inventory level reaches its target value in the buffer ($M_t = 5$). The last group of states is dedicated to the fact that an arrival may occur when the inventory level is zero and therefore block the production of product 1 at stage 2 with $M_t = 6$. Thus, $\{N_t, M_t, t \geq 0\}$ is a finite-state Markov chain that possesses a unique steady-state distribution. The probabilities of this Markov chain can be obtained in a recursive manner and provide us with an approximation for the average inventory level of product 1 in the buffer. It also provides an approximation for the idleness probabilities, based on which one can approximate the effective processing time for product 1 at stage 2.

Let us next focus on how to approximate the expected delay for each product at stage 1. Once the steady-state distribution of the Markov chain for each product is obtained, the approximate probability that the first stage is idle can be obtained by

$$P_1(I) = 1 - \sum_{i=1}^{N} \Pr(\text{Stage 1 is setting up or producing } i).$$

When product 1 initiates a request for the first processor's attention, the length of the delay that it may experience depends on the status of the first processor at the time of the request. The processor may be idle or it may be producing one of the N products, including product 1, in which case the delay is zero. Note that if the processor is producing product i at the time product 1 places a request for production, then we know that the Markov chain of the inventory process for product i must be in a state from group 5, that is, production is on but the demand arrival process is turned off. Let

$r_1(i, 1) = \Pr(\text{stage 1 is producing product } i, \text{ at the time product 1 places a request for production})$,
$q_1(i) = \Pr(\text{stage 1 is producing } i, \text{ no demand for } i)$,
$H_1(i) = $ the time from the request of product 1 at stage 1 until the inventory level of i reaches R_i^b given that stage 1 is producing i).

From the steady-state distribution of the Markov chain of each product, it is possible to obtain $q_1(i)$ and $E[H_1(i)]$. Then, an approximation for the expected delay of product 1 is given by

$$E(D_1) \cong \sum_{i \neq j} r_1(i, 1) \cdot E[H_1(i)],$$

where for $i \neq 1$ we have

$$r_1(i, 1) \cong \frac{q_1(i)}{P_1(I) + \sum_{j=2}^{N} q_1(j)}.$$

In a similar manner, one can obtain the approximate expected delay for all of the products.

It is possible to develop an iterative procedure that starts with initial guesses of the expected delays and obtains the steady-state probabilities of the Markov chains in the order of $\Omega(1, 1), \ldots, \Omega(1, N)$. It executes a few iterations (i.e., 3 to 5) among $\Omega(1, i)$s without waiting for a desired convergence. Then, with the new values of the idleness probabilities, a fixed-point iteration is carried out among $\Omega(2, i)$s until the convergence in the expected delay is obtained at stage 2. Next, the iteration among $\Omega(1, i)$s is repeated, and this way it continues back and forth between stages 1 and 2 until a desired level of convergence is obtained. The measures of performance of our interest, the average inventory and the backorder levels can be obtained from the above subsystems.

7.8. A Two-Stage, MultiProduct System   333

TABLE 7.10. The average inventory levels for products 1-4 (and 5-8) in the warehouse for various congestion levels

Example	Input	Product 1		Product 2		Product 3		Product 4	
	(R, r)	(40,15)		(20,10)		(50,30)		(10,5)	
	$\bar{x}$	3		2		4		5	
	$\bar{u}$	4		5		2		1	
	Output	Appx.	Sim.	Appx.	Sim.	Appx.	Sim.	Appx.	Sim.
1	ρ	0.06		0.05		0.05		0.05	
	$\bar{N}$	27.77	27.80	15.18	15.17	40.35	40.34	7.85	7.83
2	ρ	0.15		0.05		0.05		0.05	
	$\bar{N}$	27.19	27.32	14.94	14.95	40.24	40.25	7.74	7.71
3	ρ	0.3		0.05		0.05		0.05	
	$\bar{N}$	24.18	23.73	13.97	13.90	39.74	39.92	7.18	7.22
4	ρ	0.06		0.15		0.05		0.05	
	$\bar{N}$	27.71	27.74	14.20	13.98	40.28	40.28	7.78	7.75
5	ρ	0.06		0.05		0.15		0.05	
	$\bar{N}$	27.63	27.68	14.96	14.95	40.33	40.32	7.65	7.59

TABLE 7.11. The average inventory levels for products 1-4 (and 5-8) in the buffer

Example	Input	Product 1		Product 2		Product 3		Product 4	
	(R,r)	(20,10)		(8,5)		(15,10)		(4,2)	
	$\bar{x}$	2		1		2		3	
	$\bar{u}$	2		3		1		1	
	Output	Appx.	Sim.	Appx.	Sim.	Appx.	Sim.	Appx.	Sim.
1	$\bar{N}$	19.58	19.62	7.87	7.84	14.89	14.89	3.92	3.91
2	$\bar{N}$	19.01	19.11	7.86	7.84	14.89	14.88	3.91	3.91
3	$\bar{N}$	18.52	18.54	7.87	7.84	14.89	14.88	3.92	3.91
4	$\bar{N}$	19.58	19.62	7.66	7.64	14.89	14.88	3.92	3.92
5	$\bar{N}$	19.59	19.62	7.86	7.84	14.71	14.71	3.92	3.91

EXAMPLE 7.7. In Tables 7.10 and 7.11, we present a group of numerical examples to give the reader an idea about how accurate such a procedure can be. The approximate results were tested against their simulation counterparts. The examples are designed to report on the impact of the congestion level on the accuracy of the approximation. All of the examples involve eight products. The input in each problem consists of the values of R_i^b, r_i^b, R_i^w and r_i^w, the first moment of the processing times, and the set-up times for each product. Products 1 and 5 have the same parameters, products 2 and 6 have the same parameters, and so on. The intensity ρ of each product in stage 2 is altered by changing its arrival rate. The results for products 1 through 4 are given in Tables 7.10 and 7.11 for stages 1 and 2, respectively. The results for products 5 through 8 are almost the same as those of 1 through 4. The orders of the examples in the two tables are compatible. The output in each example consists of the average inventory level $\bar{I}$ for each product. Numerical experience shows that the preceding procedure converges in a variety of examples. As expected, it cannot be proven that it always converges. The error level in the approximation is highly acceptable. These results encourage us to consider decomposition approximations for more general networks of manufacturing systems.

7.9 The Look-Back Policy

The look-back policy dictates that each stage checks the immediate upstream buffer for a sufficient amount of WIP inventory before setting up for production. Let us consider the simplified two-stage, single-product system shown in Figure 7.24. The first stage always has raw material to process. Between stages, there is an intermediate storage that operates under the base-stock policy, that is, the continuous-review $(R_b, R_b - 1)$ policy. The second stage produces products to store in the finished-product warehouse. Let us assume exponential processing times with rates μ_1 and μ_2 for stages 1 and 2, respectively, and Poisson customer demand arrivals with rate λ. The inventory in the warehouse is controlled by a continuous-review (R_W, r_W) policy. The decision to start production at stage 2 depends on the sufficiency of the WIP level in the buffer. At the moment of a production request at stage 2, if there are not enough units in the buffer, stage 2 may delay its production until the inventory in the buffer reaches a certain threshold level denoted by r^*. In a design problem, one may associate cost figures and times with set-ups and start-ups to penalize frequent unproductive periods. Here, we omit these considerations and keep the analysis simple.

Clearly, one can study the underlying Markov chain representing the behavior of the entire system. Instead, we decompose the system into two subsystems that approximate the behaviors of the inventory levels in the buffer and the warehouse. We abbreviate the subsystems by $\Omega(1)$ and $\Omega(2)$, and the inventory levels in the buffer and the warehouse by N_b and N_W, respectively. Next, we describe the subsystems in detail.

7.9. The Look-Back Policy

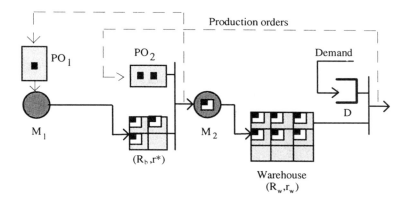

Figure 7.24. Two-stage system with the look-back policy.

Characterization of $\Omega(1)$

The first subsystem is a two-stage system with an intermediate buffer of capacity R_b. The first stage has an exponential processing time with rate μ_1 and is always busy except when the buffer is full. It starts processing when N_b drops to $R_b - 1$. Let $P(N_i = k) = P_i(k)$ for $i = b, w$, be the steady-state probability that there are k units in the buffer and in the warehouse at any point in time.

The second stage in $\Omega(1)$ is more involved. Upon process completion at stage 2 in the original system, if the number of items in the warehouse is R_W, stage 2 ceases its production. The probability of this event is

$$\Pi = P(N_W = R_W - 1 \mid \text{a departure is just to occur at node 2}).$$

Once the target level in the warehouse is reached, $R_W - r_W$ demand arrivals must occur to initiate a request for production. That is, the particular demand arrival that sees $r_W + 1$ units in the warehouse initiates a request for production. At this moment, production starts if $N_b > r^*$; otherwise, it is delayed until $N_b = r^*$. This delay may be interpreted as the set-up time for stage 2. Let U denote this set-up time and Z denote the time from the moment N_W reaches R_W until stage 2 restarts its production. Z can be interpreted as the time station 2 is blocked. Then, we have

$$\text{E}[Z] = \frac{R_W - r_W}{\lambda} + q_o \text{E}[U], \qquad (7.65)$$

where $q_o = P$ (demand arrival sees $N_W = r_W + 1$ and $N_b < r^*$) and

$$\text{E}[U] = \frac{1}{\mu_1} \sum_{k=1}^{r^*} k \frac{P(N_b = r^* - k, N_W = r_W + 1)}{P(N_b < r^*, N_W = r_W + 1)}. \qquad (7.66)$$

Let $P_2(B)$ be the steady-state probability that stage 2 is blocked. We know that stage 2 is blocked while N_W drops from R_W to r_W, and blocking persists if the inventory in the buffer is less than r when N_W down crosses r_W. Thus, $P_2(B)$ can be expressed as

$$P_2(B) = P(N_W > r_W \text{ and no production at stage 2}) + P(N_b < r^*, N_W = r_W + 1).$$

Than, Π can be obtained, similar to 7.51, by

$$\Pi = \frac{P_2(B)}{E[Z]\bar{o}_2}, \tag{7.67}$$

where $\bar{o}_2$ is the steady-state throughput in $\Omega(2)$.

The processing time at node 2 of $\Omega(1)$ is defined by

$$X_{12} = \begin{cases} X_2 & \text{wp } 1 - \Pi, \\ X_2 + Z & \text{wp } \Pi, \end{cases}$$

where the distribution of Z can be modeled using the MGE distribution shown in Figure 7.25. Notice that the branching probability a_i in Figure 7.25 is given by

$$a_k = \frac{P(N_b = r^* - k, N_W = r_W + 1)}{P(N_b < r^*, N_W = r_W + 1)}. \tag{7.68}$$

Consequently, $\Omega(1)$ can be studied as an M/PH/1/ $R_b + 1$ work station using the procedure presented in the Appendix. E[U], q_o, and $P_2(B)$ can be approximated using independence assumptions.

Characterization of $\Omega(2)$

$\Omega(2)$ is a single-stage system replicating the behavior of the inventory level in the warehouse. Once the production starts at stage 2 of the original system, it continues

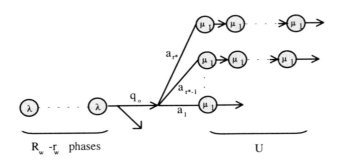

Figure 7.25. PH representation of Z.

TABLE 7.12. Approximate versus numerical results from the two-node system with the look-back policy

R_W	r_W	R_b	r^*	λ	μ_1	μ_2	$\overline{N}_{\text{Num}}$	$\overline{N}_{\text{App}}$
10	6	6	1	1	1	2	6.8232	6.7946
6	3	4	2	1	1	2	4.2378	4.4196
10	6	6	3	2	2	1	6.0147	6.5187

until $N_W = R_W$. During this period, a departure at stage 2 may leave $N_b = 0$. In this case, stage 2 waits for stage 1 to produce a unit into the buffer. The probability of this event is

$$\Delta = \Pr(N_b = 0 \mid \text{a departure is just to occur at stage 2}) = \frac{\mu_1 P_b(0)}{\overline{o}_1},$$

where $\overline{o}_1$ is the steady-state throughput of $\Omega(1)$.

Thus, the imaginary processor producing finished units into the warehouse has a processing time X_{21} defined by

$$X_{21} = \begin{cases} X_2 & \text{wp } 1 - \Delta, \\ X_1 + X_2 & \text{wp } \Delta. \end{cases}$$

The distribution of X_{21} can be modeled by an MGE-2 distribution. Notice also that at a request for production, in order to accommodate the look-back policy, stage 2 has to check the inventory level in the buffer. If it is less than r, stage 2 has to wait until it reaches r. This period can be interpreted as the start-up time for stage 2. The start-up time is an MGE random variable with the number of phases depending on the inventory level in the buffer at the moment of a demand arrival. It has exactly the same phase structure as the set-up time (U). The start-up time is zero if $N_b > r^*$.

$\Omega(2)$ can be modeled as an M/PH/1/ R_W work station with a start-up time and a threshold service policy (such as Q-policy). That is, the server waits until ($R_W - r_W$) units accumulate. It may go through a start-up time (look-back). When it starts, it continues serving until the system is cleared. The units in this M/PH/1/R_W station are the empty cells (holes) in the original warehouse. Again, the work station can be analyzed using the procedure presented in the Appendix. The above probability $P(k = r^* - j \mid j < r^*)$ for the number of phases in the start-up time may also be approximated using independence assumptions.

Again, $\Omega(1)$ and $\Omega(2)$ can be incorporated into an iterative procedure similar to the ones we have employed earlier, which iterates until $\overline{o}_1$ and $\overline{o}_2$ become sufficiently close to each other. Some results are presented in Table 7.12.

Related Bibliography

Many studies have appeared in the literature on the concept of changing the production rate according to the inventory level. In the early work, it was achieved by adding more servers to the system; see, for instance, Romani (1957) and Moder and Phillips (1962). Later, Yadin and Naor (1967) and Gebhard (1967) changed the processing rate as a function of the inventory level. Yadin and Zacks (1970) and Zacks and Yadin (1971) studied the optimal control of Markovian queues with variable service rates. There also exists a considerable work showing the optimality of a two-level control policy for situations where the server is not readily available or a set-up is involved every time the process restarts. See, for instance, Heyman (1968), Sobel (1969), and Bell (1971). Finding the cost-optimal values of the parameters R and r was studied by Gavish and Graves (1980, 1981), de Kok (1985), Lee and Srinivasan (1987), and Altiok (1989).

Yadin and Naor (1963) first studied queueing systems with removable servers (vacations) and developed relations with nonremovable servers. Miller (1964), Heyman (1968, 1977) Levy and Yechiali (1975), Lee (1984), Doshi (1985), Servi (1986), and Borthakur et al. (1987) all studied vacation queues with varying assumptions. Doshi (1986) gave an overview of vacation queues with various applications. The stochastic decomposition property (7.19) was first presented by Yadin and Naor (1963) and later generalized by Fuhrmann and Cooper (1985). Scholl and Kleinrock (1983) showed the property in terms of the time spent in the system. Altiok (1987) generalized the property to systems with group arrivals. The property was further studied by Harris and Marchal (1988) and Shanthikumar (1988), among many others.

The field of multiproduct systems with finite target levels and priorities has not attracted much attention. Avi-Itzhak et al. (1965) considered a queueing system with alternating priorities, which is equivalent to a two-product production/inventory system with (0,-1) continuous-review inventory control policy applied to both products allowing backorders. Takacs (1968), through a different approach, studied the same queueing system and obtained the distribution of the waiting time of a customer. Sykes (1970) studied an alternating priority queueing model with set-up times. Eisenberg (1972, 1979) reported on queues with periodic service and change over times as well as two queues with alternating service. Hodgson (1972) considered a two-product problem with set-ups and lost sales. Graves (1980) studied the multiproduct production cycling problem. Hofri and Ross (1987) recently studied a two-queue system with set-ups and a threshold-switching policy, which can be viewed as a two-product production/inventory system with (R,r) policy and backorders. Zheng and Zipkin (1989) considered a two-product problem with backorders and with Markovian assumptions. Altiok and Shiue (1994, 1995) studied single-stage, multiproduct systems with lost sales and backorders. The analysis of the two-product problem in Section 7.4 is due to Altiok and Shiue (1994). Similar problems have also been looked at in the area of polling systems with various communication protocols; see Fuhrmann and Wang (1988) and Takagi (1990), among others.

Comparisons of pull and push systems and related control policies can be found in Karmarkar (1986), Spearman and Zazanis (1992), Spearman et al. (1990), and Hopp and Spearman (1991), which model pull systems as closed queueing networks due to the finite number of Kanbans in the system. For instance, CONWIP (Spearman et al. (1990)) capitalizes on the fixed number on Kanban cards in the system. This is quite the same as the backorder-driven system shown in Figure 7.6.

Multistage production/inventory systems have been studied in the context of Kanban controlled pull-type systems. These are mostly single-product systems. The generalized Kanban scheme described in Section 7.5 is due to Buzacott (1989), who made comparisons with the Material Requirements Planning (MRP) systems. Various models of Kanban systems include Buzacott and Shanthikumar (1992), Tabe et al. (1980), Kimura and Terada (1981), Kim (1985), Karmarkar and Kekre (1987), and Deleersnyder et al. (1988). These works looked into the performance issues as well as the impact of demand variability on the inventory levels. Yamashita et al. (1988), Mitra and Mitrani (1990, 1991), and Altiok and Ranjan (1995) considered multistage systems with variants of the Kanban discipline and proposed approximations for performance evaluation. The generalized Kanban scheme presented in Section 7.6 is due to Mitra and Mitrani (1990). The analysis of the make-to-order systems presented in Section 7.5.3 is due to Rhee and Perros (1996). Sending production orders to multiple stages was proposed by Dallery (1995). Di Mascolo et al. (1991) developed petri-net models of various pull-type systems with the Kanban discipline. Uzsoy and Martin-Vega (1990) presented a survey and critique of the existing models of Kanban-based pull systems.

Multistage and multiproduct systems with finite target inventory levels were studied by Di Mascolo et al. (1992) and Shiue and Altiok (1993). Di Mascolo et al. (1992) developed closed queueing network approximations for production systems with the generalized Kanban scheme where the number of Kanbans is the limitation on the number of units that can be held in a stage as well as a network of stages. Shiue and Altiok (1993) proposed control policies and introduced approximations to obtain performance measures. The approximations presented in Section 7.8 are due to Shiue and Altiok (1993). Amin and Altiok (1995) developed various exhaustive and nonexhaustive production policies with various switching rules for multi-stage and multiproduct systems. Also, Onvural and Perros (1990) studied multistage systems with multiple classes of customers and proposed aggregation approximations. In their model, classes occupy the same buffer and each class can fill up the buffer. The look-back policy was proposed by Gursoy et al. (1994).

For the design and optimization of multistage systems, the multivariable search procedures that do not require taking derivatives become important. These can be found in Bazaraa and Shetty (1977), and in Reklaitis et al. (1983).

References

Altiok, T., 1987, Queues with Group Arrivals and Exhaustive Service Discipline, *Queueing Systems*, Vol. 2, pp. 307–320.

Altiok, T., 1989, (R, r) Production/Inventory Systems, *Operations Research*, Vol. 37, pp. 266- 276.

Altiok, T., and R. Ranjan, 1995, Multi-Stage Pull-type Production/Inventory Systems, *IIE Transactions*, Vol. 27, pp. 190–200.

Altiok, T., and G. A. Shiue, 1994, Single-Stage, Multi-Product Production/Inventory Systems with Backorders, *IIE Transactions*, Vol. 26, No. 2, pp. 52–61.

Altiok, T., and G. A. Shiue, 1995, Single-Stage, Multi-Product Production/Inventory Systems with Lost Sales, *Naval Research Logistics*, Vol. 42, pp. 889–913.

Amin, M. and T. Altiok, 1995, Control Policies for Multi-Product Multi-Stage Manufacturing Systems, To Appear in *Intl. J. Production Research*.

Avi-Itzhak, B., W. L. Maxwell and L. W. Miller, 1965, Queueing with Alternating Priorities, *Operations Research*, Vol. 13, pp. 306–318.

Bazaraa, M. S. and C. M. Shetty, 1977, *Nonlinear Programming*, John Wiley, New York.

Bell, C., 1971, Characterization and Computation of Optimal Policies for Operating an M/G/1 Queueing System with Removable Server, *Operations Research*, Vol. 19, pp. 208–218.

Borthakur, A., J. Medhi, and R. Gohain, 1987, Poisson Input Queueing System with Start-up and Under Control-Operating Policy, *Comp. and Operations Research*, Vol. 14, no. 1, pp. 33–40.

Buzacott, J., 1989, Generalized Kanban/MRP Systems, Working Paper, Department of Management Sciences, University of Waterloo, Ontario.

Buzacott J., and J. G. Shanthikumar, 1992, A General Approach for coordinating Production in Multiple Cell Manufacturing Systems, *Production and Operations Management*, Vol. 1, pp. 34–52.

Dallery, Y., 1995, Personal conversation.

de Kok, A. G., 1985, Approximations for a Lost-Sales Production/Inventory Control Model with Service Level Constraints, *Management Science*, Vol. 31, pp. 729–737.

Deleersnyder, J-L., T. J. Hodgson, H. Muller, and P. J. O'Grady, 1988, Kanban Controlled Pull Systems- Insights From a Markovian Model, IE Technical Report, NCSU, Raleigh, North Carolina.

Di Mascolo, M., Y. Frein, B. Baynat, and Y. Dallery, 1992, Queueing Network Modeling and Analysis of Generalized Kanban Systems, Laboratoire d'Automatique de Grenoble, Saint Martin d'Heres, France.

Di Mascolo, M., Y. Frein, Y. Dallery, and R. David, 1991, A Unified Modeling of Kanban Systems Using Petri Nets, *Intl. J. of Flexible Manufacturing Systems*, Vol. 3, pp. 275–307.

Doshi, B. T., 1985, A Note on Stochastic Decomposition in a GI/G/1 Queue with Vacations or Set-up Times, *J. Applied Probability*, Vol. 22, pp. 419–428.

Doshi, B. T., 1986, Queueing Systems with Vacations: A Survey, *Queueing Systems*, Vol. 1, pp. 29–66.

Eisenberg, M., 1972, Queues with Periodic Service and Changeover Times, *Operations Research*, Vol. 20, pp. 440–451.

Eisenberg, M., 1979, Two-Queues with Alternating Service, *SIAM J. Applied Math.*, Vol. 36, pp. 287–303.

Fuhrmann, S.W., and R. Cooper, 1985, Stochastic Decompositions in the M/G/1 Queue with Generalized Vacations, *Operations Research*, Vol. 33, pp. 1117–1129.

Fuhrmann, S. W., and Y. T. Wang, 1988, Analysis of Cyclic Service Systems with Unlimited Service: Bounds and Approximations, *Performance Evaluation*, Vol. 9, pp. 35–54.

Gavish, B., and S.C. Graves, 1980, A One-Product Production/Inventory Problem Under Continuous Review Policy, *Operations Research*, Vol. 28, pp. 1228–1236.

Gavish, B., and S. C. Graves, 1981, Production/Inventory Systems with a Stochastic Production Rate Under a Continuous Review Policy, *Comp. Operations Research*, Vol. 8, pp. 169–183.

Gebhard, R. F., 1967, A Queueing Process with Bilevel Hysteretic Service-Rate Control, *Naval. Res. Log. Quarterly*, Vol. 14, pp. 55–68.

Graves, S. C., 1980, The Multi-Product Production Cycling Problem, *AIIE Transactions*, Vol. 12, pp. 233–240.

Gursoy, M. B., T. Altiok, and H. Danhong, 1994, Look-Back Policies for Two-Stage, Pull-Type Production/Inventory systems, *Annals of Operations Research*, Vol. 48, pp. 381-400.

Harris, C. M., and W. G. Marchal, 1988, State Dependence in M/G/1 Server Vacation Models, *Operations Research*, Vol. 36, pp. 560–565.

Heyman, D. P., 1968, Optimal Operating Policies for M/G/1 Queueing Systems, *Operations Research*, Vol. 16, pp. 362–382.

Heyman, D. P., 1977, The T-Policy for the M/G/1 Queue, *Management Science*, Vol. 23, pp. 775–778.

Hodgson, T., 1972, An Analytical Model of a Two-Product, One-Machine Production/Inventory System, *Management Science*, Vol. 19, pp. 391–405.

Hofri, M., and K. W. Ross, 1987, On the Optimal Control of Two Queues with Server Set-up Times and Its Analysis, *SIAM J. Comp.*, Vol. 16, pp. 399–420.

Hooke, R., and T. A. Jeeves, Direct Search of Numerical and Statistical Problems, *J. ACM*, Vol. 8, pp. 212–229.

Hopp, W. J., and M. L. Spearman, 1991, Throughput of a Constant Work in Process Manufacturing Line Subject to Failures, *Intl. J. Production Research*, Vol. 29, pp. 635–655.

Johnson, L., and D. C. Montgomery, 1974, *Operations Research in Production Planning, Scheduling, and Inventory Control*, John Wiley and Sons, New York.

Karmarkar, U. S., 1986, Push, Pull and Hybrid Control Schemes, Working Paper Series, QM 86-14, W. E. Simon Grad. Sch. of Bus. Adm., University of Rochester, Rochester, NY.

Karmarkar, U. S., and S. Kekre, 1987, Batching Policy in Kanban Systems, Working Paper QM87-06, Graduate School of Management, University of Rochester, Rochester, NY.

Kim, T. M., 1985, Just-In Time Manufacturing Systems: A Periodic Pull System, *Intl. J. Production Research*, Vol. 23, pp. 553–562.

Kimura, O., and H. Terada, 1981, Design and Analysis of Pull Systems, A Method of Multi-Stage Production Control, *Intl. J. Production Research*, Vol. 19, pp. 241–253.

Lambrecht, M. R., J. A. Muckstadt, and R. Luyten, 1984, Protective Stocks in Multi-Stage Production Systems, *Intl. J. Production Research*, Vol. 19, pp. 241–253.

Lee, T. T., 1984, M/G/1/N Queue with Vacation Time and Exhaustive Service Discipline, *Operations Research*, Vol. 32, pp. 774–784.

Lee, H. S., and M. M. Srinivasan, 1987, The Continuous Review (s,S) Policy for Production/Inventory Systems with Poisson Demand and Arbitrary Processing Times, Technical Report 87-33, Department of Industrial and Operations Engineering, The University of Michigan, Ann Arbor,.

Levy, Y., and U. Yechiali, 1975, Utilization of Idle Time in an M/G/1 Queueing System, *Management Science*, Vol. 22. pp. 202–211.

Miller, L. W., 1964, Alternating Priorities in Multi-Class Queues, Ph.D. dissertation, Cornell University, Ithaca, NY.

Mitra, D., and I. Mitrani, 1990, Analysis of Kanban Discipline for Cell Coordination in Production Lines, I, *Management Science*, Vol. 36, pp. 1548–1566.

Mitra, D., and I. Mitrani, 1991, Analysis of Kanban Discipline for Cell Coordination in Production Lines, II, *Operations Research*, Vol. 35, pp. 807–823.

Moder, J., and C. R. Phillips, 1962, Queueing Models with Fixed and Variable Channels, *Operations Research*, Vol. 10, pp. 218–231.

Onvural, R. O., and H. G. Perros, 1990, Approximate Analysis of Multi-Class Tandem Queueing Networks with Coxian Parameters and Finite Buffers, *Proc. Performance 90*, October, Edinburg.

Reklaitis, G. V., A. Ravindran, and K. M. Ragsdell, 1983, *Engineering Optimization: Methods and Applications*, John Wiley, New York.

Rhee, Y. and H. G. Perros, 1996, Analysis of an open tandem queueing network with population constraints and constant servicetimer, *E. J. Operations Research*, Vol.92, pp. 99–111.

Romani, J., 1957, Un Modelo de la Teoria de Colas con Numero Variable de Canales, *Trabajos Estadestica*, Vol. 8, pp. 175–189.

Scholl, M., and L. Kleinrock, 1983, On the M/G/1 Queue with Rest Periods and Certain Service-Independent Queueing Disciplines, *Operations Research*, Vol. 31, pp. 705–719.

Servi, L., 1986, D/G/1 Queues with Vacations, *Operations Research*, Vol. 34, pp. 619–629.

Shanthikumar, J. G., 1988, On the Stochastic Decomposition in M/G/1 Type Queues with Generalized Server Vacations, *Operations Research*, Vol. 36, pp. 566–569.

Shiue, G. A., and T. Altiok, 1993, Two-Stage, Multi-Product Production/Inventory Systems, *Performance Evaluation*, Vol. 17, pp. 225–232.

Sobel, M., 1969, Optimal Average Cost Policy for a Queue with Start-up and Shut-down Costs, *Operations Research*, Vol. 17, pp. 145–162.

Spearman M. L., D. L. Woodruff, and W. J. Hopp, 1990, CONWIP: A Pull Alternative to Kanban, *Intl. J. Production Research*, Vol. 28, pp. 879–894.

Spearman, M. L., and M. A. Zazanis, 1992, Push and Pull Production Systems: Issues and Comparisons, *Operations Research*, Vol. 40, pp. 521–532.

Sykes, J. S., 1970, Simplified Analysis of an Alternating Priority Queueing Model with Set-up Times, *Operations Research*, Vol. 18, pp. 1182–1192.

Tabe, T., R. Muramatsu, and Y. Tanaka, 1980, Analysis of Production Ordering Quantities and Inventory Variations in a Multi-Stage Production Ordering System, *Intl. J. Production Research*, Vol. 18, pp. 245–257.

Takacs, L., 1968, Two Queues Attended by a Single Server, *Operations Research*, Vol. 16, pp. 639–650.

Takagi, H., 1990, Queueing Analysis of Polling Models: An Update, *Stochastic Analysis of Computer and Communication Systems*, (H. Takagi, ed.), Elsevier Science Publishers, Amsterdam.

Uzsoy, R., and L. Martin-Vega, 1990, Modeling Kanban Based Demand-Pull Systems, A Survey and Critique, *Manufacturing Review*, Vol. 3, pp. 155–160.

Yadin, M., and P. Naor, 1963, Queueing Systems with a Removable Service Station, *Operations Research Quart.*, Vol. 14, pp. 393–405.

Yadin, M., and P. Naor, 1967, On Queueing Systems with Variable Service Capacities, *Naval Res. Log. Quarterly*, Vol. 14, pp. 43–53.

Yadin, M., and S. Zacks, 1970, Analytical Characterization of the Optimal Control of a Queueing System, *J. Applied Probability*, Vol. 7, pp. 617–633.

Yamashita, H., H. Nagasaka, H. Nakagawa, and S. Suzuki, 1988, Modeling and Analysis of Production Line Operated According to Demand, Working Paper, Sophia University, Tokyo.

Zacks, S., and M. Yadin, 1971, The Optimal Control of a Queueing Process, *Developments in O.R.*, (B. Avi-Itzhak, ed.), pp. 241-252.

Zeng, Y. and P. Zipkin, 1989, A Queueing Model to Analyze the Value of Centralized Inventory Information, Working Paper No. 86-6 (Revised), Grad. Sch. of Management, Columbia University, New York.

Problems

7.1 Consider the single-stage, pull-type manufacturing system shown in Figure 7.3 with start-up (i.e., delay) and set-up times. The demand arrival process is Poisson with rate 5 per day, and the unit processing time is exponentially distributed with a mean of 0.18 days. The start-up and set-up times are both exponentially distributed with means 1.0 and 2.5 hours, respectively. The inventory in the warehouse is controlled using an (R, r) policy with $R = 50$ and $r = 30$ units. Obtain the long-run average inventory on-hand, the probability of backorder, and the probabilities of the processor's status when backorders are allowed. Resolve the problem when the unsatisfied demand is lost.

7.2 Consider again the single-stage system analyzed in Problem 7.1. Let us assume that the demand process is compound Poisson. That is, the demand interarrival times are exponentially distributed (say, with rate λ) and the demand size is a discrete random variable k with a probability distribution of c_k. Assuming exponentially distributed start-up, set-up, and processing times, develop a matrix-recursive algorithm to obtain the long-run joint probabilities of the system state. Consider both cases of backordering and lost sales. There are two possibilities in the case of lost sales, namely partial fill and complete fill. In the case of partial fill, it is possible that only a portion of the demand is satisfied and the rest is lost. In the complete-fill case, the customer is lost if its demand is not fully satisfied. Implement your algorithm using the data from Problem 7.1 with $c_1 = 0.4$ and $c_2 = 0.6$.

7.3 Using the flow-balance equations (7.10), prove the flow-equivalence in cuts 1 and 2 in Figure 7.10. The flow equivalence in cut 2 is given by (7.13).

7.4 Develop a mathematical reasoning for where the peak of the probability distribution P_n given by (7.15) may be. Utilize Table 7.1 and Eqs. (7.9), (7.11), and (7.12).

7.5 Consider a single-stage manufacturing system with set-up and shut-down times. The (R, r) policy is used in the warehouse. The demand process is Poisson with rate λ, and the set-up, shut-down, and unit processing times are all exponentially distributed with rates γ, μ, and β, respectively.

(a) Assume that the demand process is turned off during the shut-down period. Obtain the long-run probabilities of the processor's status.
(b) Repeat (a) in case the demand process is not turned off during the shut-down period.

Hint: First of all, the regenerative argument used in Section 7.3 can again be employed in both (a) and (b). In (b), the underlying Markov chain has states such as $(SD,R), (SD, R-1), \ldots, (SD,r), \ldots, (I, R), (I, R-1), \ldots, (I, r+1)$, among others. Using the flow-balance equations of states (SD,n) and (I, n), it is not

difficult to obtain $E[T_I]$ as a function of $E[T_c]$. Also, notice that $P_R = 1/\lambda E[T_c]$ is still valid in (b).
(c) For both (a) and (b), develop a matrix-recursive algorithm to obtain the long-run probabilities of the inventory and backorder levels.

7.6 Consider a single-stage manufacturing system with Poisson demand process with rate $\lambda = 5$ units per unit time. The start-up (delay), set-up, and unit processing times are defined through their means and the squared coefficient of variations: delay (3.0,1.5), set-up (2.0,0.5), and unit processing time (0.1,0.25). The inventory in the warehouse is controlled by using an (R, r) policy with $R = 50$ and $r = 20$. Implement the algorithm presented in the Appendix in a computer code to obtain the long-run probabilities of the inventory and backorder levels up to 25 units of backorders. Use the results of the case where all the random variables are exponential to verify your code.

7.7 In a single-stage manufacturing system with long start-up and set-up times, the production is requested when the inventory level drops to zero. Let us assume that the target level is $R = 10$ units. The unit processing time has a mean of 2.0 and a squared coefficient of variation 0.75. The start-up and set-up times are exponentially distributed with means of 15 and 10, respectively. Obtain the long-run average inventory on hand, the probabilities of the processor's status, the backordering probability, and the average backorders. Also, obtain the average outstanding production orders both while the processor is not producing and while it is producing.

7.8 Consider the single-stage system with the shut-down process in Problem 7.5. Assuming an (R, r) policy and no delays or set-ups, show the stochastic decomposition property of the outstanding production orders.

7.9 Assume a shut-down period that is Erlang distributed with 3 phases and a mean of 10 time units. If the demand process is Poisson with a rate of 4 units per unit time, what is the time-average accumulation of production orders and the total number of orders at the end of the shut-down process?

7.10 In a single-stage manufacturing system controlled using an $(R, R-1)$ policy with $R = 10$, every time the processing of an order is completed, with probability $q = 0.4$ the processor switches to another product for an exponentially distributed period with a mean of 5 time units. After this period, the processing of another order starts if there is any. If there are no orders to process, the processor remains idle. There are no delays or set-ups before the process starts. The unit processing time is exponentially distributed with a mean of 1.0 time unit. The demand process is Poisson with rate $\lambda = 0.75$ units per unit time.

(a) Obtain the long-run average inventory and backorder levels.
(b) Repeat (a) for a compound Poisson demand arrival process with $\lambda = 0.4$ and the demand distribution of $c_1 = 0.2$, $c_2 = 0.5$, and $c_3 = 0.3$.

7.11 In the following single-stage system, the unit processing time, the set-up time and the demand interarrival time are all defined through their means and the squared coefficient of variations as given in the following table:

	Mean	Cv²
Processing time	2.50	0.75
Set-up time	2.00	1.50
Demand interarrival time	2.00	2.00

The unsatisfied demand is lost. An (R, r) policy is implemented with $R = 5$ and $r = 2$.

(a) Define the state of the system, and write the flow-balance equations.
(b) Develop a matrix-recursive algorithm and implement it in a computer code to obtain the long-run marginal probabilities of the inventory level. We encourage you to test your algorithm for large values of R and for stability using exponential random variables.
(c) What is the processor utilization, and what percentage of the demand is not lost?
(d) What is the average number of orders when the processor is not processing?

Note: Your algorithm can easily be implemented to obtain the probabilities of the subsystems in the decomposition of the multistage pull-type systems.

7.12 Show that the expression in (7.52) for $\Pi(K)$ of subsystem $\Omega(K)$ is the same as the expression in (7.9) for P_R.

7.13 Starting from $\overline{B} = \sum_{n=R+1}^{\infty} (n - R)q_n$, show that $\overline{B} = \overline{L} + \overline{N} - R$, as given by (7.26).

7.14 *Design Problem:* Consider the single-stage system given in Problem 7.1. On separate graphs, show the behavior of the cost-minimizing values of R and r for the ranging values of $K_s \epsilon (100, 500)$, $h \epsilon (.1, 5)$ and $g \epsilon (0.1, 5)$. In each graph, keep the other cost parameters at their mean values. Utilize (7.27) with $\pi = 0$ in evaluating the expected total cost.

7.15 Consider a two-product manufacturing system as shown in Figure 7.13. The arrival processes are Poisson with rates $\lambda_A = 0.5$ and $\lambda_B = 0.6$. The unsatisfied demand is backordered. The mean and the squared coefficient of variation of the set-up and unit processing times are given in the following table:

Products	Set-up Time	Processing Time
A	(5,1.0)	(0.5,0.5)
B	(15,1.0)	(0.75,0.5)

The inventory of each product in the warehouse is controlled by an (R, r) policy with $R = 5, r = 2$ for product A and $R = 8, r = 3$ for product B.

(a) Using the algorithm in Section 7.4.2., approximate the mean delay each product type experiences in getting the processor's attention. Show the path of convergence as shown in Table 7.4.

(b) Obtain the approximate average outstanding production orders of each product during their delay periods.

7.16 Consider the following two-stage manufacturing system:

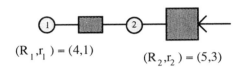

$(R_1, r_1) = (4,1)$ $\qquad (R_2, r_2) = (5,3)$

The processing times are both exponentially distributed with rates 4 and 2 units per unit time. The demand process is Poisson with rate 0.75 per unit time. The set-up times are also exponentially distributed with rates 0.1 and 0.2 at stages 1 and 2, respectively. Implement the decomposition method to obtain the approximate values of the performance measures such as the average inventory levels in both storage locations and the average backorders in the warehouse.

Note: Each subsystem can be analyzed as an M/MGE/1 work station with set-ups. Remember that the queue capacity has to be augmented by one in the analysis of $\Omega(1)$.

7.17 Consider a three-stage, pull-type P/I system with a Poisson demand process with rate λ. The unit processing times are of MGE-2 type with parameters μ_1^i, μ_2^i and a_i at stage i. The set-up process is also of MGE-2 type with parameters γ_1^i, γ_2^i and α_i at stage i. The inventory in the storage locations are controlled by using (R, r) policies with $(R_1, r_1) = (4, 1)$, $(R_2, r_2) = (5, 4)$ and $(R_3, r_3) = (6, 3)$. Explain the details of the decomposition method with its pictorial representation similar to Figure 7.16 as well as the expressions for all the parameters such as Δ_i, Π_i, etc.

Note: The implementation of the decomposition method in a computer code to analyze manufacturing systems with three or more stages is a bigger project than a homework problem. We encourage the students who are interested in semester projects to work in this problem. The input to such a code may be in the form of the mean and the variance of all the random variables. The code would convert them to MGE variables using the moment approximations of Chapter 3. Then, the algorithm in Section 7.5.2 may be implemented with the analysis of Problem 7.13 as being a subroutine. The results may be in the form of probability distributions of the interstage inventories and the measures of the warehouse. At every opportunity, we encourage the use of 3-moment approximations to reduce a PH distribution to its MGE-k equivalent.

7.18 Implement the generalized Kanban scheme in the two-stage system of Problem 7.16. Assume that stage 1 has 2 cards and stage 2 has 7 cards. Approximate and compare the performance measures of the average inventory and backorder levels with those in Problem 7.16.

7.19 Consider the following four-stage manufacturing system with a compound Poisson demand process with rate $\lambda = 0.2$ and the demand probabilities of $c_1 = 0.4$, $c_2 =$

0.4, and $c_3 = 0.2$ ($\bar{c} = 1.8$). The unit processing and set-up times are defined through their means and the squared coefficient of variations given in the following table:

	Stages			
	1	2	3	4
Proc. Time	(0.25,0.5)	(0.2,0.7)	(0.15,0.30)	(0.25,0.20)
Set-up Time	(2.0,1.0)	(4.0,0.75)	(2.0,1.5)	(3.0,1.0)
(R, r)	(6,3)	(8,6)	(5,4)	(20,10)

Using computer simulation (with any simulation language of your preference), obtain the performance measures of the system under the policies of Problems 7.16, 7.18, and 7.19 and compare the results. Furthermore, reflect the finished-product demand to stages 1 and 2 and stages 1 and 3, and observe the results.

7.20 Consider the single-stage manufacturing system shown in Figure 7.21 where the raw material is procured using an (r, Q) pure inventory control policy. A raw material order is placed as soon as Q units accumulate. It takes an exponentially distributed period with rate ξ for the order of Q units to come back to the input buffer. Let us assume that the unit processing and demand interarrival times are both exponentially distributed with rates μ and λ. Explain the details of how one would analyze the system to obtain the approximate values of the performance measures. The probability of zero raw material inventory at any time may be considered as a measure of the delivery performance. Also, the probability of zero raw material at the time a production order is just processed may be the measure of how well the control policy is operating.

7.21 Consider a two-stage, two-product system. The unit processing and set-up times of both products are exponentially distributed with rates μ_i^j and γ_i^j for product i at stage j. The demand process is assumed to be Poisson with rate λ_i for product i. Assuming $R_i^w - r_i^w > R_i^b$ as we have done in Section 7.8, explain how you would decompose the system to be able to approximate the values of the measures of interest.

7.22 The following pull-type manufacturing system has 3 stages. Assume that the processing time at each stage is exponentially distributed with rates 2, 3, and 1 unit per hour. The customer arrivals are according to a Poisson process with rate 0.5 customers per hour. The space available between stages to store WIP inventory is given below.

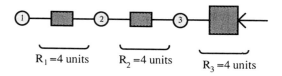

$R_1 = 4$ units $R_2 = 4$ units $R_3 = 4$ units

(a) Implement the local control policy assuming $r_i = R_i/2$ for each buffer in the system. Apply the decomposition algorithm of Section 7.2.2 to obtain approximate average WIP levels at each storage. Obtain the customer service level.

(b) Now implement the generalized Kanban scheme, and obtain the corresponding approximate average WIP levels and the customer service level.

(c) Compare the systems in (a) and (b) with appropriately chosen values for C_ks.

Appendix A Single-Stage P/I System with PH Delay, Set-up, and Processing Times

In this appendix, we are going extend the analysis presented in Section 7.3.1 to be able to study single-stage production/inventory systems with more general assumptions. In particular, we allow the delay, set-up, and processing times to be of mixtures of generalized Erlang type, as explained in Chapter 3. The processing time can also be thought of as the process completion time (including failures), as we have done in Chapter 4. Moreover, the demand arrival process can be compound Poisson;[1] however, we will retain our Poisson assumption for simplicity. We keep the same (R, r) continuous-review policy to control the inventory level in the warehouse. The behavior of the production/inventory system remains to be the same as the one shown in Figure 7.9. Therefore, it is still possible to develop a tractable procedure to obtain the stationary probabilities of the inventory and backorder levels.

Let us assume that the delay period is of MGE type with a $(\underline{\nu}, Z)$ representation with K_D phases; the set-up time has an $(\underline{\eta}, W)$ representation with K_S phases; and the processing time has an $(\underline{\alpha}, S)$ representation with K_P phases. Also, let β_i, γ_i, and μ_i be the rates of the ith phases in the delay, set-up, and processing times, respectively. Due to the special structure of the MGE distribution, $Z(i, j)$, $W(i, j)$, and $S(i, j) > 0$ only for $j = i + 1$, and $\underline{\nu}$, $\underline{\eta}$ and $\underline{\alpha}$ are vectors of type $(1, 0, \ldots, 0)^T$. Here, we again assume that a vector is a column vector unless its transpose is indicated. Finally, let $\underline{Z}^o$, $\underline{W}^o$ and $\underline{S}^o$ be the vectors of the departure rates from the phases of the delay, set-up, and processing times.

Let us consider the stochastic process $\{N_t, D_t, S_t, J_t, t \geq 0\}$, where N_t represents the inventory level in the warehouse (positive or negative) and D_t, S_t and J_t represent the phases of the delay, set-up, and processing times at time t, respectively. Thus, D_t, S_t and J_t are disjunctive variables of which only one can be positive at a time while the others are zero. $\{N_t, D_t, S_t, J_t\}$ constitutes a continuous-time, discrete-state Markov chain with an infinite number of states. Let (n, i, j, k) be a particular state of the process and $P(n, i, j, k)$ be its associated steady-state probability. Again, $(R, 0, 0, 0)$ is the state initiating the regeneration cycles with the idle state. Notice that $P(R, 0, 0, 0) = P_R$, given by (7.9), and consequently the steady-state probabilities of the processor's status are insensitive to the higher moments of the delay, set-up, and processing times. Therefore, (7.6)–

---

[1] The compound Poisson demand arrival process can be visualized as the Poisson customer arrivals each demanding a random number of units.

(7.9) are still valid for $P(I)$, $P(D)$, $P(S)$, and $P(P)$ in the system studied here. Also, again $\rho_p < 1$ is the necessary and sufficient condition for the existence of the steady-state probabilities.

We will not go through the flow-balance equations of the Markov chain here. It is left as an exercise to the reader (see Problem 7.6) for some specific MGE distributions. Ahead, we summarize a matrix-recursive algorithm to obtain the joint probabilities of $\{N_t, D_t, S_t, J_t\}$.

Let us define the probability vector $\tilde{\underline{P}}_P(n)$ as

$$\tilde{\underline{P}}_P(n) = \begin{bmatrix} P(n, 0, 0, 1) \\ \vdots \\ P(n, 0, 0, K_P) \end{bmatrix}, \qquad n = R-1, \ldots, r+1,$$

representing the probabilities of the production states for $n > r$. Then, $\tilde{\underline{P}}_P(n)$ is given by

$$\tilde{\underline{P}}_P(n) = \lambda \mathbf{A}_P^{-1} \underline{\mathbf{G}}_P(n+1), \qquad n = R-1, \ldots, r+1, \qquad (A.1)$$

where

$$\mathbf{A}_P = \begin{bmatrix} S^o(1) & S^o(2) & \cdot & \cdots & S^o(K_P-1) & S^o(K_P) \\ -S(1,2) & (\lambda + \mu_2) & 0 & \cdots & 0 & 0 \\ 0 & -S(2,3) & (\lambda + \mu_3) & \cdots & 0 & 0 \\ \vdots & \vdots & \vdots & \vdots & \vdots & \vdots \\ \cdot & \cdot & \cdot & \cdots & (\lambda + \mu_{K_P-1}) & 0 \\ 0 & \cdot & \cdot & \cdots & -S(K_P-1, K_P) & (\lambda + \mu_{K_P}) \end{bmatrix},$$

and

$$\underline{\mathbf{G}}_P(n+1) = \begin{bmatrix} P_{n+1} \\ P(n+1, 0, 0, 2) \\ \vdots \\ (n+1, 0, 0, K_P) \end{bmatrix}, \quad \text{with} \quad \underline{\mathbf{G}}_P(R) = \begin{bmatrix} P_R \\ 0 \\ \vdots \\ 0 \end{bmatrix}.$$

P_{n+1} is the marginal probability of $N_t = n+1$ at any time t.

7. Pull-Type Manufacturing Systems

Next, for $n = r, r - 1, \ldots$, let us define the following probability vector:

$$\underline{\tilde{\mathbf{P}}}(n) = \begin{bmatrix} P(n, 1, 0, 0) \\ \vdots \\ P(n, K_D, 0, 0) \\ \cdots \\ P(n, 0, 1, 0) \\ \vdots \\ P(n, 0, K_S, 0) \\ \cdots \\ P(n, 0, 0, 1) \\ \vdots \\ P(n, 0, 0, K_P) \end{bmatrix},$$

representing the delay, set-up, and production states for $n < r + 1$. Then the matrix recursion progresses in a similar manner, that is,

$$\underline{\tilde{\mathbf{P}}}(n) = \lambda \mathbf{A}^{-1} \underline{\mathbf{G}}(n + 1), \qquad n = r, r - 1, \ldots \qquad (A.2)$$

where

$$\mathbf{A} = \begin{bmatrix} \mathbf{A}_D & & \\ \mathbf{B} & \mathbf{A}_S & \\ & & \mathbf{A}_P \end{bmatrix},$$

with

$$\mathbf{A}_D = \begin{bmatrix} \lambda + \beta_1 & 0 & \cdots & 0 & 0 \\ -Z(1, 2) & \lambda + \beta_2 & \cdots & 0 & 0 \\ 0 & -Z(2, 3) & \cdots & \cdots & \cdots \\ \vdots & 0 & \vdots & \vdots & \vdots \\ \cdot & \cdot & \cdots & \lambda + \beta_{K_D - 1} & 0 \\ 0 & \cdot & \cdots & -Z(K_D - 1, K_D) & \lambda + \beta_{K_D} \end{bmatrix}_{K_D \times K_D},$$

and similarly

$$\mathbf{A}_S = \begin{bmatrix} \lambda+\gamma_1 & 0 & \cdots & 0 & 0 \\ -W(1,2) & \lambda+\gamma_2 & \cdots & 0 & 0 \\ 0 & -W(2,3) & \cdots & \cdot & \cdot \\ \vdots & 0 & \vdots & \vdots & \vdots \\ \cdot & \cdot & \cdots & \lambda+\gamma_{K_D-1} & 0 \\ 0 & \cdot & \cdots & -W(K_S-1, K_S) & \lambda+\gamma_{K_S} \end{bmatrix}_{K_S \times K_S},$$

$$\mathbf{B} = \begin{bmatrix} -Z^o(1) & \cdots & \cdots & -Z^o(K_D) \\ 0 & \cdots & \cdots & 0 \\ \vdots & \vdots & \ddots & \vdots \\ 0 & \cdots & \cdots & 0 \end{bmatrix}_{K_S \times K_D}.$$

B is the matrix of the rates of transitions from delay to set-up phases. Finally, $\underline{G}(n+1)$ is given by

$$\underline{G}(n+1) = \begin{bmatrix} \tilde{\mathbf{P}}_D(n+1) \\ \tilde{\mathbf{P}}_S(n+1) \\ \underline{G}_P(n+1) \end{bmatrix}, \quad \text{with} \quad \underline{G}(r+1) = \begin{bmatrix} \underline{G}(R) \\ 0 \\ \underline{G}_P(r+1) \end{bmatrix}.$$

The preceding matrix-recursive procedure does not create any numerical instability problems. However, it should be understood that very small probabilities, which are close to the smallest number acceptable by the computer being used, may assume negative values due to round-off errors and subtractions involved in the recursive procedure. For practical purposes, these probabilities are assumed to be zero.

Once, $\tilde{\mathbf{P}}(n)$ is known, the marginal probabilities of the inventory and backorder levels are given by

$$P_n = \begin{cases} P(n,0,0,0) + \sum_{i=1}^{K_P} P(n,0,0,i), & n > r, \\ \sum_{j=1}^{K_D} P(n,j,0,0) + \sum_{k=1}^{K_S} P(n,0,k,0) + \sum_{\ell=1}^{K_P} P(n,0,0,\ell), & n \leq r. \end{cases}$$

Index

Arbitrary configurations, 6, 268
Assembly/Disassembly Systems, 7, 267
Assembly systems, 7, 243

Backorder
   average, 296
   probability, 289
Backward pass, 154, 224, 316, 325
Base-stock policy, 274
Blocking, 5,
   blocking-after-processing, 120
   blocking-before-processing, 120
Blocking probability, 119, 127, 131, 144, 196, 218, 243, 245
Bottleneck, 5
Bounds,
   lower, 162, 165
   upper, 163, 165
Bowl Phenomena, 10, 169
Buffer allocation, 169
Buffer clearing policies, 197
Bulk-service queues, 258
Buzacott-type models, 234

Chapman-Kolmogorov equations, 28
Coefficient of variation, 19, 95
Compound Poisson distribution, 348
Conditional distribution, 16
Continuity property,
   transform functions of, 20, 35
Continuous-review (Q, r) policy, 326
Continuous-review (R, r) policy, 276, 283, 297, 300, 309, 329, 334, 348
Conwip, 339

Correlation coefficient, 19
Covariance, 19
Coxian distributions, 10, 42
Cumulative distribution function, 15
Cycle time, 185

Decomposition, 143, 217, 231, 255, 311, 328, 334
Density function, 15
Design problems, 168, 213
Deterministic processing times, 157
Distributions,
   bernoulli, 24
   binomial, 24
   exponential, 22
   gamma, 23
   geometric, 24
   normal, 23
   poisson, 25
   uniform, 22
Down time probability, 82, 85, 196, 202, 218, 223
Dynamic order priority, 278

Effective input rate, 109
Effective output rate, 109
Empty cells, 140, 276, 292
Equivalence relations, 140, 252, 291
Erlang distribution, 10, 41, 44, 45
Excess jobs, process, 249, 241
Exhaustive production, 278
Exponential distribution,
   generalized, 45
   phase-type representation, 51

354   Index

Exponential distribution (*cont.*)
  shifted, 65

Failure-repair policies, 93
  process restarts after repair, 96
  process resumes after repair, 93
  scrapping, 100
  single-failure assumption, 99
Failures,
  operation-dependent, 82, 85, 91, 186
  operation-independent, 82, 85, 92, 186
Flow-balance equations, 22, 32, 34, 78, 123, 125, 199, 202, 208, 246, 287
Flow line, 1, 120
Forward pass, 154, 224, 316, 325

Generalized Kanban systems, 282, 323, 339
Geometric solution, 36
Gershgorin's theorem, 135
Globally controlled systems, 280

Holder's inequality, 19
Hooke and Jeeves search procedure, 172
Hyperexponential distribution, 46
  weighted, 55

Indicator functions, 21
Input buffer, 6
Interarrival time, 66
Inventory control, 38, 62, 326
Inventory systems, 3
Iterative algorithm, 151, 223, 234, 256, 307, 316, 325
Iterative search, 9

Job-shop, 1
Joint density function, 16

Kanban system, 8, 276, 319, 323, 339
Kraemer and Langen Bach Belz, 74, 162

Laplace-Stieltjes transform, 20, 42, 94, 95, 97, 99
Lexicographic order, 134
Line behavior, 137, 210
Little's formula, 69, 150, 256, 315
Local balance equation, 32

Locally controlled systems, 279
Look-back policy, 334
Lost sale, 289

Machine-repairman (interference)
  problem, 31, 62, 65, 104, 117
Make-to-order systems, 319, 339
Markov chains, 26
  discretization, 28
  stationary probabilities, 27
Matching buffers, 243, 260
Matrix-geometric solution, 80, 201, 209, 288
Matrix-recursive procedure, 80, 130, 247
Memoryless property,
  exponential distribution, 22
  geometric distribution, 25
Message passing, 8
MGE distributions, 42, 95, 124, 154, 348
Moment generating function, 19, 191
Moment matching, 52
Moments of random variables, 18, 42, 52, 94, 97, 99, 101
MRP, 339
Multiproduct systems, 277, 300
Multistage, multiproduct systems, 328
Multistage systems, 309, 319, 322, 339

Nonexhaustive approach, 278
Numerical approach, 133

Optimization, 39, 117, 170, 214, 297
Order quantity, 327
Order-up-to-R policy, 275
Output buffer, 6
Output rate, 69, 121, 157, 165, 194, 202, 248
  variance of, 215, 231

PASTA property, 37, 38, 72, 73, 84, 294
Performance evaluation, 3
Phase-type distributions, 40, 42
Pollaczek-Khintchine formula, 72
Power method, 135
Priorities, 75
Process completion time, 5, 91
Production control policies, 274
Production/inventory system, 7

Profit maximization, 171, 216
Pseudo processing times, 147
Pull systems, 8, 118, 309
Push systems, 8, 118, 273

Queues
  M/M/1, 70, 73, 257
  M/M/1/N, 124, 139
  M/M/m, 74
  M/D/1, 162, 163
  M/G/1, 70, 71, 73, 107, 294
  M/G/ , 106
  M/MGE-2/1, 256, 335
  G/G/1, 73, 320
  with different customer types, 75
  with SPT rule, 76
  with vacations, 293, 338
  quasi-reversible, 106
Queueing analogy, 291
Queueing networks, 174, 264, 268, 339

Regeneration points and cycles, 29, 68, 188, 285
Regenerative processes, 29
Reorder point, 275, 284, 300, 326, 334
Residual life, 30, 72, 75, 76, 84, 92
Reversibility property, 122, 138

Scrapping, 189, 100

Search, 9, 171
Set-up time, 284, 300, 309, 328, 334, 338
Single-stage systems, 275
Starvation, 5, 144
  probability, 144, 256, 312, 337
Stochastic decomposition property, 293
Stochastic processes, 26
Synchronous lines, 185
System time, 70, 124

Target inventory level, 275, 284, 300, 326, 328, 334
Throughput, 215 (*see* output rate)
Transfer lines, 2, 185, 188, 195, 228

Uncertainty in manufacturing systems, 5, 98, 101
Utilization, 69, 124, 202, 218, 220, 285

Warehouse, 276, 281, 296, 300, 312, 328, 325
Waiting time, 72, 75, 84
Work-in-process inventory (WIP), 77
  probability distribution, 79, 87
Work load allocation, 168
Work station, 71
  subject to failures, 81
  job arrival process turned off, 88

LANCHESTER LIBRARY

3 8001 00215 9303

£34.00

LANCHESTER LIBRARY, Coventry University
Much Park Street, Coventry CV1 2HF   TEL. 0203 838292

This book is due to be returned no later than the date stamped above.
Fines are charged on overdue books